A Scientist's Voice in American Culture

A Scientist's Voice in American Culture

Simon Newcomb and the Rhetoric of Scientific Method

Albert E. Moyer

UNIVERSITY OF CALIFORNIA PRESS
Berkeley • Los Angeles • Oxford

University of California Press
Berkeley and Los Angeles, California

University of California Press
Oxford, England

Copyright © 1992 by The Regents of the University of California

Library of Congress Cataloging-in-Publication Data
Moyer, Albert E., 1945-
    A scientist's voice in American culture : Simon Newcomb and the rhetoric of scientific method / Albert E. Moyer.
      p.   cm.
    Includes bibliographical references and index.
    ISBN 0-520-07689-3
    1. Newcomb, Simon, 1835–1909.   2. Science—Methodology.
3. Science—United States—History.   4. Scientists—United States—Biography.   I. Title.
Q143.N49M68   1992
509.2—dc20
    [B]   91-48271
                                                                               CIP

Printed in the United States of America

1 2 3 4 5 6 7 8 9

The paper used in this publication meets the minimum requirements of American National Standard for Information Sciences—Permanence of Paper for Printed Library Materials, ANSI Z39.48-1984 ∞

For Lynette, Holly, Emily, and Rebecca

Contents

Illustrations x

Preface xi

Acknowledgments xvii

PART I. INTRODUCTION

1. Method, Rhetoric, and Newcomb 3
    The Rhetorical Aspects of Scientific Method 6
    Newcomb and Methodological Rhetoric 9

PART II. NEWCOMB'S LIFE AND THOUGHT

2. Formative Years 19
    Growing Up in the Maritime Provinces 20
    Encountering Joseph Henry 26
    Scientific Cambridge 30
3. Influences of Comte, Darwin, and Mill 36
    Auguste Comte 37
    Charles Darwin 38
    John Stuart Mill 41
    A Textbook Statement on Method 45

4. Interactions with Wright and Peirce — 52
   Chauncey Wright — 52
   Charles Sanders Peirce — 58
   Views of Wright and Peirce — 62
5. Midcareer — 66
   Taking Charge of the Almanac Office — 68
   The Planets and the Moon — 71
   Professional Involvements — 78
6. American Science, Scientific Method, and Social Progress — 82
   Institutional Deficiencies and Public Indifference — 83
   The Nation's Need for Scientific Method — 86
   Social Progress and the Scientific Use of Language — 90
   The Languages of Physics, Business, and Philosophy — 94
7. Political Economics: Old versus the New School — 98
   Newcomb as Political Economist — 100
   Partisan Liberal Policies — 105
   Defending the Old School — 113
   Confronting Richard Ely and Edmund James — 120
8. Religion: A Clash with Gray, Porter, and McCosh — 128
   The Saint Louis Speech — 130
   Asa Gray, Noah Porter, and James McCosh — 135
   Newcomb's Personal Religious Skepticism — 139
9. Physics and Mathematics: Public Understanding and Educational Reform — 146
   Response to John Stallo's Critique of Physics — 148
   Clarifying Scientific Terminology — 155
   The Teaching of Introductory Mathematics — 158
10. Mental and Psychical Sciences: Challenging Current Beliefs — 166
    Scientific Materialism — 167
    Psychical Research — 169
    Address to the Psychical Society — 175
11. Later Years — 183
    Facts and Fiction — 184
    Reflections on Method — 188
    Old Issues Rejoined — 194

PART III. COMMENTARY

12. Newcomb and American Pragmatism	205
Pragmatists in Historical Context	207
The Philosophical Core of Pragmatism	214
Newcomb as a Catalyst	221
13. Pragmatism and Methodological Rhetoric	224
Pragmatists as Apostles of Scientific Method	227
Pragmatism's Emergence in the Late Nineteenth Century	230
Notes	239
Select Bibliography	279
Index	293

Illustrations

Simon Newcomb in 1857 at age twenty-two
Pages from Newcomb's diary during his Cambridge years
Simon and Mary Newcomb shortly after their marriage in 1863
"Professor" Newcomb at the Naval Observatory's telescope
Astronomers at the Naval Observatory preparing for the 1874 transit of Venus
Newcomb in the 1870s
The superintendent of the Nautical Almanac Office and president of the Washington Philosophical Society
A draft, in Newcomb's hand, of his "Religious Autobiography," ca 1879–1880
The elder statesman of American science, Newcomb in 1903 at age sixty-eight

Preface

If we judge a scientist's influence by recognition from his contemporaries, then Simon Newcomb stands out as the most influential American scientist of the late nineteenth century. The American and international scientific communities repeatedly honored this mathematical astronomer for his comprehensive studies of the motions and positions of the sun, moon, planets, and stars and his supportive studies in mathematics and physics. Conducted primarily during the decades following the Civil War, these investigations culminated in definitive sets of astronomical constants, tables, and computational methods. The general public and the nation's leaders joined in recognizing Newcomb's contributions, in part because he addressed popular issues as well as technical topics. He eagerly discussed the nature of science and its relation to religion, philosophy, and society, especially political economy. Thus, he earned acclaim in the United States not only as a scientist but also as a commentator on science.

Despite such success, Newcomb slipped from prominence during the half century following his death in 1909. Apparently, the contributions of this "last of the great masters" of classical, Newtonian astronomy—to use Albert Einstein's epithet—were overshadowed by the new astronomy of spectroscopic observations and relativistic theories.[1] Though Newcomb lifted classical astronomy to a new level of refinement, few nonastronomers could appreciate the significance of his precise measurements and complex calculations, and unfortunately, he lacked that single major discovery or breakthrough to which nonastronomers could readily attach his name. Add to this that he worked not at his own easily distinguishable observatory but at the U.S. Naval Observatory and then at the Nautical Almanac Office—modest government agencies soon dwarfed by the increasingly conspicuous laboratories of expanding universities, corporations, and private foundations.

Only in recent decades have historians delved into Newcomb's contributions to astronomy as well as to mathematics, physics, economics, and the growth of American scientific institutions. While now more attentive to this polymath's place in particular disciplines and organizations, investigators continue to overlook his role in broader historical developments of the late nineteenth century. In fact, Newcomb both reflected and helped shape the quickening interplay between, on the one hand, scientific modes of thought and action and, on the other hand, wider intellectual, cultural, and social currents within the United States. He was in the vanguard of Gilded Age "scientism," as Americans increasingly embraced not only the practices of science but also its perceived forms and values—practices, forms, and values that would come to even fuller flower in the Progressive era of the early twentieth century.

A particular image of science colored Newcomb's view on its relationship to other areas of life and inquiry. He believed that scientists' successes derived mainly from adhering to the "scientific method." That is, they succeeded because they governed their studies using a definite set of procedural rules, primary of which was the rule to employ only those concepts that can be defined in terms of concrete experiences. Moreover, comparable successes awaited others who chose to apply the same rules to investigations outside natural science. Scientific method, in other words, provided the key not only to scientific but to all intellectual progress. Thus, in his writings and speeches, Newcomb repeatedly called for nonscientists to adopt the method; at the same time, he tried to bring method to bear on issues in fields such as political economy, religion, philosophy, and even psychical research.

Though he saw himself as an unbiased scientist addressing issues objectively, Newcomb actually used pronouncements about method for persuasive effect. That is, he used method rhetorically. In some instances, to produce the desired rhetorical effect, he did not actually apply the method but merely extolled its virtues. For example, to elevate the status of science—and hence of scientists—in the United

States, he set out to convince politicians, economists, educators, and other uninitiated Americans of the utility of scientific method for their fields.

At other times, he applied the method in a direct manner, turning its rules and criteria into working rhetorical tools. Thus, in an effort to consolidate the institutional gains of American scientists and encourage the further professionalization of the scientific community, he employed method to demarcate the boundaries of science, variously distinguishing it from and, when useful, associating it with theological, philosophical, commercial, and other traditionally nonscientific realms of culture. Not limiting himself to the broader social relations of science, he also used method on a more personal plane. In particular, he relied on the authoritative language of method to justify his own political, economic, pedagogic, religious, and philosophical positions. In other words, this prominent scientist legitimated components of his personal ideology, including his religious skepticism and political liberalism, by invoking scientific method.

The insights of historians vary according to the particular person or group they choose to study. Studies of men and women who, for one reason or another, are lauded today but had essentially no following in their own times tell us as much about our present interests as about the past. On the other hand, studies of individuals such as Newcomb who, though less well known today, were esteemed in their own times inform us more directly about the concerns and aspirations that prevailed in prior societies. In this book, I attempt to clarify late nineteenth-century American history, particularly its scientistic strand, by examining both the context in which Newcomb formulated his outlook on scientific method and the actual rhetorical uses he made of the method. In particular, I locate Newcomb against a backdrop of heightened interaction between science and general thought, culture, and society. As an advocate of science, he helped to mold, and was molded by, the intensifying dialogue concerning the place of science in the United States. As an advocate of scientific method, he contributed to a faith in method that, for better or for worse, took firm hold and persists within American culture to this day.

This book, being a first attempt to examine the sweep of Newcomb's career, is preliminary in the sense of clearing the way. Confronted by a mass of archival and published documents that impinge on a wide range of topics, I seek to evoke basic patterns, or at least tendencies, in Newcomb's intellectual pursuits and interactions. Ideally, this initial ordering of Newcomb's career will move him further into the mainstream of scholarship, where other historians of science or of American thought and culture will refine his story.

I intend the book to work on four levels. On the most detailed level—the level that underpins the entire book—it provides a historical analysis of Newcomb's views on the nature of science and science's wider ties. In effect, this analysis constitutes a partial intellectual biography: it draws on Newcomb's unpublished diaries, letters, and manuscripts as well as his published writings to document his evolving views on science and its province in human affairs. On a second, broader level, the book furnishes insight into intellectual and cultural developments in the United States during the Gilded Age. It illuminates the harmonies and tensions between natural science and areas such as religion, philosophy, pedagogy, and social science. In addition, it reveals late nineteenth-century perceptions about the nation's assimilation of scientific attitudes and support of scientific research.

Third, and more generally, the book furthers our understanding of one of the most penetrating cultural currents in the United States during the decades bracketing 1900—pragmatism, with its emphasis on the "practical" and "scientific." Through his methodological pronouncements, Newcomb joined Chauncey Wright, Charles Peirce, and other Americans in contributing to the formulation and transmission of basic pragmatic outlooks. He played this catalytic role, as will be explained in a concluding section of commentary, not merely by insisting that concepts be definable through sensory experiences but also by aligning himself with a characteristic cluster of beliefs and hopes about American culture in a scientific age. Finally, on the most general level, the book adds to our awareness of the ways in which scientists use methodology, often unconsciously, for political and rhetorical purposes and the degree to which their ideas on method are historically contingent in both origin and application. A final chapter of commentary suggests that this broader rhetorical tradition even subsumes aspects of American pragmatism.

In structure, the book divides into three sections: an introduction to the rhetoric of scientific method, a core of ten central chapters on Newcomb's life and thought, and the concluding commentary on pragmatism and scientific method. The account of Newcomb's life and thought combines chronological and topical approaches. It begins with a discussion of Newcomb's intellectual meanderings in his younger years and his first encounters with Joseph Henry and other practicing scientists (chap. 2), his eventual exposure to the ideas of Comte, Darwin, and Mill (chap. 3), and his interactions as a young man with Wright and Peirce (chap. 4). After an interlude reviewing his professional attainments during midcareer (chap. 5), the account continues by examining Newcomb's perception of the institutional deficiencies of American science and his resulting call for wider public support of science; seeking to convince the citizenry of the benefits of science, he argues that fuller public adoption of the method of science would engender social progress (chap. 6). This leads to a series of cases in which Newcomb uses the language of scientific method as a rhetorical resource, bringing method variously to bear on controversial issues in: political economics (chap. 7), religion (chap. 8), physics and mathematics (chap. 9), and mental and psychical research (chap. 10). The biographical account ends with a brief look at his later years (chap. 11).

When Newcomb died in 1909, Robert S. Woodward, president of the Carnegie Institution in Washington, singled out his former colleague for not only his "unrivaled productivity" but also the "unusual clearness" of his prose. "He enjoyed the distinction amongst his fellow-countrymen," Woodward explained, "of being able to write the clearest English produced by any man of science in America."[2] To date, scholars have tapped only a small portion of Newcomb's vast, lifetime output of unpublished and published writings. The Simon Newcomb Papers at the Library of Congress number approximately 46,200 items and occupy sixty-two linear feet of shelf space. His published writings extend to almost 550 works, including many books and lengthy monographs. Because the archival documents are virtually unknown and because even his published writings do not have wide currency, I quote liberally in the following chapters from Newcomb's own words. My intention is not only to display the texture of the archival and published sources, but also to restore the voice to one of America's most vocal scientists.

Acknowledgments

In trying to understand Newcomb's career, I turned for help to many colleagues and friends over the past decade. Providing substantial assistance were David Hollinger, Edward Madden, and Arthur Norberg. I also learned much from referees, all anonymous except for one, Theodore Porter. Many other persons provided specific advice or information; this group included Peter Barker, James Beichler, Robert Bruce, Paul Conkin, Michael Dennis, James Fleming, Craufurd Goodwin, Richard Hirsh, Jon Hodge, Burton Kaufman, William Keith, Larry Laudan, Rachel Laudan, David Lux, Seymour Mauskopf, Roy North, Duncan Porter, Nathan Reingold, Margaret Schabas, Daniel Siegel, Eugene Taylor, and Richard Yeo.

For archival materials, I relied heavily on the staff of the Manuscript Division at the Library of Congress. Also helping me flesh out the archival record were: Susan Bluhm at the Smithsonian Institution Archives; Stuart Campbell at the Clark University Archives; Brenda Corbin at the U.S. Naval Observatory Library; Clark Elliott at the Harvard University Archives, and the staff at Harvard's Houghton Library; Joan Grattan at the Eisenhower Library of the Johns Hopkins University; James Matlock at the Library of the American Society for Psychical Research; David Ment at the Milbank Memorial Library at Teachers College Columbia University; and Philip Weimershirch at the Burndy Library.

My wife, Lynette, used her literary skills to help knock my prose into a more intelligible form. My daughters, Holly, Emily, and Rebecca, used their fresh minds and bright faces to return me to the joys of fatherhood after long hours of converting fading documents into computerized text.

I thank all of these persons.

PART I

Introduction

CHAPTER I

Method, Rhetoric, and Newcomb

The words "scientific method" call to mind a set of rules, procedures, and criteria that enable scientists to govern and direct their study of nature. That is, methodological rules are thought to be the means to scientific ends.[1] In the United States during the closing decades of the nineteenth century, Simon Newcomb took a firm public stand on methodological issues. He presented his position with authority, being the nation's most eminent astronomer as well as one of its most respected mathematicians, physicists, and political economists. In fact, Newcomb was the most prominent American natural scientist of his day to address methodological matters seriously. His pronouncements on method were neither isolated initiatives nor momentary diversions in his career, but were integral and recurrent components of his life as a practicing scientist and scientific spokesman. And he presented his views not merely in his formative or retirement years, but also during the midcareer decades of the 1870s and 1880s, his most productive and celebrated period of research. Through commentaries in his own books, articles in widely read journals, and speeches to influential groups, he delivered his message on method with clarity and conviction. Indeed, for the historian, scientific method emanates as a unifying theme in Newcomb's multifaceted career; as such, method provides a window into his life and times.

Newcomb's understanding of how method operates and develops among scientists matches that held by most scientists and nonscientists over the past few centuries. According to this traditional view, scientific method functions autonomously. That is, methodological rules operate and develop independently of the scientists' broader cultural and social environment, their particular professional and institutional setting, and even their specific theoretical and experimental framework. Supposedly, this autonomy of method allows scientists to achieve objectivity in their research. By embracing the same set of unequivocal rules, scientists acquire the means of generating new knowledge through consensus while at the same time isolating themselves from the quirks and prejudices of popular opinion and peer sentiment. And the rules, being autonomous, are equally applicable to all fields of scientific inquiry. They function universally.

In recent decades, historians, philosophers, and sociologists of science have been challenging the long-standing belief in an autonomous scientific method. One historiographer comments that just as historians of science have taught us "that the positivist belief in a value-free, culturally independent science is a myth," so too have they taught us "that *the* scientific method, perceived as an absolute, canonized doctrine, is an artifact."[2] Scholars now agree that scientists' methodological rules, rather than being autonomous, actually exist in a symbiotic relationship with elements in the scientists' cultural, institutional, and conceptual environment. Scientific method, in other words, is historically contingent. Admittedly, different scientists sometimes hold in common particular investigational precepts, but they seldom share the same total set of precepts—a set that might seem to constitute a monolithic, universally applicable method. Rather, methodological rules appear much more localized than previously thought; they seem much more dependent on the particular circumstances of a group or an individual. And these circumstances—cultural, institutional, and conceptual—impinge to such an extent in real scientific life that the scientists' formal methodological statements often do not correspond to their concrete actions. Historical actuality mitigates abstract methodology.

While agreeing that scientific method is historically contingent, the revisionists disagree on the true role of methodological rules. Some historians, philosophers, and sociologists of science go so far as to maintain that methodological rules do not contribute to the generation of new knowledge. Rules regarding the evaluation of evidence, for example, are often so inherently ambiguous that they do not allow scientists to choose unequivocally the best theory from among alternative theories that seem to fit the evidence. This problem of "underdetermination" is compounded when different scientists do not agree on the same rules or do not interpret shared rules uniformly. Other scholars still grant that methodological rules contribute to the generation of new knowledge, but that they contribute only in a partial manner. Proponents of this moderate position acknowledge that agreement on methodological rules does not ensure agreement on rival factual claims or hypotheses. They emphasize, however, that even when a shared rule does not enable scientists to make an unequivocal choice between rival theories, the rule often permits them to make useful comparisons and to identify preferences. In this limited sense, scientific method does contribute to the constitution of new knowledge.

Part of the problem confronting all of the revisionist scholars is the challenge of identifying or defining scientific method. Some even question whether abstract method can be separated or extracted from the substantive content or actual operation of the sciences. To help clarify matters, Larry Laudan makes a heuristic distinction between three avenues of study open to students of scientific methodology: "(a) a study of the methods which are *implicitly* utilized by working scientists in making their appraisals of theories, experiments, etc.; (b) a study of the *methodological pronouncements* of scientists about the methods they use; and (c) a study of the methods which rational scientists *ought* to use." Noting that the three realms are interrelated, he labels them descriptive methodology, ideological methodology, and normative methodology.[3] Descriptive methodology falls within the purview of scholars with an anthropological or ethnographic bent or the knack to discern often unspoken investigational beliefs and their possible ties to scientists' actions and practices. Inquiry into normative methodology requires primarily the evaluative and analytic talents of philosophers. Ideological methodology—scientists' pronouncements about the methods they use—lends itself to scrutiny by historians, sociologists, and rhetoricians. In studying Newcomb, I focus on ideological methodology. Specifically, noticing the extent to which Newcomb relied on method throughout his career to champion or legitimate particular views and interests, I present a historical analysis of his methodological pronouncements. Because such pronouncements are likely conditioned by a scientist's personal and professional circumstances, the study of ideological methodology offers the possibility of illuminating the human and cultural dimensions of science. In Newcomb's case, his professed methodology is interwoven with his personal political, economic, philosophical, religious, and pedagogic convictions and his perceptions of the institutional and conceptual status of science in late nineteenth-century America. And as with his colleagues Chauncey Wright and Charles Sanders Peirce, his views on method connect to the broader cultural tradition of American pragmatism.

THE RHETORICAL ASPECTS OF SCIENTIFIC METHOD

Of the few recent studies that deal exclusively with ideological methodology, one stands out for providing an overview of the topic: *The Politics and Rhetoric of Scientific Method*, edited by John Schuster and Richard Yeo. This collection of essays includes an editors' introduction that establishes a historiographic framework for analyzing the political and rhetorical — or what Laudan calls the ideological — aspects of scientific method. In the introduction, Schuster and Yeo first inventory the recent challenges to the traditional belief, dating back to at least the seventeenth century, that there exists "a single, transferable, efficacious scientific method." Drawing on the findings of historians, philosophers, and sociologists of science beginning with Alexandre Koyré, Gaston Bachelard, and Thomas Kuhn and extending through Hilary Putnam, Paul Feyerabend, John Ziman, Larry Laudan, Barry Barnes, and Bruno Latour, they note that while the old belief is now discredited, there remains the problem of understanding "the actual sociocognitive functions of method doctrines within the practice and social organization of the sciences." They propose that historians make a twofold examination of the problem, where each of the two inquiries informs the other. First, historians should attempt to reconstruct the cultural and intellectual context in which a scientist formulated his method claims. They should examine not only the scientist's relation to "the grand tradition of theorizing about method" but also how this relation is conditioned by the scientist's "biography, social location, institutional affiliations and perceived interests." Conversely, they should avoid the past tendency to a "Whiggish" interpretation of the history, in which a scientist's methodological claims are prejudicially judged using present methodological criteria. They should also avoid a strict "internalist" account of the history, in which a scientist's methodological pronouncements are portrayed as disembodied dictates developing in isolation from all outside circumstances.[4]

Second, historians of methodology should, while bearing in mind the historical context, attempt to identify and analyze the actual political and rhetorical uses of method claims. Schuster and Yeo distinguish three sometimes overlapping levels of possible use: "(1) the 'internal' level of technical debate and argument where knowledge claims are initially framed, negotiated, and evaluated; (2) the level of institutional and disciplinary organization and politics; and (3) the level of the 'public politics' of the scientific community." What they mean by the political and rhetorical aspects of scientific method shows most plainly on levels two and three. At the second level, that of institutional and disciplinary organization, methodological pronouncements permit members of scientific societies, for example, to construct and enforce guiding ideologies. Similarly, methodological arguments and appeals enable practitioners to demarcate the boundaries of distinct "socio-cognitive groupings" either inside or between disciplines. They also allow scientists to sanction particular disciplinary outlooks or techniques in which they have vested interests.

At the third level, that of "public science," methodological pronouncements serve to present a positive image of science to the broader society and to enhance the political and cultural standing of the scientific community. For example, statements about method allow scientists to portray scientific inquiry as a consensual activity; in particular, scientists tend to claim that consensus on method, rather than disagreement over specific theories, is the defining characteristic of their field. Similarly, representatives of the natural sciences find it useful to portray this method, with its putative objectivity and power, as accessible to nonscientists and transferable to other fields. Furthermore, method serves in the public arena variously to associate science with or distinguish it from "other intellectual discourses such as theology or philosophy." (This is an outer-directed consolidation or demarcation of scientists and lay people as contrasted to the innerdirected consolidation or demarcation, on Schuster and Yeo's second or disciplinary level, of purely scientific groupings of practitioners.) Finally, method enables scientists and their apologists to associate science with or distinguish it from "broader social and political ideologies."[5] Whereas riding on ideological coattails can foster the collective aims of a scientific community, such linkages also can reflect the personal biases of individual scientists. Thus, while Schuster and Yeo focus on public initiatives serving the scientific collective, we must remember that such initiatives often entail private agendas stemming from personal social and cultural beliefs. Though scientists often curb their individual moral judgments in deference to the collective enterprise,[6] private opinions, we will see, also find ready justification in the language of scientific method.

Schuster and Yeo borrow the label for their third level, "public science," from historian Frank Turner. Turner explains that because science exists in a dialectical relationship with the broader society and culture, scientists must justify their pursuits to the political leaders and other persons who control essential authorizations and resources. "The body of rhetoric, argument, and polemic produced in this process may be termed *public science*, and those who sustain the enterprise may be regarded as *public scientists*." Though Turner does not concentrate only on methodological statements and though he confines his study to Britain from 1800 to 1919, he still provides support for Schuster and Yeo's analysis of methodological rhetoric and, in turn, an analysis of Newcomb in his American setting. He does this by detailing the wide range of activities in which public scientists engage. The activities include "lobbying various nonscientific elites, persuading the public or government that science can perform desired social and economic functions, defining as important those public issues that scientists can address through their particular knowledge or expertise, stressing professional standards among scientists, and defining the position of scientists vis-à-vis other rival intellectual or social elites, such as the clergy." This latter activity, the public demarcation of science from rival fields, has also caught the attention of sociologists of science. Thomas Gieryn and his sociological associates concentrate on the "boundary-work" done by American scientists at the science-religion interface and how such efforts at demarcation contribute to the scientists' collective professional development. These scientists, to legitimate their claims to public support and "cognitive authority," depend on a rhetoric that presents delimiting but self-serving images and ideologies of science, including ones involving method. That is, they depend on such rhetoric to further two professional goals: "to justify enlarged investments in scientific research and education, and to monopolize professional authority over a sphere of knowledge in order to protect collective resources of scientists."[7]

The first level on which scientists make methodological pronouncements, the internal level of technical argument, is more complicated — and controversial. Schuster and Yeo must explicitly confront the issue, mentioned earlier, of if and how method contributes to the generation of new knowledge and, thus, possibly plays more than either a public or an institutional and disciplinary role. They grant the possibility that "methodological discourse can be partially constitutive of knowledge claims in science." Nevertheless, they insist that "scientific argument is essentially persuasive argument and therefore is rightly termed *rhetorical* in the sense defined by students of 'the new rhetoric,' where 'rhetoric' denotes the entire field of discursive structures and strategies used to render arguments persuasive in given situations." In other words, even if the use of method contributes to the actual generation of scientific knowledge, it remains a rhetorical use to the extent that all scientific discourse involves persuasion and advocacy. Indeed, among the "new rhetoricians" who are studying the conduct of scholarly inquiry in general, there is a circle whose members assign a universal epistemic role to rhetoric. They maintain that all knowledge claims are rhetorical constructs, even supposedly objective claims involving "logic" and "facts."[8] Scientists' rhetorical structures and strategies have ranged from imagistic and metaphorical modes of expression to appeals, at least through earlier centuries, to seemingly irrefutable rules of logic such as in induction as well as to supposedly indisputable sages such as Francis Bacon, deferentially cited as *Lord* Bacon. To say that such forms of argumentation, expression, and appeal are rhetorical is neither to disparage them as superficial oratory nor to deny their value within the community of practicing scientists; it is merely to highlight the interpersonal, persuasive aspect of method-based scientific inquiry. The language of method, with its appeals to logic and lords, fills rhetorical functions by providing scientists with authority of voice.

NEWCOMB AND METHODOLOGICAL RHETORIC

In this book, aiming to gain entry into Newcomb's life and time, I adopt Schuster and Yeo's twofold approach to studying the history of methodology. Bearing in mind that the two avenues of inquiry intertwine, I examine both the historical context in which Newcomb devised his particular method and the actual rhetorical uses he made of the method on the internal, disciplinary, and public levels. I thus further illustrate the degree to which methodological rhetoric is historically contingent in both origin and application. In the course of this process, I also further illuminate three other topics of increasing specificity: the roots of American pragmatism, intellectual and cultural developments pertaining to science in the Gilded Age, and Newcomb's career as an active advocate of science. In other words, a study of the origins and rhetorical uses of Newcomb's methodological views provides a vehicle for examining not only scientific method but also pragmatism, late nineteenth-century American culture, and ultimately Newcomb's life and thought—the latter being the initial stimulus and main focus of the entire study.

About the origins of his view, I conclude that Newcomb did not arrive at his view by simply tapping into a universal body of methodological rules that somehow existed independently of outside influences. To be sure, he had links to an ongoing philosophical tradition of empiricism and its latest offshoots, particularly positivism, but the tradition was diverse and constantly shifting and, as I show, professional and personal circumstances of a passing nature tempered his response to it. He ended by emphasizing and elaborating one strain of the various empirical and positivist philosophies of science: the strain concerned with method and language. For Newcomb, the essence of scientific method lay in the appropriate use of language, of terms and propositions. In turn, the proper use of language entailed adherence to the basic precepts of empiricism. This combined linguistic and empirical focus resulted in one central methodological rule: scientists, in probing nature, should use only those concepts that can be derived from or explicated in terms of sense experience. More specifically, this rule translated into what Newcomb called the "scientific use of language," meaning in its simplest form that scientists must be able to relate all their concepts ultimately to particular sensory objects.[9]

Though Newcomb insisted that scientific concepts show empirical warrant, he avoided strict adherence to the empirical method commonly attributed to Francis Bacon; instead, he favored a hypothetico-deductive approach to scientific inquiry. That is, like most practicing scientists in the second half of the nineteenth century, he had renounced a Baconian insistence on following inductive rules of discovery and had moved on to a concern with the assessment and verification of scientific hypotheses, whatever their origin. Unlike the Baconians, he granted a large role for hypothetical and unobservable entities in science, requiring only that the concepts specifying the entities somehow relate (preferably through quantitative measurements) to sense experience.[10] As we will see, this vision of scientific method—based on a linguistic, empirical foundation within a hypothetico-deductive framework—both joined Newcomb to and separated him from prevailing empirical methodologies, specifically the positivist outlooks of Europeans such as Auguste Comte and John Stuart Mill and the nascent pragmatic outlooks of Americans such as Chauncey Wright and Charles Peirce.

Regarding rhetorical use of the rules, I conclude that Newcomb's particular perceptions of his professional environment in the United States during the late nineteenth century led him to a particular application of the rules. A scientist's professional environment includes both conceptual and institutional aspects. For example, a researcher must contend with ideas on theory, experiment, instrumentation, and method as well as arrangements regarding education, publication, and funding. These two aspects are, in turn, tied to the broader culture in which a scientist lives, including religious, philosophical, political, and economic elements. Newcomb's perceptions of his conceptual and institutional setting, although sometimes embellishing reality, were based on existing circumstances in the United States, especially in astronomy, physics, and mathematics, between the late 1850s when he began his formal scientific studies and the late 1890s when he officially retired. Newcomb, and most other American physical scientists whose careers peaked during this period punctuated by the Civil War, believed that the community's conceptual foundation was secure while the institutional framework was insecure. That is, as I have discussed elsewhere, they agreed—generally—on the conceptual core of physical science. While sometimes differing on particulars, most recognized that research programs based on various elaborations of classical mechanics offered reasonably sure opportunities for making important breakthroughs.[11] Yet, as historians from Daniel Kevles through Robert Bruce have demonstrated, these American physical scientists felt themselves burdened by an institutional framework that not only was embryonic but also had been shaken by the Civil War. They sensed that their fledgling university programs, professional societies, scientific journals, and other institutional supports were deficient but promising, as were arrangements for funding and for the recruitment and retention of scientists. "In science as in other matters," Bruce comments in his overview of American natural science, "the nineteenth century was a time for organizing."[12] This likewise was true, perhaps even more so, for the social sciences, including Newcomb's area of intermittent interest, economics. As the century waned, American social scientists were also groping toward professionalization.[13]

Newcomb's sense of his conceptual and institutional setting led him to use his linguistic, empirical methodology on primarily the public level, that is, on Schuster and Yeo's third rhetorical level. He was anxious to strengthen the institutional infrastructure of American natural science so that practitioners could take fuller advantage of the opportunities offered by the flourishing research programs of the day; consequently, he tried to cultivate deeper public support of science in the United States by convincing politicians and other influential citizens of the civic value not merely of the science itself but of its method. He sought to achieve this expansion of science's cognitive authority through precept (simply by extolling the merits of method and its potential public contributions) as well as example (by actually applying the method to public issues). Attempting to solidify institutional gains, foster the further professionalization of the scientific community, and advertise the distinctiveness of scientists' contribution to American society and thought, he also used method to demarcate the external boundaries of science; he distinguished it from potential competitors such as philosophy, theology, and lay social analysis, and, when useful, associated it with business and other nonscientific realms of culture. Similarly, he turned to method to rebuff philosophical pundits who, through their reappraisals of fundamentals in fields such as physics, seemed to be tarnishing the objectivist image of the natural sciences.

While taking these various public stands, Newcomb also used the rhetoric of method to associate science with broader social and cultural ideologies. In addition to bolstering the public image of science, such ideological linkages often promoted Newcomb's personal beliefs. Indeed, Newcomb's personal convictions and biases complicate our understanding of his initiatives on the public level. Often in his methodological pronouncements, he was speaking not only for "American science" but also for himself, thereby mixing two agendas: one for science as a collective enterprise (Schuster and Yeo's primary concern) and the other for himself as an individual citizen-scientist seeking to serve society at large. That is, he frequently linked his efforts to improve the social relations of science with his efforts to promote personal political, economic, religious, philosophical, and pedagogic convictions. Outspoken and often openly partisan, he inserted himself in this dual manner into a variety of public debates. For example, while demonstrating method's usefulness in clarifying political issues, he also was using method to justify his liberal ideology. That Newcomb often had a personal stake in manifestly collective campaigns alerts us to a related point: though he sometimes followed an intentional rhetorical strategy, this dedicated and ingenuous advocate of scientific method often seemed to be introducing the rhetorical tactics into his discourse unconsciously—or, at least, without explicit acknowledgment. This applies equally to his initiatives as a spokesman for the scientific collective and as an individual citizen-scientist attempting to promote personal beliefs. Whether using the tactics knowingly or unknowingly, he displayed no cynicism and employed no chicanery in his various and often intermixed initiatives, believing fully in the virtues of scientific method and in the rightness of what he was doing.

Only occasionally did Newcomb use methodology on Schuster and Yeo's second rhetorical level—the disciplinary. When he did use method as a means to achieve the end of "institutional control and disciplinary hegemony," he usually restricted his attention to the social sciences and other research areas outside what was considered the mainstream of science. Seeking to maintain scientific rigor in the study of political economy, for example, he tried through the criterion of method to distinguish those who supposedly conducted legitimate scientific inquiries from those meddlers who did not.[14] Likewise, he tried to discriminate between those "scientists" committed to the Christian tenets of natural theology and those who, in his opinion, more appropriately eschewed any connection between natural phenomena and God's possible design in nature. Even less often did he deploy methodological rhetoric on Schuster and Yeo's level of internal, technical argument—the first level. Generally confident of the conceptual core of natural science, he seldom turned to method to advance or appraise contemporary scientific ideas or programs. When he did, he deployed the rhetoric mainly in areas that were at the fringe of conventional scientific practice, areas such as psychical research.

Though we find examples of Newcomb carrying method beyond the public level to both the disciplinary and internal levels, we realize that even his activities on the latter two levels resemble "public science." That is, we can equally construe his occasional disciplinary and internal initiatives as being variants of his more common public involvements. In particular, his methodological incursions into fields such as political economy and natural theology served as much to disassociate scientists from nonscientists affiliated with those fields (an outer-directed demarcation on Schuster and Yeo's public level) as to disassociate one faction of practicing scientists from another (an inner-directed demarcation on the disciplinary level). After all, to charge that certain practitioners were not doing legitimate science (an inner-directed demarcation) was, in effect, to charge that they were either pseudoscientists or nonscientists (an outer-directed demarcation). By expelling outsiders from a field such as political economy, he was gaining a greater public role for "science" as he construed it; said differently, he was extending the cultural and cognitive sway of those scientists having the requisite credentials. Similarly, his faultfinding forays into marginal sciences such as psychical research served as much to protect the popular image of science (on the public level) as to clarify the fundamentals of the fields (on the internal level).

In this manner, the combination of confidence in the conceptual foundation and ambition for the institutional framework gave Newcomb's rhetoric a scientistic or evangelical flavor. Certain that the successes of contemporary physical science reflected the use of sound method, he proclaimed the doctrines of scientific method with missionary zeal, directing his exhortations primarily outward to lay people rather than inward to the community of believers. Of course, though fancying himself to be proclaiming the true method of science, he was actually proclaiming his notion of scientific method. We need to remember the warning that scientism is not a campaign to colonize the humanities and social sciences with the methods of the natural sciences but a campaign to colonize them with what are construed to be the methods.[15] And regardless of the degree to which Newcomb's campaign ultimately benefited the American scientific community—and perhaps even the public well-being, as he also intended—it did serve to carry scientific thinking further into the mainstream of American culture. Indeed, Newcomb contributed to a broader cultural embrace of the values and methods of science that ultimately helped catalyze American pragmatism.

A word of caution is in order regarding the historical reconstruction of the origins and rhetorical uses of Newcomb's methodological views. Whereas throughout his career Newcomb emphasized the centrality of method in science, he seldom achieved in his pronouncements the consistency and coherence of academic philosophers. As a practicing scientist, he was somewhat eclectic in his statements, incorporating ideas from various sources. Similarly, he seldom employed the distinctions or categories of formal philosophy. He usually relied on lay language and appealed to common knowledge in his discussions of method. And frequently, he mixed methodological matters with either ontological, epistemic, or axiological issues. Although methodology is intimately related to these other realms of thought, Newcomb usually discussed the various realms without using or observing the analytic distinctions of academic philosophers. Again as a practicing scientist, he laced his discussions of procedural rules with statements concerning, for example, the ultimate constituents of the physical world, the nature of scientific knowledge, and the aims of scientific inquiry.

That Newcomb was primarily concerned with methodological rules, procedures, and criteria becomes evident, however, from his specific writings and speeches. Thus, when he invoked the seemingly comprehensive phrase "scientific philosophy," his accompanying discussion reveals that he was referring to a viewpoint ultimately grounded on method.[16] As we will see, he presented critiques and analyses that usually could be translated into investigational imperatives and directives. The origins of and uses to which he put these imperatives and directives tell us much about not merely Newcomb but also late nineteenth-century culture, American pragmatism, and the rhetoric of scientific method. They inform us of his broader cultural and intellectual ties because, though he probably possessed certain innate propensities, he developed his particular "scientific philosophy" only over time and only through dialogue and experience. As Chauncey Wright commented in 1873, "Positivists, unlike poets, become—are not born—such thinkers."[17]

PART II

Newcomb's Life and Thought

CHAPTER II

Formative Years

Simon Newcomb grew up in a region where, even in the mid nineteenth century, books were scarce enough to be treasured possessions. Years later, he could still rattle off the titles and recount the texts of the two dozen books that he chanced on as a child. Of course, not all the children who were reared alongside Newcomb in the small towns and villages of Nova Scotia placed such a premium on the few books around them. Young Newcomb was, as he later bemoaned, something of a *lusus naturae*, a sport of nature, an eccentric given over to the life of the mind.[1] In his autobiography, he titled the chapter on his earliest years "The World of Cold and Darkness," an allusion to the discomfort he felt as a lone bookworm in a rustic society of farmers and laborers. Only in his late teens, when he discovered the communities of scientists working in Washington, D.C., and Cambridge, Massachusetts, did his sense of oppression lift. Only then did he enter, as he titled a later chapter in his autobiography, "The World of Sweetness and Light."

Though he felt that he had escaped the benighted villages of his childhood and reached cosmopolitan havens of learning, he never seemed to overcome a sense of not belonging. Whereas in Nova Scotia he had been a highbrow among rustics, in Washington and Cambridge he found himself a rustic among highbrows. Even after rising to a station that permitted him to move with ease through the "World of Sweetness and Light," he still carried a burden: the self-doubt of his younger years. He constantly needed to prove himself. Whether making another exhaustive analysis of a planetary orbit or assuming another exhausting presidency of a learned society, he seemed compelled to achieve.

GROWING UP IN THE MARITIME PROVINCES

Newcomb was born in Wallace, Nova Scotia, in 1835, a year notable for the return of Halley's Comet. (In a fitting coincidence, the astronomer died in 1909 on the eve of Halley's next visit, two months before his colleagues sighted the comet's blur in their telescopes.) Though born outside the United States, Newcomb was, as he later expressed it, "of almost pure New England descent."[2] Indeed, on both his mother's and father's side, he could trace his New England forebears back four or five generations, to well before the Revolutionary War. While this was old Yankee stock, it was not particularly distinguished stock. Certainly, his ancestors were respectable: his paternal grandfather (the fourth in a line of Newcombs named Simon) owned a stone quarry, while his maternal grandfather (Thomas Prince) was a magistrate. Nevertheless, Newcomb began life in modest circumstances. His father, John Newcomb, eked out a living as a country schoolteacher, forced to move every year or two to find pupils in the thinly populated outlands of Nova Scotia and Prince Edward Island. Consequently, young Newcomb spent his first sixteen years without secure community roots. In addition, to help support the family, which eventually included six younger brothers and sisters, he lived away from home for part of each year doing chores on neighboring farms or in households. At age sixteen, he left home permanently and apprenticed himself to a local doctor. Also about this time, his mother, Emily Prince Newcomb, became seriously ill; a woman apparently of strong intellect but weak health, she died soon thereafter, at age thirty-seven.

Newcomb was not merely a precocious child but an obsessive one when it came to learning. When his father and other relatives introduced him to reading, writing, geography, and arithmetic, he immersed himself in the subjects with the intensity that most other four- to six-year-old children reserve for play. His father recalled that for a period around age four and a half, Simon spent hours on his own each day doing arithmetic. Of course, his father, a teacher whom Newcomb later described as "the most rational and the most dispassionate of men," contributed to this fixation on learning. "One result of my father's occupation was," Newcomb recalled, "that I breathed, in early childhood, an atmosphere which had at least the scent of learning. The spelling book was more familiar than the plow, and the idea that there was a correct way of using language was acquired at as early an age as if we had lived in cultivated society." Besides receiving this informal education, Newcomb also periodically attended his father's school, with his father personally drilling him on the day's lessons. The upshot was that by about age six and a half, he was able to read the Bible and calculate simple cube roots.

All this study at an early age apparently took its toll. Although he did not remember the incident, Newcomb had a "mental breakdown" at age six and a half. John Newcomb confirmed this, decades later, in a long letter of reminiscences that he sent to his son: "You had lost all relish for reading, study, play, or talk. Sat most of the day flat on the floor or hearth.... From the time you were taken down until you commenced recovery was about a month.... [Then] after a few weeks I began to examine you in figures, and found you had forgotten nearly all you had ever learned."[3] Supposedly, this breakdown convinced John Newcomb to shelter Simon from formal schooling with its rigid procedures, assignments, and examinations. The fear that his son might again lose himself in overstudy led him to find work for the boy on neighboring farms, an arrangement also intended to strengthen his son's frail body and improve the family's financial standing. For years to come, Newcomb remained without ongoing, formal schooling. "I attained the age of twenty-one," he was fond of pointing out, "without meeting a college professor, or any one else who could give me help or advice in the pursuit of my studies beyond the point where parental guidance ceased."

If Newcomb's rationalistic father left a mark on his early development, an equally strong but apparently antithetical mark was left by his pious mother and other religious authorities. Baptized as an Episcopalian but raised as a Baptist under the doctrinaire eye of his mother, young Newcomb learned what he later loosely called "the old Calvinistic orthodoxy in its gloomiest form." As he elaborates in one of the drafts of an unpublished "Religious Autobiography": "I was born in the country before the leaven of 'liberalism' had been felt far outside the great cities, and bred in a church which neither glazed over nor softened down the beliefs of the New England Puritans from whom it derived its strength." More specifically, he recounts: "Preachers dwelt with especial emphasis upon the doctrines of the innate depravity of man, the dreadful future of the unbelievers, the futility of mere good works, and the necessity of being born again. The most familiar texts of scripture were those which described weeping, wailing and gnashing of teeth." Taking these teachings literally, but yet profoundly aware that he had not personally experienced a "conversion" and been "born again," he developed deep anxieties about an imminent Judgment Day and his consignment to hell. He relates that, from about age ten through fourteen, "I suffered untold terrors before my prospective fate in the next world. The only hope of escape rested upon the chance of ultimate conversion." Try as he might, he remained unconverted. Concluding that God showed favor and charity to others but not to him, he experienced an "overwhelming force of self-condemnation."[4]

Around age fifteen, he began to distance himself from orthodox Christian doctrines and to lose his fear of divine judgment. His attitude toward Christian teachings turned from "active belief" to "indifference." The shift reflected in part a callousness developed through years of exposure to the severe and frightening doctrines. It also reflected his growing awareness of differences between the spheres "of sacred and profane history." It dawned on him "that Heaven, instead of being a visible place, of which the sky was the floor, was a place from which no news had been received for eighteen centuries." In other words, he began to realize that he did not have to respond to biblical teachings with the same literal-mindedness that he responded to facts about, for example, geography.

Books kindled Newcomb's adolescent awakening. Two in particular that triggered "qualms of conscience" regarding his coarse religious training were also favorites of John Newcomb. Being prized by his father, they likely reinforced the tension between his mother's Christian piety and the views of his father—"the most rational and the most dispassionate of men." The two books were phrenology texts: George Combe's *The Constitution of Man Considered in Relation to External Objects* and Orson Fowler's *Phrenology*. Newcomb later explained the impact that these texts had on him when he first studied them between the ages of ten and fourteen: "It may appear strange to the reader if a system so completely exploded as that of phrenology should have any value as a mental discipline. Its real value consisted, not in what it taught about the position of the 'organs,' but in presenting a study of human nature which, if not scientific in form, was truly so in spirit. I acquired the habit of looking on the characters and capabilities of men as the result of their organism." Man, Newcomb read, was a product of natural law. Particularly influential in reinforcing this naturalistic message was the *Constitution of Man*, by Combe, an early nineteenth-century Edinburgh lawyer and phrenologist. Extremely popular, this wide-ranging work combined a rationalistic view of mental action with nineteenth-century liberalism—espousing, as Newcomb would in later years, individual freedoms, property rights, and free trade.[5] Newcomb knew that his father respected this book so much that he had based his philosophy of life on it: indeed, John Newcomb had courted and married Emily Prince as the culmination of a systematic search for a mate, all in accord with phrenological principles. Simon himself took from the book the lesson that "all individual and social ills were due to men's disregard of the laws of Nature, which were classified as physical and moral. Obey the laws of health and we and our posterity will all reach the age of one hundred years. Obey the moral law and social evils will disappear." As thinkers from at least the time of the Enlightenment had found, and as Newcomb's own father and mother likely demonstrated, this affirmation of natural law was potentially at odds with the orthodox Christian contention that God was directly responsible for provident care of his creations.

Though the books that came his way were sparse and sometimes suspect, Simon persisted with each one until he had drained it of useful information. Doggedly, he extended his intellectual grasp. The farmers with whom he lived sometimes loaned him books, which he recalled reading "by the light of the blazing fire in Winter evenings." His main source of books, however, was his father, who in turn tapped his own father's modest library. These "ancestral volumes" included three books relevant to mathematics and science: an eighteenth-century popularization of algebra, an edition of Euclid's *Elements*, and a standard treatise on navigation. Using the first book, his father started teaching him algebra when he was about twelve; this led by the next year to study of the Euclid volume. John Newcomb recalled that his son was "enraptured" with algebra and geometry: "The pleasure of intellectual exercise in demonstrating or analyzing a geometrical problem, or solving an algebraic equation, seemed to be your only object." So delighted was Simon with his first exposure to Euclidean geometry that he immediately lectured his younger brother Thomas on the proof of the Pythagorean theorem. As for the third volume, on navigation, John Newcomb further recollected that his son had an "almost intuitive knowledge of geography, navigation, and nautical matters in general."

Astronomy and physics also lured Newcomb. When he was about ten, he "read with avidity" an old astronomy book that his father had obtained, astronomy being one of John Newcomb's long-standing interests. (Soon thereafter, during an evening lecture on astronomy that his father was giving at their home, the young boy amused the audience by correcting his father on a point; John Newcomb, not amused, consequently consigned his son to the household of an Anglican rector, where he was to do chores and tutor the rector's two sons.) And while his father tried to teach him the rudiments of physics, Simon ended by instructing himself through a chance encounter with a book that belonged to one of his father's students; he recalled that, as a twelve year old,

> one day after school I saw lying on the desk of one of the scholars, an unusual-looking book, which proved to be Mrs. Marcet's "Conversations on Natural Philosophy." I devoured it in a very few days, by stealthily making my way into the school house after hours. Never since have I tasted such intellectual pleasure as was offered by this first insight into the mysteries of nature.

He continued his education in physical science by reading other books that his father brought home and by listening to occasional traveling lecturers who demonstrated marvels such as electricity.

John Newcomb remembered his son as being a stickler for truth. "I never knew you to deviate from it in one single instance, either in infancy or youth." The elder Newcomb also remembered him as being thoroughgoing and punctilious in his pursuit of new knowledge. Thus, in a description that foretold Simon's adult preoccupation with exact definitions of terms, John Newcomb wrote: "An extraordinary peculiarity in you was never to leap past a word you could not make out. I certainly never gave you any particular instructions about this, or the fact itself would not at the time have appeared so strange to me." In a telling sentence from an unpublished reminiscence, Newcomb himself observed that as a teenager he already displayed "an inborn tendency to interpret statements with literal exactness." Newcomb's oldest daughter, Anita Newcomb McGee, would concur with her father's self-appraisal. After his death in 1909, while reading the manuscript containing the appraisal, she marked the sentence with a marginal bracket and penciled in: "How characteristic of Papa!"[6]

As devoted as he was to learning, young Newcomb was discontented with his life. Years later, he explained why: "Notwithstanding the intellectual pleasure ... my boyhood was on the whole one of sadness. Occasionally my love of books brought a word of commendation from some visitor, perhaps a Methodist minister, who patted me on the head with a word of praise. Otherwise it caused only exclamations of wonder which were distasteful." In an even blunter comment, he attested that he would not want to live his childhood over again.

At age sixteen, Newcomb went to live with his maternal grandfather in Moncton, New Brunswick, and faced the choice of becoming either a carpenter or the apprentice to a locally prominent practitioner of "the botanic system of medicine." Although Newcomb lacked a prior interest in medicine, he was impressed by the doctor's charming manners and intellectual airs. He agreed to become the doctor's assistant until age twenty-one in exchange for room, board, clothing, and medical training. The doctor lived in the nearby village of Salisbury when Newcomb moved in with the doctor's family.

Newcomb was soon disillusioned with the herbalist, whose true personality turned out to be cold, and whose actual knowledge of medicine was nil (his remedies were typically laced with opium). Whereas he had believed that he would serve as the doctor's student and assistant, he discovered that his actual role was one of drudge: "I cared for the horse, cut wood for the fire, searched field and forest for medicinal herbs, ordered other medicines from a druggist in St. John, kept the doctor's accounts, made his pills, and mixed his powders." Among the few books that the doctor grudgingly supplied to his apprentice was Fowler's *Phrenology*—a text that Newcomb had already mastered. Newcomb did, however, salvage time to carry his own education into new areas; he privately studied medicine, Latin, and chemistry (using the textbook by John Draper) as well as participated in local cultural groups.

"As time passed on," Newcomb further recalled, "the consciousness that I was wasting my growing years increased." Eventually, after enduring the apprenticeship for two years, he decided to run away. Leaving a letter explaining that the doctor had not fulfilled their agreement, he slipped away before daybreak on a September day in 1853. He walked until after dark—some fifty miles through New Brunswick—wanting to be sure to elude the doctor, whom he correctly feared would pursue him. Whereas in the process he lost "all the books of my childhood which I had, as well as the little mementoes of my mother," he later characterized this action as the "most momentous" and this day as the "most memorable" of his life. Traveling with little money and no possessions, he slowly made his way to the United States. Here he reunited with his father, who after the recent death of Simon's mother had settled in Maryland.

ENCOUNTERING JOSEPH HENRY

Newcomb celebrated his nineteenth birthday in March of 1854 as a teacher in a Maryland country school, "a log school house the floor of which was very near the ceiling, and the four walls very near each other." His father found him this job, which paid an annual salary of $250, in the rural community of Massey's Cross Roads. The following year he went on to a better school a few miles away in Sudlersville. He moved yet again during his third and final year in Maryland, assuming the position of private tutor for the children of a planter in Prince Georges County. Still lacking any outside guidance in his own education, he continued in his free time to read voraciously and eclectically. From his father (who in late 1856 moved with two of Simon's younger brothers to Minnesota), he received a copy of William Cobbett's *A Grammar of the English Language*, a book designed to help common folk, "Soldiers, Sailors, Apprentices, and Plough-boys," improve themselves intellectually. John Newcomb based his philosophy of education on the teachings of Cobbett, an outspoken, early nineteenth-century British journalist and social reformer who lived briefly in the United States. Like his father before him, Simon must have noticed Cobbett's denunciation of rote learning: "Never attempt to *get by rote* any part of your instructions. Whoever falls into that practice soon begins to esteem the powers of *memory* more than those of *reason*; and the former are despicable indeed when compared to the latter." And he must have read about the critical importance of clear expression, advice that he would later incorporate into his distinctively linguistic notion of scientific method. According to Cobbett, grammar "teaches us *how to make use of words*; that is to say, it teaches us how to make use of them in a proper manner." Specifically, grammar "enables us, not only to express our meaning fully and clearly, but so to express it as to enable us to defy the ingenuity of man to give to our words any other meaning than that which we ourselves intend them to express. This, therefore, is a science of substantial utility."[7] To polish his foreign language skills, Newcomb also turned to another of Cobbett's self-help books, *A French Grammar, Or, Plain Instructions for the Learning of French.*

A volume by Jean-Baptiste Say, an early nineteenth-century French exponent of the economic doctrines of Adam Smith, provided Newcomb with his initial taste of political economy. Newcomb recalled that "Say's [A Treatise on] Political Economy was the first book I read on that subject, and it was quite a delight to see human affairs treated by scientific methods." What particular method did Say follow? He developed his analysis, he told his readers, in strict accord with classic Baconian inductivism: "The excellence of this method consists in only admitting facts carefully observed, and the consequences rigorously deduced from them; thereby effectually excluding those prejudices and authorities which, in every department of literature and science, have so often been interposed between man and truth." As Cobbett had done regarding grammar, Say also reminded his readers of the importance of clear expression in the "science" of political economy: "The utmost precision must be given to the phraseology we employ, so as to prevent the same word from ever being understood in two different senses."[8]

While in Maryland, Newcomb also continued to read books on Christianity that reinforced his awareness of the differences between the sacred and profane worlds—the latter including the world of natural science and mathematics, in which his interest was burgeoning. Analytic geometry and calculus were among the technical topics that he studied, turning to both English and French textbooks. Concluding that "mathematics was the study I was best fitted to follow," he went so far as to obtain an English translation of Isaac Newton's *Principia* and an issue of the *American Journal of Science.* Though he struggled through both, he soon realized they were beyond his ken.

Always a ravenous reader, Newcomb developed during his Maryland years into an equally prolific writer. While teaching at Massey's Cross Roads, he drafted a lengthy manuscript, intended to be his first book. "Essay on Human Happiness" is a multichapter, didactic tract on the nature of happiness and how it is acquired, beginning with advice on child rearing. Though neither completed nor submitted to a publisher, the manuscript reveals that Newcomb at age nineteen was already a skilled writer. It also reveals glimmerings of his later methodological imperative that language must be grounded in sensory experience. Thus, in a section of the book on educating children, he presents this pedagogic maxim: "Every rule and every principle a knowledge of which is to be acquired by the pupil should, if possible, be *illustrated* by some suitable method." "I regard this as one of the most important means of education," he proceeds to explain, "yet in our common schools it is almost entirely neglected. The element of all our knowledge is in the first place obtained through the medium of the senses, and knowledge which we ourselves derive through this medium is far more durably impressed upon our minds than if we received it from another person."[9] In other words, the best way to teach a principle is to relate it to concrete, sensory experiences.

The next year, while at Sudlersville, Newcomb also tried his hand at mathematical and scientific writing. Drawing on his private studies, he produced a paper titled "A New Demonstration of the Binomial Theorem." Having by this time visited Washington, D.C., and learned of the Smithsonian Institution, Newcomb sent his mathematical demonstration to no less a figure than Joseph Henry (1797–1878), Smithsonian secretary and laureate of American physical science. Newcomb sought Henry's advice on the merits of the paper and its suitability for publication. Henry, after asking the opinion of a colleague in mathematics, responded with both reserve and encouragement. "Though the opinion expressed is not in all points as favorable as you may have expected," Henry demurred, "still it is sufficiently so to assure you that you may do something more important by a persevering application of your talents."[10] During the same year that he corresponded with Henry, Newcomb also applied his talents to answering the author of a letter in a Washington periodical, the *National Intelligencer*, who purported to refute Copernicus's theory. Submitting a rebuttal and subsequently seeing his name in print — "in large capitals, in a newspaper" — Newcomb gained further encouragement to pursue his mathematical and scientific interests.[11]

After moving to Prince Georges County to serve as a private tutor, Newcomb began to frequent the library of the Smithsonian Institution in nearby Washington. Increasingly drawn to mathematical astronomy, he located in the library one of the classics of the field. "Here I was delighted to find the greatest treasure that my imagination had ever pictured,—a work that I had thought of almost as belonging to fairyland. And here it was right before my eyes—four enormous volumes,—'Mécanique Céleste, by the Marquis de Laplace, Peer of France; translated by Nathaniel Bowditch, LL.D., Member of the Royal Societies of London, Edinburg, and Dublin.' " Soon thereafter, he personally encountered another treasure of the Smithsonian, Professor Henry, with whom he previously had merely corresponded.

Of his first interactions with the young tutor, Henry later stated: "I was not only impressed with his remarkable intellectual abilities, but, also with the correctness of his moral deportment. He, evidently, knew little of the world and was struggling manfully in the battle of life."[12] These interactions were the beginning of an enduring friendship with a man who, Newcomb would say in a later autobiographical sketch, "encouraged and promoted my advancement in a way which will make his name ever remembered by me and my children." In the course of the friendship, this veteran scientist also certainly promoted Newcomb's understanding of the character of scientific inquiry. Though he had established himself as an experimental scientist through studies on magnetism and electricity, Henry valued and encouraged the use of mathematics in science. He also disdained a naive Baconian method, advocating instead the use of hypothesis in research.[13] Thus, looking just at Henry's public speeches and writings from the 1850s, we find him repeatedly downplaying "the mere accumulation of facts" and heralding hypotheses as "the great instruments of discovery." "Indeed, in the investigation of nature," he explains in a typical passage, "we provisionally adopt hypotheses as antecedent probabilities, which we seek to prove or disprove by subsequent observation and experiment; and it is in this way that science is most rapidly and securely advanced." We also find him stressing the fixed and invariable character of natural law, which he attributes to the governance of all phenomena by "a Supreme Intelligence, who knows no change." Even when pondering seemingly random phenomena, such as meteorological events, Henry discerns "a permanency and an order." Though convinced of nature's regularity and the power of hypotheses to enlarge our understanding of the regularity, he warns that absolute truths—God's truths—are ultimately unknowable to mere mortals. Human faculties are simply too limited to comprehend the full complexity of nature.[14]

Henry advised young Newcomb to contact the U.S. Coast Survey about the possibility of obtaining a suitable technical job. Now devoting his private time to refining the Cavendish method of determining the earth's density and using the tables of an ephemeris to make astronomical calculations, Newcomb eventually met geophysicist Julius E. Hilgard at the Washington office of the Coast Survey. While unable to employ Newcomb at his own office, Hilgard arranged for the bright, twenty-one-year-old tutor to assume the position of "computer" under the direction of astronomer Joseph Winlock at the Nautical Almanac Office in Cambridge, Massachusetts. Congress had established this government agency, under the control of the secretary of the navy, seven years earlier. Becoming a computer meant joining a small group that performed the routine mathematical calculations for various lunar and planetary tables useful in navigation and astronomy—tables published in the *American Ephemeris*. Newcomb traveled to Cambridge in late December, 1856, and soon began his new work: "I date the fruition of my hopes, my actual citizenship of the world of my childish dreams and youthful aspirations, from one frosty morning in January, 1857, when I took my seat before a blazing fire in the 'Nautical Almanac' office." Indeed, Newcomb had chanced onto the nation's leading scientific metropolis, Boston-Cambridge, the nation's largest telescope, the fifteen-inch refractor at the Harvard Observatory, and the nation's leading mathematical astronomer, Benjamin Peirce.[15]

Pleased that Newcomb had taken the Cambridge job, Henry wrote to assure him that he would certainly develop into "a distinguished as well as a good man." "Do not however be too anxious to obtain reputation," Henry went on to caution in words that Newcomb would often later paraphrase, "this comes most surely and pleasantly to those who deserve it by labouring unselfishly to enlarge the bounds of knowledge from the love of truth and the desire to be useful to their fellow men." Henry closed this letter by adding, "I shall ever feel a warm interest in your welfare." Newcomb reciprocated this warmth, and in succeeding years, as we will see, the two men remained close. In fact, Newcomb increasingly used Henry as a role model, particularly in his perception of himself as a spokesman for and advocate of American science.[16]

SCIENTIFIC CAMBRIDGE

Newcomb quickly distinguished himself at the Almanac Office. Within half a year of his arrival, according to a report submitted by Superintendent Winlock, the young computer showed "evidence of Mathematical talent and knowledge very unusual for his age and limited opportunities.... With his love for mathematics and his industry, he will in a short time be one of the most suitable assistants engaged in our work."[17] Happily, like his fellow computers who performed the routine calculations for the astronomical tables, Newcomb had much free time in his daily schedule of work at the Almanac Office. He allocated some of the hours to his favorite pastime, chess. (In a diary entry written after the office was visited by two eminent Harvard scientists, Newcomb sheepishly revealed: "Profs Peirce and Agassiz came into the office while I was playing chess with Edmunds." Eight months later, Newcomb recorded that the lax work schedule was being tightened slightly: "An order issued in the N. Alm. Office that the computers should hereafter work six hours per day."[18]) He also found amusement in philosophical discussions with his co-workers, especially Chauncey Wright.[19] In addition, he had enough open hours to enroll as a student of mathematics at Harvard's Lawrence Scientific School. Studying primarily under Benjamin Peirce, Newcomb was in a loosely structured program that required little formal course work.[20]

Benjamin Peirce (1809–1880) was, as Newcomb stated, "the leading mathematician of America." He had published complex theoretical studies of the positions and motions of the planets, moon, and comets, including mathematical analyses of errors of observation. Besides serving as a professor at the Scientific School, he used his skills in mathematics and astronomy as theoretical advisor for the Almanac Office. Newcomb had initiated a relationship with Peirce when, while still in Maryland, he sent the eminent Harvard professor a calculation concerning the earth-moon system; Peirce had asked to see the calculation (a foretoken of Newcomb's lifelong fascination with lunar motions) following a visit from Newcomb's father, John. After Newcomb became better acquainted with Peirce both in the classroom and at gatherings at the professor's home, he realized with pleasure that Peirce was "simple and unreserved as a farmer, and as cordial as if he had never known what dignity was." Peirce's relaxed style of teaching mathematics and astronomy, which included frequent digressions on the theological import of mathematics, reflected these personality traits. "The student was advised to attend the mathematical lectures that the professor delivered to the senior undergraduates," Newcomb recalled, "and to read La Place; outside of that he did what he pleased till he came up for final examination for a degree." Peirce's unrestrictive approach to teaching apparently worked for Newcomb, a young man well accustomed to directing his own studies; in the summer of 1858, he was "examined for the degree of S. B. and got passed for summa cum laude." To perform so well on this examination, Newcomb had first labored through numerous works on mathematics and physical science, including Peirce's own *System of Analytic Mechanics* and *Elementary Treatise on Curves, Functions, and Forces*. After graduating, he continued to draw on Peirce for scientific counsel.[21]

Remaining in Cambridge, Newcomb soon demonstrated that his skills in mathematical astronomy were developing well beyond the requirements for an assistant at the Almanac Office. Building on studies that he began in 1858, he crafted a precise analysis of the orbital motions of the asteroids, the small bodies orbiting between Mars and Jupiter. He demonstrated that the asteroids could not have originated, as was commonly believed, from the shattering of a single planet. Though he published his main results in 1860 in the *Memoirs* of the American Academy of Arts and Sciences, he rehashed the findings two years later in the German journal *Astronomische Nachrichten*. He later remarked that the asteroid study was the first of his research projects "to attract especial notice in foreign scientific journals."[22]

With free time at the Almanac Office and liberty to follow his own interests at the Scientific School, Newcomb developed his intellectual skills in areas besides physical science and mathematics. Pleased to discover a more liberal attitude toward a much more benevolent Christian faith than he had experienced as a youth, he began to delve deeply into Christian theology. "I had quite a taste for philosophical studies and discussions," he recalled in one of his unpublished religious reminiscences, "and became much interested in the evidences of christianity." He regularly attended Sunday services at the college chapel and at various Cambridge churches, ranging from the Swedenborgian to the Episcopal and Methodist. He especially frequented the services of college chaplain Frederick D. Huntington, a Unitarian minister who shifted to Episcopalianism about 1860, and Huntington's replacement, Andrew P. Peabody, whom Newcomb described as "nominally a unitarian, and doubtless a liberal." A serious student of religion, Newcomb systematically recorded the theme of each sermon he heard. (This practice, to his eventual amusement, allowed him to pinpoint each time a minister repeated a sermon over the course of a few years.) Eventually, he formed a close personal relationship with Peabody, often visiting his home. Hoping to gain Peabody's approval, he even "went so far as to construct a refutation of Hume's argument against the credibility of miracles." Newcomb also discussed religious issues with his fellow students; indeed, during most of his five years in Cambridge he roomed in Harvard's Divinity Hall (since there were not enough theological students to fill all the quarters).[23]

Books continued to bolster Newcomb's religious education. According to his diary, by spring of 1860, he counted 172 volumes in his private library, of which 21 fell in the category of "Logical and Religious." His personal "Alphabetical Catalogue of Books," dating from this period, reveals that the religious titles included classics such as Joseph Butler's *The Analogy of Religion, Natural and Revealed, to the Constitution and Course of Nature* and William Paley's *Natural Theology; or Evidences of the Existence and Attributes of the Deity*, both gifts to the aspiring scholar. So serious was his interest that, when he traveled into the wilds of the upper Saskatchewan River in 1860 on an eclipse expedition, the only book that he took with him was Butler's *Analogy*—"a choice which excited some merriment on the part of my companions." The upshot of all these various religious involvements during his Cambridge years was an increased sympathy for the Christian faith coupled to a residual inability to give himself over to the faith. Evidently, he was still trying to resolve the tension between his mother's Christian piety and his father's rationalistic naturalism. He had to admit, as he later expressed it, that even his intricate refutation of Hume's argument against miracles was a sham: "I must confess that my conscience pricked me when the question arose how much faith I put in my own argument. It said:—you do not construct this argument because you see into its soundness, but only to please Dr. Peabody as well as yourself, and to help yourself to become a religious believer." Similarly, though part of him wanted to join the college church, "the creed of which was of the most innocent kind," another part resisted: he had to acknowledge that "the aversion was still unconquerable." As for the books, although intrigued by their arguments and wanting to be a believer, he was unable to share the foundation of faith on which the authors built their cases. "These writers did not become believers through the evidence which they cited," he complained, "but they made up their minds beforehand that christian doctrines were true and then constructed their arguments to fit the case."[24]

Whereas Harvard's liberal Unitarianism seemed more benign than the strict Calvinistic Protestantism of Newcomb's childhood, it still sprang from the same root: New England Puritanism. And though Puritan theological doctrines ultimately deterred Newcomb from making a formal religious commitment, cultural values and intellectual dispositions associated with Puritanism might have triggered, or at least reinforced, certain tendencies in the young man. Admittedly, it is difficult to distinguish the elusive "Puritan temperament," but two of Newcomb's lifelong traits recall traits of the Puritan mind. He displayed an almost compulsive tendency for intense study and hard work. Also, he exhibited a moralistic but down-to-earth fixation on honesty and actuality—what George Santayana later aptly described in his novel *The Last Puritan* as "hatred of all shams, scorn of all mummeries, [and] a bitter merciless pleasure in the hard facts."[25] In many ways, Newcomb's life would be a Puritan pilgrimage of hard work and hard facts.

Besides extending his grasp of Christian religion during these Cambridge years, early 1857 until late 1861, Newcomb added to his understanding of traditional and contemporary thought on the nature of science and its relationship to philosophy and society. He recorded in his "Alphabetical Catalogue of Books" that, along with scientific and religious texts, he had purchased volumes such as George Boole's *Laws of Thought*, Richard Whately's *Logic*, Francis Bowen's *Political Economy*, and Thomas Reid's *Intellectual Powers of Man*.[26] Bowen was on the Harvard faculty, and Harvard president James Walker had edited the American edition of Reid's *Intellectual Powers*.

Reid's books along with those of his younger Scottish countryman Dugald Stewart had dominated American philosophy, especially that branch stressing mental or "psychological" processes. Thus, Walker had observed in the 1850 foreword to his edition of Reid's *Intellectual Powers*: "The psychology generally taught in England and this country for the last fifty years has been that of the Scotch school, of which Dr. Reid is the acknowledged head."[27] But for Americans such as Newcomb, Reid's Scottish Realism carried with it more than a "common sense" trust in sensory experiences and the reality of objects of the external world. It also carried an emphasis on Francis Bacon's inductive and empirical conception of science. Reid and Stewart had integrated Bacon's philosophy into their writings, retaining many of the master's precepts, but altering others. Thus, they held that the term "cause" lacks metaphysical import and signifies nothing more than an invariable connection between physical events; similarly, they felt that the goal of science is merely to describe phenomena and their connections.[28] Looking at early to mid nineteenth-century Protestants in the United States, it becomes evident that many Americans had embraced the resulting package, thus indirectly establishing Baconianism.[29] In addition, Baconian philosophy as mediated by Reid and Stewart, or at least lip service to the philosophy, came to prevail among American scientists active in the decades leading to mid century.[30] The vogue of Scottish Realism extended to the Harvard community of Newcomb's day. The community leaned particularly to the Scottish philosophy as joined to the theological values of Unitarianism.[31]

With a blend of Scottish Realism and Baconianism pervading thought on the national and local Cambridge levels, Newcomb would have repeatedly encountered the tenets of the common-sense and empirical philosophies. But, being exposed to Henry and other practicing scientists who rejected raw Baconianism, he also looked beyond these accepted philosophies to consider the heterodox teachings and theories of Auguste Comte, Charles Darwin, and John Stuart Mill.

CHAPTER III
Influences of Comte, Darwin, and Mill

In the years bracketing 1860, many professors, ministers, and other illuminati in the Harvard community were opposed to Comte's positivism, Darwin's evolutionary principles, and Mill's skeptical empiricism. The community was, after all, predisposed to the philosophical precepts of Scottish Realism as intertwined with the theological premises of Unitarianism. Nevertheless, some members of the community, especially among the growing contingent of scientifically minded thinkers, were sympathetic to Comte, Darwin, and Mill. That a young Cambridge scholar, Newcomb, could become so well versed in the provocative views of the three Europeans was due to the coupling of normal Cambridge opportunities with a twofold scientific exposure: his dual tenure at Harvard's Lawrence Scientific School and the Almanac Office. Through his contacts at the Scientific School, he would hear some of the first discussions in North America of Darwin's controversial theory of evolution as well as explications of Mill's more familiar but also controversial "philosophy of experience." And during a period when even Harvard was threatening undergraduate John Fiske "with dismissal from college if he was caught talking Comtism to anybody,"[1] Newcomb could listen at the Almanac Office to Chauncey Wright discourse on Comte—as well as on Darwin and Mill.[2]

AUGUSTE COMTE

Newcomb had an early interest in Comte and his positivist philosophy. Three months after receiving his bachelor of science degree, but while still on the Harvard rolls as a "resident graduate" and working as a computer at the Almanac Office,[3] he wrote in his diary: "I think I shall write a dissertation on Auguste Comte for the Bowdoin prize." On the very next day, he adds: "Got Comte Philosophe [sic] from the Library." Harvard's Bowdoin Prize committee had included the topic of "Auguste Comte" as one of two possibilities on which resident graduates (not impressionable undergraduates like Fiske) might compose dissertations during the academic year 1858–59.[4] It was in his *Cours de philosophie positive* that Comte (1798–1857) rejected the theological or metaphysical quest for first or final causes of phenomena; instead, he embraced the positivistic search for invariable natural laws of phenomena as revealed in relations of succession and similarity. Especially after Mill called attention to Comte's writings on positivism, the French philosopher's readership grew to include Anglo-American thinkers, including practicing scientists such as Darwin.[5]

Newcomb did not win the fifty-dollar prize in the Bowdoin competition; the money went to a resident graduate in the Divinity School. It remains uncertain whether he even completed the dissertation and submitted it.[6] Tucked away among his papers, however, there do exist what appear to be three different partial drafts of the dissertation.[7] In each of these fragments, Newcomb reveals a firm understanding of Comte's philosophy. In all three, he also expresses general agreement with Comte's overall outlook but dissatisfaction with specific aspects. In one of the drafts, for example, he seeks to evaluate Comtian positivism from a historical perspective. After sarcastically deriding the French philosopher for his smug skepticism toward all metaphysical systems, Newcomb adds in carefully qualified words that "it must be confessed that the assertion that man has no knowledge of anything above the realm of the sensuous and material world is not without some foundation."

In another draft, Newcomb contends that positivists are guilty of harboring a metaphysical bias and thus of advocating a "species of partial and inconsonant skepticism." He begins this critique by surveying positivists' avowed disdain for discussions of first causes. A positivist, according to Newcomb, sees planets moving, bodies falling, seeds germinating, plants growing, animals walking, and men acting, but he sees no God, no Nature, no vital, life forming principle. It is impossible he declares to arrive at any knowledge of the causes of these operations of nature; they are of necessity beyond the possible limit of human investigation. The utmost that we can do is to observe the laws of succession of phenomena, and to express them in a few and simple formulae.

Using language reminiscent of John Stuart Mill,[8] Newcomb proceeds to argue that even this austere philosophy is in fact metaphysical since it goes beyond the realm of experienced sensations and illegitimately presupposes the existence of a universe external to ourselves.

If it comes to this, confining the term see to that which we are immediately conscious of [that which we immediately experience] I deny that you see planets moving, or seeds germinating, or men acting or any other natural operation! You experience sensations, and sensations only. You experience hardness, softness, heat, cold, colors, sensations of sound, and taste and smell, but you do *not* experience planets moving or bodies falling or seeds germinating or plants growing or animals walking or men moving. But these sensations of feeling and hearing and seeing you desire to account for, and you do it by erecting this vast fabric you call a universe, in which the qualities you experience reside, and which is thus regarded as the cause of your sensations. But how do you know that your sensations may not proceed from some other cause than this? Positivism gives no reply.

In other words, when a positivist claims to see "planets moving," the positivist exceeds the bounds of raw sensory experience and is, in fact, engaging in a metaphysical speculation about a world external to the viewer—a world that is the cause of the raw sensations. Newcomb further explains that only a short step separates the metaphysician's belief in final cause from what Newcomb claims to be the positivist's belief in an external universe beyond sensations. In later years, he would continue to reject various details of Comte's positivist program while remaining sympathetic to his fundamental goal of describing phenomena without the crutch of metaphysics.

CHARLES DARWIN

Four months after deciding to write a prize dissertation on Comte, Newcomb encountered the controversial ideas of a second major scientific thinker of the nineteenth century, Charles Darwin (1809–1882). In presenting his theory of evolution in *Origin of Species* and other writings, Darwin did more than espouse the novel idea of variation within species through natural selection. He also espoused—at first implicitly, later explicitly—an empirically cautious, even positivistic approach to scientific inquiry. Feeling that it was sufficient for a scientist to describe phenomena, he professed to refrain from speculating on, for example, the ultimate cause of "natural selection" or the metaphysical import of "specie." Repeatedly, Darwin revealed this positivist tendency by claiming merely to report uniform laws of nature involving only natural, not supernatural, causes. As a result, contemporary commentators often linked the names of Darwin and Comte.[9]

Newcomb's diary reveals the specific—and, in retrospect, historic—avenue of his exposure to Darwin's views. On Monday, 21 February 1859, he wrote, "Prof. Peirce got me permission to attend the special meetings of the A. A. S." Then on the next day, Tuesday, he recorded, "Went into Boston, and heard a discussion at the A. A. S. on the origin of plants and animals. Walked out with Bartlett and Prof. Bowen, whom I attacked on the subject of fatalism."[10] Benjamin Peirce had arranged these now-famous meetings of the American Academy of Arts and Sciences to provide a fuller airing of the growing controversy during the past two months between Asa Gray and Louis Agassiz regarding Darwin's recently announced theory of evolution. Peirce also was the one who had arranged for this historic debate between these eminent naturalists to be opened to nonmembers of the Academy such as Newcomb; also present were Newcomb's fellow student from Harvard, William Bartlett, and Newcomb's co-worker at the Almanac Office, Chauncey Wright.[11]

Francis Bowen, the Harvard professor of philosophy and economics with whom Newcomb walked home after the Tuesday meeting and "attacked on the subject of fatalism" (Newcomb did not believe in traditional notions of free will), would himself soon join the somewhat beleaguered Louis Agassiz on the Academy's podium as an opponent of Darwinian Asa Gray. Bowen appeared in a second series of meetings held the next year as the debate quickened following dissemination of *Origin of Species*, just published in 1859.[12] Again, Newcomb was present. His diary for 27 March 1860 reads: "Attended the Academy. Darwin's origin of species was the only subject discussed." His interest, however, began to wane by the last session of the second series: "Attended a special meeting of the Academy called for the purpose of finishing up the Darwinism discussion, which I hope was done."[13]

With primary interests in physical science and mathematics, Newcomb did not pursue the technical aspects of Darwin's theory of evolution. Subsequently, he was concerned mainly with the theory's religious implications, especially as reported in weekly religious journals such as the *New York Observer*. In an unpublished manuscript of his "Religious Autobiography," he reconstructed his involvement with evolutionary and related ideas beginning in his Cambridge years and extending into the 1860s, when he first lived in Washington:

> Darwin, Huxley, Spencer, Clifford, Tyndall, and Haeckel came one by one into note, and I read the discussions of their views with gradually increasing interest, though my occupations left me no time for a careful reading of their books.... So much discussion of these men [in religious journals] gradually led to a desire to know what they had to say for themselves, and I began now and then to glance at their shorter essays, and to dip into their books.

Though he was not convinced by the arguments of all these writers, he was impressed by "the spirit with which they took up their work." That is, "whether they were right or wrong, their spirit was that of earnest searchers after truth, and their method of search was to make the largest possible collection of facts and draw their conclusions from them. Particularly striking was the readiness of each one to point out and correct any hasty conclusion which he himself or any one else had reached." Reacting to their spirit and method rather than their particular doctrines, he found that his "sympathies had gone over" to these scientists rather than to their religious critics.[14]

In a related passage in another draft of his religious autobiography, he further explained that because of his limited exposure to the writings of evolutionists and other advocates of "modern thought," such writings had not been a cause of his own skepticism toward orthodox Christian doctrines. "What I have read of Darwin, Huxley, Tyndall, and Spencer," he added, "has been in a critical rather than a sympathetic spirit, and I have always had a repugnance to wholesale attacks on Christianity."[15] Although he was only a casual student of evolutionary theory with reservations about anti-Christian assertions made in the name of evolution, Newcomb would become an advocate of Darwin's views on natural selection as well as Darwin's empirically cautious approach to scientific inquiry.

JOHN STUART MILL

Of the nineteenth-century European thinkers who affected Newcomb's outlook on science and method during his formative years, the most influential was John Stuart Mill (1806–1873). Mill regarded his "philosophy of experience" as not only a scholarly contribution to logic but also a practical response to broader political, economic, moral, and religious issues. A leading exponent of what has come to be known as British empiricism, he used this philosophy to underpin his polemics supporting British utilitarianism and philosophical Radicalism. For example, when he affirmed the experiential foundations of knowledge in his influential 1843 text *A System of Logic*, he was consciously reacting against contemporary thinkers such as William Hamilton; he portrayed men of this ilk as impeding social progress by perpetuating the misguided idea that beliefs could be certified through "intuition." He later explained: "The notion that truths external to the mind may be known by intuition or consciousness, independently of observation and experience, is, I am persuaded, in these times, the great intellectual support of false doctrines and bad institutions."[16]

Aware that the intuitionists had justified their philosophical perspectives by taking examples from mathematics and the physical sciences, Mill sought to counteract their influence by demonstrating the actual empirical and inductive nature of scientific method and, hence, of logic in general. The logic of science is "the universal Logic, applicable to all inquiries in which man can engage," he contended in his 1843 textbook, with the full title *A System of Logic, Ratiocinative and Inductive; Being a Connected View of the Principles of Evidence and the Methods of Scientific Investigation*. He specifically argued that all knowledge, even mathematics, derives ultimately from inductions based on sensory observation. In claiming that inductive inferences can lead to the establishment of natural laws, he recognized that he was making two key assumptions. First, he was presupposing that the course of nature is uniform—"namely, that there are such things in nature as parallel cases; that what happens once, will, under a sufficient degree of similarity of circumstances, happen again, and not only again, but always." Second, he was assuming that causation is universal—"every event, or the beginning of every phenomenon, must have some cause; some antecedent, upon the existence of which it is invariably and unconditionally consequent."[17]

Again, Newcomb pinpointed in his diary the date of one of his earliest exposures to the English philosopher. In January 1860, he wrote: "Went into Boston, bought Murphy's games, Peirce's Trigonometry, and Mill's Logic." Then, a few months later, he reported calling on Harvard professor Francis Bowen, an opponent of Mill's views, "to see Mill's Logic." Newcomb further recorded in his "Alphabetical Catalogue of Books" for the period from 1855 to 1862 that about the time when he paid $1.12 for Mill's *A System of Logic*, he also purchased for $2.00 a companion volume on which Mill had relied and to which he had reacted, William Whewell's *History of the Inductive Sciences*.[18]

The philosophical debate between Mill and Whewell, particularly as summarized by Mill in the *Logic*, provided Newcomb with a point of departure in an essay that appears to date from these early years.[19] Although only a fragment of this essay exists, it sheds light on both Newcomb's understanding of Mill and his willingness to go beyond Mill. Where the fragment begins, Newcomb is discussing the antithetical views regarding man's knowledge of nature held by "the idealist or theorist on the one side, and the sensationist or empiricist on the other." After summarizing several general disagreements between these two types of philosophers, Newcomb focuses on the disparate ideas of Whewell and Mill. He points out that Whewell believes in the a priori nature of the axioms of geometry and of the laws of motion, whereas Mill argues that the axioms and laws are inductions from experience.[20]

Newcomb, having characterized the views of "idealist" Whewell and "sensationist" Mill, next outlines his own purpose in writing this essay. Rather than enter the murky debate over the origin of knowledge (mind or experience), he ignores questions of origin and discusses only the "necessity" of the particular propositions regarding geometry and physics.

> Our present purpose is rather to consider the *necessity* of these results, admitted on both sides as truths, than to consider the source of our knowledge of them. The latter question from its very nature, does not admit of a definite and satisfactory answer . . . . But the *necessity* of a proposition or law is something which admits of being considered with much more rigor, as the conceptions with which we have to deal are more primitive and simple in their nature, and the discussion need not occupy so much indefinite ground.
>
> It may be well to begin with two remarks respecting the method by which we are to proceed in our examination, for great confusion of ideas

has arisen from neglecting certain necessary precautions and conditions in discussing questions which lie on the boundary between physical and metaphysical science.

1. Before any discussion respecting the truth or falsity of a proposition, it should be distinctly ascertained that both parties attach the same ideas to the words in which the proposition is expressed. Otherwise the discussion will simply be one of terminology.

2. When one asserts that a certain act or effect is possible, he should also state distinctly the conditions under which it is possible.

Finally, Newcomb goes on to analyze the specific propositions confounding Whewell and Mill, beginning with Newton's first law of motion. "No doubtfulness attaches to the meaning of any term used in this proposition," Newcomb points out, "except perhaps the word 'force.' " This leads to a detailed discussion of definitions of the word. At this point, unfortunately, the fragment of the essay terminates. It is evident from the above excerpts, nevertheless, that Newcomb was engaged in an elaboration of Mill's empirical views. As both he, Chauncey Wright, and later pragmatic thinkers such as William James would do in formal publications, he was directing his attention away from the origins of scientific knowledge toward the warrant for specific propositions and the meaning of terms.[21] Mill was aware of a distinction between the psychological source and logical proof of an idea, but he had at times blurred the two issues.[22] In general, it was only in the 1860s and 1870s that leading scientists and methodologists abandoned traditional empirical rules of inductive reasoning in favor of the hypothetico-deductive method with its emphasis on verification as opposed to discovery.[23]

In another early paper, this time definitely datable to 1860, Newcomb further revealed his familiarity with and willingness to develop Mill's ideas as presented in *A System of Logic*. In May of 1860, he addressed the American Academy of Arts and Sciences on objections that Mill had raised in *Logic* to Pierre Laplace's view on the logical foundation of the theory of probability.[24] (A year before, he had purchased Laplace's *Théorie Analytique des Probabilités*; the theory of probabilities emerged during this period as another of Newcomb's lifelong preoccupations, essential to his studies of the positions and motions of stars, planets, and other celestial bodies.[25]) In this address, essentially an elaboration of Mill's analysis, Newcomb begins with a fundamental, philosophical inquiry concerning the definition of probability. Specifically, he discusses the impossibility of defining "the probability of a proposition" in terms of "the *amount* of our belief in the truth of that proposition." He explains that "neither belief, nor any other affection of the mind, admits of being measured as a *quantity*." What does admit of measurement—and statistical computation—are the actual "events" and "combinations of circumstances" that underpin a belief. Newcomb also grapples with another problem identified by Mill, the relationship between "law and chance." While acknowledging that some events seemed to behave according to chance, Newcomb agreed with Mill that ultimately every event in nature is governed by exact laws.[26]

As we will see later, Newcomb also had a deep knowledge of Mill's philosophy as it extended beyond the natural sciences and mathematics to issues involving society. As early as 1865, in a book published at his own expense on the government's financial policies during the Civil War, he endorsed Mill's reservations regarding the issuance of paper notes as currency. In subsequent years, Newcomb regularly defended Mill's "celebrated definition" of political economy as a delimited science that "considers mankind as occupied solely in acquiring and consuming wealth." Newcomb's commitment to Mill's perspective on society, politics, and economics also appeared generally in his repeated endorsements of *A System of Logic* for study by public leaders.[27]

Newcomb's admiration for Mill was assured when in 1870 at age thirty-five he had the good fortune to meet with the sixty-four-year-old philosopher. During a trip to Great Britain, he recorded the following entry in his diary: "This morning went to Greenwich. Left wife at observatory while I went and made a call on John Stuart Mill at Black Heath, with whom I had a very pleasant conversation." In his *Reminiscences*, he later explained that he had wanted to see Mill, "to whom I was attracted not only by his fame as a philosopher and the interest with which I had read his books, but also because he was the author of an excellent pamphlet on the Union side during our civil war." Regarding his meeting with Mill, Newcomb further recalled: "The cordiality of his greeting was more than I could have expected."[28]

While in Edinburgh during the prior week, Newcomb spent an evening visiting with both the widow and daughter of another eminent philosopher, William Hamilton. Mill had targeted Hamilton, "the great fortress of the intuitional philosophy," in an 1865 supplement to *A System of Logic* titled *An Examination of Sir William Hamilton's Philosophy*. Newcomb remembered: "I talked with Miss Hamilton [herself a philosopher] about Mill, whose 'Examination of Sir William Hamilton's Philosophy' was still fresh in men's minds. Of course she did not believe in this book, and said that Mill could not understand her father's philosophy."[29] Newcomb was probably able to converse intelligently about intuitionist Hamilton as well as the more familiar Mill. Two years earlier, according to his diary, he had "read Sir William Hamilton on Metaphysics, with a view to a paper on the limits of Natural Law for the N.A. Review." And while still at the Nautical Almanac Office in 1858 he had listened to his co-worker, Chauncey Wright, expound on Hamilton's newly published *Lectures on Metaphysics*.[30]

While commenting in his *Reminiscences* on his chat with Mill, Newcomb took the opportunity to summarize his opinion of the value and relevance for science of Mill's philosophy and especially his treatise *A System of Logic*. He judged that the incomplete state of mid-century science had inhibited Mill from becoming a major voice in "the new philosophy." Nevertheless, he believed that *A System of Logic* remained until the end of the century the definitive work on scientific method.[31] Indeed, Mill's *Logic*—which Newcomb first purchased in 1860 at age twenty-five and which he then endorsed throughout his career—provided Newcomb with a schema for much of his own thought and work in the natural and social sciences. The entire first section of *Logic*, about 100 pages, develops the foundational subject "Of Names and Propositions," while later sections amplify this topic; in like manner, the mature Newcomb placed a concern with language and definition at the very base of his scientific method. In the middle portions of his treatise, Mill furnishes a detailed examination of scientific reasoning and, in particular, induction. Again, as we will see, these are topics that Newcomb regularly stressed, though he tempered Mill's inductivism. Finally, the closing section is titled "On the Logic of the Moral Sciences"—a section in which Mill, himself a respected economist, calls for the extension of the methods of physical science to the realms of human nature and society. Newcomb installed the premise of the social relevance of scientific methods at the base of his own extensive efforts in political economy.[32]

A TEXTBOOK STATEMENT ON METHOD

Although the allure of mathematical astronomy was strong for Newcomb while working and studying in Cambridge, he also was drawn to the closely related field of physics or, as it was still sometimes known, natural philosophy. At Harvard, he studied physics as well as mathematics and celestial mechanics under Benjamin Peirce. He also maintained close contact with his sponsor and mentor, physicist Joseph Henry. Newcomb's diary entries and letters during his Cambridge years are laced with references to his investigations of heat, light, electricity, and the kinetic theory of gases, some of which he reported to the American Academy of Arts and Sciences in Cambridge. He even sought a professorship in physics at Washington University in Saint Louis; though he went for an interview at the new school in 1860, he lost the position to a more experienced West Point instructor. He later commented that, had he obtained the position, it "would have changed the whole course of my life."[33]

During the following year, something did change the course of his life: intervention by a group of leading scientists who called themselves "the Lazzaroni." An informal network built around a core of about half a dozen scientists, the Lazzaroni held a modicum of sway over the nation's science policies and scientific appointments from the late 1840s through the early 1860s. Newcomb was well connected to the network, whose core included Henry, Benjamin Peirce, and a younger Cambridge astronomer, Benjamin Gould. In 1861, Gould alerted Newcomb to an opening for a "Professor of Mathematics" at the U.S. Naval Observatory in Washington. Founded three decades earlier, ostensibly to meet the navy's navigational needs, this agency had evolved into a major research observatory; the navy relied on a commissioned corps of "professors" to provide technical expertise and instruction at the observatory and the U.S. Naval Academy. Only a few months before Gould's overture to Newcomb, the Civil War had broken out and the superintendent of the observatory, Matthew Maury, had resigned to enlist with the Confederacy. Disliked by the Lazzaroni all along, Maury had revealed himself to be only a lukewarm supporter of astronomical research. Gould was delighted to see him replaced by one of the observatory's founders, James Gilliss. "I think," Gould wrote to Newcomb, explaining the observatory opening for a professor, "that an active effort on the part of your friends would secure the place for you." Listing as references such Lazzaroni "friends" as Peirce, Henry, and Gould, Newcomb obtained the position. Thus, with a letter of appointment from President Lincoln, he actually began his post-Cambridge career as an observational astronomer. He would spend the remainder of the 1860s mainly performing basic observations of stars, pinpointing their right ascension (celestial longitude) using a transit instrument and their declination (celestial latitude) using a mural circle. Though he came to Washington inexperienced in observational work, he would soon take the lead in organizing and unifying the Naval Observatory's methods and then, using a new transit instrument, take on the challenge of disclosing systematic errors in the right ascensions of stars that plagued leading observatories around the world. As we will see, Newcomb once again would find himself in a lax work regimen that allowed time for other research. He recalled that whenever he or his fellow observer tired of their late-night vigils, "we could 'vote it cloudy' and go out for a plate of oysters at a neighboring restaurant." Incidentally, in 1858 while still in Massachusetts, Newcomb had begun the procedure to become a citizen of the United States and he became naturalized six years later in the District of Columbia.[34]

Toward the end of his Cambridge years, before tilting toward a career in astronomy, he began to draft a physics textbook for college students. By the late spring of 1861, he could note in his diary: "Commenced the final revised composition of the introductory chapters of physics." While he never published these or later chapters of the textbook, many of the sections have survived, including an opening statement on scientific method.[35] This statement, intended for callow students and written in the ordered pedagogic style of the day, is somewhat simplified and didactic. Nevertheless, it contains at least a summary of Newcomb's thoughts on method as his Cambridge years drew to a close.

Both Newcomb's language and the location he assigned to the methodological statement convey to the student the importance of method in science. The statement begins on the first page of the textbook, after a preliminary definition of physics. Newcomb immediately warns students that, unless they want to waste their time "in blind and random endeavors," they must "lay out some exact *plan*, or *method* of procedure" in making scientific investigations. Adopting a traditional, empiricist line of argument, Newcomb further advises that the essence of science and the reason for its progress lie in method, particularly the method articulated by Frances Bacon in the early 1600s. While asserting the centrality of empirical method, Newcomb tacks on important caveats indicating that he was not a doctrinaire disciple of Bacon. Perhaps with Comte and certainly with Mill in mind, he points out that "numerous [or 'some'] improvements have been made in Bacon's method." Probably also with contemporary scientists such as Henry in mind—seasoned practitioners who recognized the limitations of abstract, Baconian methodology—he adds that Bacon's method "is not, when applied in its purity, as successful as he hoped." Nevertheless, Newcomb underscores his empiricist preference by stressing that Bacon's method "still furnishes the basis of every sound system of investigation in Natural Science." This preference is understandable when we recall Americans' professed allegiance to Baconianism (and Scottish Realism) through the first half of the nineteenth century.[36]

Having completed this testimony to the primacy and efficacy of method, Newcomb turns to his main task: providing a pedagogic summary of four different methodological "processes, or systems of processes, which are necessary in the construction of a science." Proceeding in a Baconian mode, he labels the first process "the collection of facts." Scientists begin their investigation of a subject by determining as exactly as possible the relevant facts. They do this through either observation or experiment, two procedures that Newcomb, like Mill, sharply distinguishes. Moving to the next stage, the scientists compare the facts that have been collected to determine relations and connections, thus arriving at laws. "The inference of the truth of these general laws from the particular phenomena in which we find them exhibited is termed *induction*," Newcomb explains, adding, probably with Mill's *Logic* in mind, "and exact rules may be laid down for carrying on the process." In depicting the inductive process, Newcomb joins step with Mill and Comte—as well as the Scottish Realists—and avoids any suggestion that scientists are actually arriving at metaphysical insights regarding underlying causes. Choosing his words carefully, he states that scientists merely attempt to compare "the facts of experience with a view to determine the relations between the phenomena themselves, and to find what circumstances are essential to the production of any particular phenomena. It needs very little examination to see that there are certain circumstances or causes which are invariably followed by certain events or effects; and that certain properties in bodies are invariably associated with certain other properties." Later, in the third chapter of the textbook, Newcomb reinforces this point when discussing force as a cause of motion. Scientists study the "effects," not the "essence," of force, "for, as remarked in the introduction, the law of the phenomena, and not the nature of the cause is the subject of inquiry in physical science. The latter belongs to the domain of metaphysics."[37]

Though these first two procedures, collection of facts and induction, follow the empirical recipe of Bacon as leavened by the Scottish common-sense philosophers, as well as Comte and particularly Mill, Newcomb again interjects a caveat that reveals his taste for more recent, non-Baconian trends. Specifically, the caveat indicates his affinity for a hypothetico-deductive methodology. Recall that Henry, along with other practicing scientists confronted by the contingencies of actual research and disenchanted with orthodox Baconianism, was encouraging the use of hypotheses rather than mere gathering of empirical facts. Even Mill acknowledged that hypotheses were "absolutely indispensable in science"; while Mill had focused on the three procedural steps of "induction, ratiocination, and verification," he insisted that the formulation of a hypothesis provided a legitimate alternative to the usual first step of induction.[38] Similarly, Newcomb identifies two types of guiding principles in science: "our assumed laws of connection between phenomena and qualities, or our hypothetical mechanism by virtue of which the phenomena are brought about, and the connection established." And the latter type of hypothetical approach has a heuristic value lacking in the former type of inductive approach. With contemporary mechanical, atomistic physics in mind, Newcomb specifically cautions that it is insufficient for a scientist merely to determine inductively an empirical connection; the scientist must also specify, through the use of a hypothesis, "the hidden mechanism by which such connection is established."

The third process is "deduction." Having formulated either an empirical law or a mechanical hypothesis and working now on the theoretical level, scientists attempt to determine the practical consequences of the law or hypothesis. Reflecting again the views of contemporary scientists such as Henry and Benjamin Peirce, Newcomb prefers that the deductions point to quantitative experiments. "We find, by mathematical reasoning, if possible, what will be the result of a great variety of experiments (untried experiments if possible); and we must find not only of *what kind* will be the result, but also express in exact numbers *how much*." Fourth and finally, scientists seek "to determine whether the theory is verified." That is, they carry out the experiments or observations that they deduced from their empirical law or mechanical hypothesis and compare the actual results with the predictions.

In outlining the process of verification, Newcomb inserts another critical caveat, again in general accord with commentator Mill and practitioner Henry, this time regarding the epistemic status of hypotheses. A positive comparison between theory and experiment does not mean that a hypothesis is true in any ultimate sense, but only in a probable sense. "If the agreement, in *quantity* as well as in *kind*, is perfect within the unavoidable error of observation, and if the verifying experiments have been quite diverse," Newcomb explains, "there will be a high probability, or a reasonable certainty, of the truth of the hypothesis." Furthermore, a contradiction between theory and experiment does not necessarily invalidate the hypothesis. "If the agreement is not perfect, it by no means follows that the hypothesis must be unconditionally rejected. There may be some imperfection in the conduct of the experiment; or some flaw in the reasoning by which we found the results of the hypothesis; or finally, the latter may be true in itself [only part of the whole cause], but combined with some other unknown cause which alters the results." In following decades, he would express the latter point by saying that scientists must be aware that hypotheses involve stipulations of simplified conditions or idealized circumstance under which the hypotheses hold. Similarly, in the fragment of the essay in which he contrasted Whewell's idealism and Mill's sensationism, he had touched on the same point, insisting that a researcher "state distinctly" the conditions under which a certain result occurs.

Though he interjects qualifications throughout this introduction to method, Newcomb seems to present the student with a firm, sequential, Baconian procedure for automatically generating scientific knowledge: observation, induction, deduction, verification. After reviewing the four steps, he concludes: "Whenever we have discovered a system of laws to which some or all of the phenomena are subjected, such a system constitutes a *Science*. If the laws account perfectly for all the phenomena, so that we can predict the latter with the same certainty that we can observe them, it becomes an exact science." But again, Newcomb adds a final, general qualification to his comments on constructing a science through the algorithm of method—a qualification that reflects his own hands-on research experiences and the down-to-earth practices of scientists such as Henry and Peirce. He cautions: "Although every step in the constitution of an exact science can be placed under one of the four heads preceding, it must not be supposed that each operation is undertaken and completed in its regular order. Nor is equal prominence given to each step." In actual performance, Newcomb warns, scientists deviate from the textbook pattern.

Influences of Comte, Darwin, and Mill

In later publications pertaining to method, Newcomb would both build on and diverge from the simplified characterizations of this unpublished textbook. While maintaining his empiricist stance as mitigated by using hypothesis, he would increasingly emphasize the appropriate use of language, of terms and propositions. One methodological rule would come to loom over others: all scientific concepts, whatever their origin, must ultimately be relatable to particular sensory objects. Comte, Darwin, and especially Mill contributed to this emerging view, as did mentors Henry and Peirce. Indeed, working scientists such as Henry illustrated through their everyday practices the necessity of moving beyond crude Baconianism, while philosophical commentators such as Mill added through their learned discourses parallel formal analyses. As we will now see, Chauncey Wright and perhaps Charles S. Peirce, Americans who combined scientific practice and philosophical analysis, also influenced Newcomb in the formulation of his eventual linguistic, empirical methodology.

CHAPTER IV

Interactions with Wright and Peirce

It is difficult to distinguish the influence on Newcomb of Comte, Darwin, and Mill from the influence of his colleague and friend, Chauncey Wright (1830–1875). The confusion results because Wright advocated elements from the outlooks of all three of the European thinkers. Charles Peirce once characterized Wright, for example, as being "one of the most acute of the followers of J. S. Mill."[1] The issue of influence is further complicated in that ideas did not always flow from the older and more experienced philosopher, Wright, to the unseasoned newcomer. Although never a match for Wright in philosophical breadth and subtlety, Newcomb contributed, at least on one occasion, to the developing ideas of the two men.

CHAUNCEY WRIGHT

For the first years after beginning work in 1857 at the Nautical Almanac Office in Cambridge, Newcomb maintained a formal, professional relationship with his co-worker, "Mr. Wright," who was five years older.[2] By 1859, however, their relationship became more personal. In his diary for April of that year, for example, he reports a philosophical talk with his senior colleagues John Runkle (later president of the Massachusetts Institute of Technology) and Wright: "The latter had Hamilton's lectures, just published." During 1861, he records two long walks with Wright and again Runkle, first "around mount Auburn, stopping for a sherry cobbler while returning," and second "to Camp Cameron . . . to see Porter's artillery."[3]

During January of the same year, 1861, Newcomb also describes the possibility of collaborating with Wright on a rebuttal to some opinion espoused by Francis Bowen. "Wright and I think of cooking up a reply to Bowen." Three days later, a Sunday, he added: "After dinner called on Chauncey Wright, and talked with him most of the afternoon."[4] There was much in Bowen's beliefs that might have sparked the ire of Wright and Newcomb. Bowen, the Alford professor of natural religion, moral philosophy, and civil polity at Harvard, opposed the views of Comte, Mill, and Darwin.[5] Perhaps Bowen's polemics against Darwin in the American Academy meeting of March 1860, as later published in the April issue of the *North American Review*, had prompted Wright and Newcomb to consider a reply.[6]

The Academy meetings in which Bowen participated did help coalesce Wright's views on natural selection. In later recalling the meetings, Newcomb wrote: "Wright was a Darwinist from the very beginning, explaining the theory in private conversation from a master's point of view." Perhaps because he had shared in Wright's early exposure to the theory of evolution, Newcomb in later years admired Wright's explications of Darwinian thought, even elevating Wright's works to the level of Asa Gray's. "In philosophic comprehension, scientific accuracy, and clearness of thought," Newcomb wrote in 1876, "the essays of Wright and of Gray might well head the list in a competition among those of all nations."[7]

Newcomb summarized his experiences as one of Wright's Cambridge "disciples" in a letter to James B. Thayer, a Harvard professor of law and friend of Wright. Thayer, following Wright's death in 1875, was compiling a volume of Wright's correspondence. Newcomb wrote:

> My acquaintance with him began in 1857, when I became a computer for the Nautical Almanac, and hence a sort of scientific colleague. He had then an abominable habit of doing his whole year's work in three or four months, during which period he would work during the greater part of the night as well as of the day, eat little, and keep up his strength by smoking. The rest of the year he was a typical philosopher of the ancient world, talking, but, so far as I know, at this period, seldom or never writing. His disciples were his fellow-computers on the almanac. He regarded philosophy as the proper complement of mathematics,—the field into which a thinking mathematician would naturally wander. Philosophic questions were our daily subjects of discussion.

Newcomb presented similar recollections in an unpublished, handwritten manuscript titled "Autobiography of My Youth":

> In some respects the lax organization [of the Nautical Almanac Office] was favorable to scientific progress, as it gave an opportunity to those so disposed to improve their mathematical knowledge, and general culture. One feature of our work which I remember with great interest was the philosophical discussions with Chauncey Wright—which were sometimes of almost daily occurrence. About the first I heard from him on the subject was in connection with Hamilton's Lectures on Metaphysics[,] the appearance of which seemed to start him on a new line of thought.[8]

Newcomb's appointment in 1861 to the Naval Observatory in Washington did not signal the end to his relationship with Wright. Through letters and Newcomb's occasional visits back to Cambridge, they maintained their philosophical, scientific, and personal discourse.[9] A flurry of letters from 1865 in which they called on each other for personal favors—and exchanged congratulatory comments on the North's victories in the closing campaigns of the Civil War—illustrates the continuing depth of their friendship. Newcomb begins the exchange by asking Wright to review his first book, on American financial policy, for the *North American Review*. "I don't ask for a puff," Newcomb explains, "or want you to agree with all my doctrines. It will satisfy me if you can conscientiously say that the book is worth reading."[10] Although Wright begged off reviewing a book outside his area of competency, he in turn asked his own favor of Newcomb. He inquired whether Newcomb would finish an article Wright had begun on the Nautical Almanac Office. Newcomb agreed, but then Wright did an about-face and asked to reclaim the project. In case this loss of a publication in the *North American Review* (and presumably the associated loss of remuneration) upset Newcomb, Wright offered consolation: "Perhaps some other subject may interest you more. If any does and you feel inclined to write it for the N. A. R. I will secure for it a favorable consideration from the Editors." Three months later, Newcomb took up Wright's offer.[11]

A few weeks after the initial flurry of letters, Newcomb instigated a more circumscribed, philosophical dialogue—a dialogue in which he appeared as the originator of substantive ideas. The ideas were an elaboration of John Stuart Mill's rejection of traditional notions of free will in favor of what has come to be called a "compatibilist" position; in particular, the English philosopher affirmed the compatibility of two seemingly antithetical doctrines, free will and determinism. Mill, troubled by the ambiguity of traditional terminology, carefully explained that, while human actions can be "free" in the sense of not coerced, they are still "determined" in the sense of being part of an invariable sequence subject to the all-inclusive law of causality.[12] Newcomb later submitted this philosophical correspondence to James Thayer for inclusion in the volume of Wright's letters published in 1878. In a cover letter to Thayer, he explained the circumstances of the exchange.

> My favorite subject [in philosophy] was that to which the enclosed correspondence relates,—the compatibility of free-will with absolute certainty regarding human acts, and the absence of any reason for supposing that human actions are any less determinate than the operations of nature. Wright was at first inclined to claim, in accordance with popular notions of free-will, that these propositions were not well founded, but at length was led to maintain that, considered simply as phenomena, they were correctly formulated; that is, that we have no reason to believe human acts, considered simply as phenomena, to be any less determinate than the operations of nature. This is the ground which you will see that we agree upon in the enclosed correspondence.[13]

In other writings, Newcomb was more emphatic and explicit about influencing Wright's view on free will during their Cambridge days. In a passage about Wright published in an 1891 autobiographical note, he commented, "I shall always remember him as the only man, so far as I know, whose theories I ever changed by argument. After much wrestling on the subject of the freedom of the will, he was led to accept the logic of Jonathan Edwards and John Stuart Mill." Similarly, in the unpublished "Autobiography of My Youth," he wrote: "The only question [in philosophy] in which I took especial interest was that of the subjection of the will to the general law of causation. For a long time Wright held out against the affirmative view, which I always maintained; but one morning he suddenly informed me that I had convinced him, and from that time he maintained the view with as sound arguments as could be desired."[14]

The initial paragraph of Newcomb's 1865 letter to Wright merits full quotation to correct a possible misconception. Thayer, in reprinting only Wright's reply, left the impression that Newcomb functioned as an inconsequential partner in the exchange.[15] This was not so.

> Dear Wright,
> I want to draw upon your classical knowledge, or rather upon your knowledge of the Greek and Latin languages for words expressive of certain philosophical ideas which have never been analysed, so far as I am aware,

but which it is extremely desirable should be. You may rember [sic] that at the [Nautical Almanac] office I used to argue that all words such as *may, can, power, possibility, liberty, free agency*, all words in fact which express or imply the *attribute of potentiality*, in contradiction to the attribute of *actuality*[,] have a threefold application; that in ordinary speech and ordinary thought we make no distinction between the three meanings, and that for want of making such distinctions men who have written on free agency have fought on verbal quibbles, and been guilty of the crime of ignoratio elechi [answering to the wrong point] to a wholloy [sic] unnecessary extent.

Newcomb then proceeds to the core of his analysis, specifying and discussing the three meanings of propositions involving potentiality. When we say that an event *may* happen, we variously mean that: (1) we are ignorant of any reason it should not happen; (2) there is nothing now in nature to cause it to happen, but something entirely new, for example an act of the will, could arise to induce it to happen; or (3) we are conscious of having liberty of choice regarding its occurrence. "The second and third meanings," Newcomb remarks, "seem to be so continually confounded that it is often hard to tell what writers on the freedom of the will really mean to assert." While acknowledging the feeling of liberty of choice (as described in the third meaning), he suggests that acts of the will (as described in the second meaning) are not independent of but, like other phenomena, actually subject to the law of causality. This is what he meant in the letter to Thayer when, in the spirit of Mill, he wrote of "the compatibility of free-will with absolute certainty regarding human acts, and the absence of any reason for supposing that human actions are any less determinate than the operations of nature." Newcomb explains to Wright that this compatibility of "freedom" with "determination" becomes obvious "if one will only think from examples, e.g. as you pass Cambridge bridge you are every moment at perfect liberty to jump over into the river. But, this fact does not interfere with the other fact that you are conscious of a good and sufficient reason why you never will jump over." He ends by asking Wright if he can "coin some words" to express the three meanings. "If you do so," he further remarks, "you will add greatly to men's power of correct thought and bring the question of human liberty into a very narrow compass."[16]

Wright begins his response to this request with a reservation: "Much more thought and care than I have yet given to it would be necessary to a final and valuable decision on so important a matter as the invention of a nomenclature, which is to ordinary metaphysics what the construction of a machine is to the working of it." Then, after generally agreeing with Newcomb "that the ideas you propose for baptism have never before been analyzed, or at least signalized, with any distinctness," he suggests appropriate Latin phrases for Newcomb's three meanings. Wright also presents his own commentary on the issue of free will. He winds up in overall accord with Newcomb (and "Mr. Mill"), stating that there is no evidence that human actions are distinct from other natural phenomena which fall under "the law of causation." He concludes that "if the terms in which the problem of philosophic liberty is discussed be freed from ambiguity and metaphor, there will be little or nothing left to discuss." That is, in line with Newcomb and Mill, he dismisses the freedom-determinism issue as being in some sense a pseudo-problem arising from ambiguity in the metaphorical use of language.[17]

Apparently, Newcomb carefully pondered Wright's reply. A week after receiving the reply, and then misplacing it, he wrote to his vacationing wife: "I read over again this morning Mr. Wright's long metaphysical letter which I received Monday. It was in my old coat pocket at the observatory all the time." Two months later, however, he sent Wright a tardy and rather nit-picking response: "I beg pardon for not acknowledging the receipt of your philosophical letter, which was very satisfactory. But don't you think it would have been preferable if the ideas of potentiality could have been expressed by single words instead of phrases?" Fifteen years later, this correspondence was still fresh in Newcomb's mind; in an 1880 address that partially dealt with meanings of the phrase "freedom of the will," he cited his philosophical exchange with Wright, "one of the most acute thinkers of the country."[18]

Although Newcomb guided this dialogue on free will, he usually assumed the role of Wright's follower. In 1868, for example, Newcomb acknowledged Wright's direct influence on his opinion regarding natural law. He made this comment while proposing an article to the editor of the *North American Review*, Charles E. Norton: "I have a set of ideas on the nature and limits of natural law, which were formed in discussion with Mr. Chauncey Wright eight or ten years ago, which have since been maturing, but which have never been commutted [sic] to paper. These I may commut [sic] to paper and offer you in the course of the present year." Although Newcomb immediately began preparing for this article by reading William Hamilton's views on metaphysics, he published nothing of substance on the topic of natural law until a decade later. When he did, it would bear the imprint of Wright.

William James could have had Newcomb in mind when he wrote his eulogy of Wright, a man who published little: "His best work has been done in conversation; and in the acts and writings of the many friends he influenced[,] his spirit will, in one way or another, as the years roll on, be more operative than it ever was in direct production."[19]

Editor Norton was an old Cambridge friend of Wright. When Wright died in 1875 at the age of forty-five, Norton arranged to collect and reprint his main writings. Newcomb was on the original subscription list of patrons who agreed to help underwrite the project.[20] About this same time, Thayer set out to compile and publish Wright's letters. In following years, Newcomb praised both of these volumes, *Philosophical Discussions* (1877), edited by Norton, and *Letters of Chauncey Wright* (1878), edited by Thayer. He felt that these posthumous volumes provided "a deserved reputation in the narrow sphere of pure philosophies" to a close colleague who previously had "remained unnoticed."[21] Newcomb also had a more immediate reaction. The issuing of these two volumes of hitherto scattered articles, reviews, and letters rekindled his appreciation for Wright's views. In particular, the volumes provided Newcomb with easy access to stimulating ideas for his own works on the nature and method of science. During the years around 1880, Newcomb borrowed heavily from Wright's writings.

CHARLES SANDERS PEIRCE

Historians of American thought often link the name of Chauncey Wright to that of his younger Cambridge colleague, Charles Sanders Peirce (1839–1914).[22] Wright served as one of Peirce's philosophical mentors, although later Peirce diverged from various of Wright's views. Peirce, remembered today primarily for his general theory of signs and secondarily for his pragmatic theory of meaning, was also an associate of Newcomb. Whereas Newcomb's relationship with Peirce seemed more formal and impersonal than that which either of them had with Wright, it was definite and long-standing.

In their younger years, around 1860, Newcomb and Peirce shared a similar professional training. Both benefited from the scientific riches of Harvard and of the government agencies located in Cambridge. Just as Newcomb worked in the Almanac Office while also attending the Lawrence Scientific School, Peirce worked for the U.S. Coast Survey while attending the Scientific School. Peirce first obtained regular degrees from Harvard in 1859 and 1862 before graduating from the Scientific School in 1863; Newcomb graduated from the school in 1858, continuing for the next three years as a resident graduate. During his Cambridge years, Newcomb profited also from the formal tutelage of Benjamin Peirce—Charles's father and professor of mathematics, astronomy, and physics at Harvard and the Lawrence Scientific School as well as an associate of the Almanac Office and Coast Survey. Newcomb shared too in the personal hospitality of Benjamin Peirce, being a frequent visitor to the professor's home. And as might be expected, Newcomb and Charles Peirce enjoyed many mutual friends and colleagues in the Cambridge scientific community. Besides Chauncey Wright, these included Alexander Agassiz, Charles Henry Davis, and Benjamin Gould. Sharing all of these common educational and professional roots, Newcomb and Charles Peirce developed similar interests in, among other topics, mathematics, astronomy, physics, and philosophy.[23]

Despite this shared background and these like interests, Newcomb and Peirce always remained only impersonal colleagues, never intimate friends. One Peirce scholar, in analyzing the Newcomb-Peirce relationship especially as manifested in their animated scientific correspondence around the 1890s, suggests: "Perhaps the fact that Peirce was four years younger than Newcomb and the fact that Newcomb lacked social roots in Cambridge accounted to a large extent for the surprising lack of evidence of a personal friendship."[24] Perhaps also there was an element of jealousy in that Peirce's lackluster scientific career paled in comparison to Newcomb's success[25]—a success acknowledged, as we will see, by even Peirce's father. Diaries and letters allow us to glimpse the tenor of this stilted but enduring relationship.

Newcomb's earliest recorded mention of Charles Peirce, although brief, reveals that Newcomb offered Peirce advice on his philosophical writings. In his diary for January 1868, seven years after moving from Cambridge to Washington, Newcomb made the following note: "Wrote a letter to Charles Sanders Peirce acknowledging receipt of his logic papers, and making comments thereon." These papers were likely the first three of five that Peirce read before the American Academy of Arts and Sciences between March and November of 1867, including his now well-known "On a New List of Categories." After having offprints of the early papers bound, Peirce distributed this package of "Three Papers on Logic" to colleagues; during the following year, the Academy published the full series in its *Proceedings*. Newcomb, presumably at Peirce's invitation, was drawing on his own considerable knowledge of logic to review some of the seminal philosophical writings of Peirce, a man today recognized for his contributions to logic. He had a chance to scrutinize Peirce's complex and evolving views on George Boole's use of mathematics to analyze logic; the logical structure of "argument," including induction and hypothesis; and the possible "categories" of conceptions. The latter topic embraced an analysis of the symbolic nature of terms, propositions, and arguments, and their relation to "the manifold of sensuous impressions."[26]

The paths of Newcomb and Peirce crossed again in a more literal sense two years later, in 1870, when both participated in a solar eclipse expedition to the Mediterranean. Newcomb was stationed with a party of observers in Gibraltar, however, and apparently had little contact with Charles Peirce, who was positioned in Sicily with his father, the director of the American expedition.[27] During the following year, Newcomb and Peirce both became active participants in the meetings of the recently founded Washington Philosophical Society, a "society for the advancement of science" presided over by Joseph Henry. With Peirce employed in Washington at the Coast Survey Office—he spent his middle years there as a gravitational specialist—and Newcomb at the Naval Observatory, they occasionally exchanged technical information, for example, regarding maps.[28]

They also once exchanged points of view on economics. During December of 1871, through mailings between Washington and Cambridge, Peirce was helping his father, Benjamin, prepare a talk on the application of mathematics to aspects of political economy. Toward the middle of the month, presumably at Peirce's invitation, Newcomb visited his Washington colleague to consult on economics. Peirce noted the visit in a letter to his wife, who also was in Cambridge: "Simon Newcomb came to see me today.... I have been quite interested in political economy which I generally spend my evenings in studying." The only surviving record of the actual discussion between the two men is a follow-up comment that Peirce sent to Newcomb later in the same day. In this letter, which opens with the formal greeting of "Dear Sir," Peirce provides a technical clarification of an earlier statement of his on the law of supply and demand, a clarification rooted in A. A. Cournot's extension of calculus to economics.[29] Whereas in 1868 Newcomb had been offering advice to Peirce on one of Peirce's specialties, logic, Peirce was now offering advice to Newcomb on one of his, economics. And whereas the earlier exchange provided Newcomb with an opportunity to delve into Peirce's views on logic, Peirce now had a chance to probe Newcomb's thoughts on political economy.

For a few years around 1880, Newcomb and Peirce both supplemented their government jobs by teaching part-time at Johns Hopkins University. While Newcomb served as a lecturer from 1876 to 1883 (and a professor of mathematics and astronomy from 1884 to 1893), Peirce lectured on logic within the Mathematics Department from 1879 to 1884.[30] A letter during this period from Newcomb to Peirce exemplifies their continuing scholarly interaction. Newcomb sought Peirce's opinion on the originality of his realization that it was theoretically impossible for spectroscopic observations to produce blatant evidence that the earth was moving through the universe relative to a pervasive ethereal medium.[31] Revealingly, Newcomb sent an almost identical inquiry to another of his colleagues at Johns Hopkins, physicist Henry Rowland; though he had opened his letter to Peirce, as was their pattern, with the formal greeting, "My Dear Sir," he addressed Rowland with the more familiar, "My Dear Rowland."[32]

Newcomb's distant but amicable relationship with Peirce began to deteriorate while they were both at Johns Hopkins. Late in 1883, Newcomb learned and passed on some personal information about Peirce that hastened his dismissal from the Hopkins faculty the following year. It seems that Newcomb inadvertently alerted a university trustee, who in turn alerted President Daniel Gilman, about Peirce's having earlier lived out of wedlock with the widow of a French count. In a letter to his wife, Newcomb described this prior relationship as a "great scandal" that had even caused the couple to have been "expelled from hotels." "It is sad to think," Newcomb further commented, "of the weaknesses which may accompany genius."[33]

Open animosity developed later in the decade. Peirce repeatedly complained, for example, that Newcomb as editor-in-chief of the *American Journal of Mathematics* (from 1885 to 1893) thwarted the publication of Peirce's papers. In addition, in 1889, a disagreement flared up between the two men in the pages of the *Nation*. This public squabble followed Newcomb's criticism of Peirce's definitions in the new *Century Dictionary* (1889) of certain crucial terms in astronomy and experimental physics. Newcomb found the definitions to be "insufficient, inaccurate, and confused in a degree which is really remarkable." Through the 1890s and up to Newcomb's death in 1909, Newcomb and Peirce settled into a pattern of scholarly dialogue broken by occasional bickering. From 1890 to 1894, for example, they had an especially creative and emotionally charged exchange of letters on the technical and conceptual foundations of mathematics and astronomy.[34] Despite the discord, Newcomb expressed guarded respect for Peirce's views on philosophy of science. In particular, Newcomb replied positively in 1894 to Peirce's request for advance financial support of a proposed twelve-volume set of his writings: "I am persuaded that whatever you might write on the subject of scientific philosophy would be provocative of thought and discussion, and therefore interesting, whether one accepted your conclusions or not." Toward the end of this reply, seemingly hoping to minimize further involvement with the multivolume project, Newcomb added a self-effacing comment in which he questioned his ability to evaluate Peirce's proposed work: "I am sorry to say that you greatly overestimate the value of any expression from me on your subject. My experience leads me to believe that people have very little confidence in my views on subjects outside of mathematics and astronomy. The general subject of the greater number of your volumes is one on which people already have their minds made up."[35]

Apparently, in seeking Newcomb's advice on his philosophical writings from at least as early as 1868 up through 1894, Peirce was one person who respected Newcomb's views both in and out of mathematics and astronomy. As Peirce, however, frequently lacked financial and institutional support for his projects, this respect partially reflected a desire to cultivate the influential Newcomb. Later, in an unsigned review from 1904 of Newcomb's *Reminiscences of an Astronomer*, Peirce had an opportunity to express more candidly his opinion on Newcomb's scientific contribution. Writing anonymously, he characterized his colleague of almost half a century as "quite the most distinguished man of science in this country to-day, as well as one of the most eminent in the whole world." "His name will remain upon the page of scientific history, and eventually take its place," Peirce added, unable to resist tempering his praise with a slight snub, "high in the second rank, distinctly above Leverrier's or even Hansen's, because of the breadth of his work."[36]

VIEWS OF WRIGHT AND PEIRCE

The 1870s loom large in the history of American thought and culture—especially, the foundations of pragmatism. Charles Peirce published his only early pragmatic essay in 1878. During the prior year, Chauncey Wright's friends reissued his main essays in the posthumous volume *Philosophical Discussions*. Then in 1878, Wright's friends published his *Letters*. It was also during the mid to late 1870s that Newcomb began regularly publishing essays on the nature and method of science and its philosophical, theological, and social implications. Newcomb's midcareer ideas on language and meaning paralleled those of his two colleagues, especially Wright. The three men agreed, for example, that the success of recent scientific inquiry reflected the scientists' reliance on concepts having clear sensory definitions. And all three men sought to extend this "practical" method of inquiry to issues in areas such as philosophy, theology, and economics. Before going on to Newcomb's midcareer writings, it is worthwhile to summarize the main parallel writings by Wright and Peirce.

Wright's writings reflected the positivism of Comte and Mill and presaged, but did not explicitly prefigure, the pragmatism of Charles Peirce and William James. As early as 1865, Wright revealed such tendencies when he published a lengthy attack on English philosopher Herbert Spencer for misappropriating and muddling the clear, experiential concepts of physics and biology. In the Spencer article, Wright endorses the positivists' criterion for establishing the meaning of "ideal or transcendental elements" in scientific discourse; such elements "must still show credentials from the senses, either by affording from themselves consequences capable of sensuous verification, or by yielding such consequences in conjunction with ideas which by themselves are verifiable."[37] Just before his death ten years later, moreover, Wright—who had remained unappreciated except by a circle of Cambridge colleagues and protégés—published one of his most precise statements on meaning. He presented the statement in a muted manner without special emphasis as part of an extended book review in the *Nation* titled "Speculative Dynamics." In evaluating a recent monograph on Newtonian mechanics, he sharply criticized the author, a nonscientist, for using scientific terms to serve philosophically speculative ends. Regarding, for example, the physicists' term "force," Wright insisted:

> All its uses in mathematical language, or the equivalent terms, acceleration, mass, momentum, and energy, refer to precise, unambiguous definitions in the measures of the phenomena of motion, and do not refer to any other substantive or noumenal existence than the universal inductive fact that the phenomena of all actual movements in nature can be clearly, and definitely,

or intelligently analyzed into phenomena, and conditions of phenomena, of which these terms denote the measures.

Similarly, he explained that physicists had in mind particular "sensible properties" and "sensible measures" when they referred to the seemingly metaphysical concept of "stores of energy," that is, potential energy.[38]

Charles Peirce expressed views similar to those of Wright, but with some significant differences. His meaning criterion, while better focused than Wright's musings on meaning, appeared originally as just one briefly used tool in what was his growing and ever changing philosophical program. Following William James's lead around 1898, most scholars seeking the American roots of pragmatism point to a single paper by Peirce, "How to Make Our Ideas Clear." Peirce published the paper in 1878 but supposedly formulated it during the early 1870s while meeting in the informal "Metaphysical Club" with Wright, James, and other Cambridge thinkers. (Newcomb was living in Washington by then.) The paper, published in *Popular Science Monthly*, was the second of a series of articles that Peirce had written to provide, as the title of the series indicated, "Illustrations of the Logic of Science," by which he specifically meant illustrations of "the method of scientific investigation."[39] Embedded within this second paper were two sentences that later became known as Peirce's "pragmatic maxim": "Consider what effects, which might conceivably have practical bearings, we conceive the object of our conception to have. Then, our conception of these effects is the whole of our conception of the object." For example, "by calling a thing hard," we mean that "it will not be scratched by many other substances."[40]

Peirce had introduced the maxim by examining the religious doctrine of transubstantiation, particularly the elusive and highly metaphysical notion of the conversion of wine into blood. After insisting that both a "sensible perception" and a "sensible result" are essential components of a conception, he drew the hard-nosed conclusion that "we can consequently mean nothing by wine but what has certain effects, direct or indirect, upon our senses." He further concluded "how impossible it is that we should have an idea in our minds which relates to anything but conceived sensible effects of things." Later, in going on to illustrate his pragmatic maxim, Peirce specifically echoed Wright in rejecting metaphysical interpretations of scientific concepts. Thus, regarding the term "force," to which commentators often attributed intangible qualities, Peirce wrote: "The idea which the word force excites in our minds has no other function than to affect our actions, and these actions can have no reference to force otherwise than through its effects. Consequently, if we know what the effects of force are, we are acquainted with every fact which is implied in saying that a force exists, and there is nothing more to know."[41] Whereas Peirce paralleled Wright in applying his meaning criterion to expressly abstract concepts such as "force," only Peirce further applied the criterion to ostensibly simple concepts such as "hard." In other words, as we will see later, Wright's meaning criterion is less general than that of Peirce and later pragmatists.[42]

To recap, Peirce's essay appeared early in 1878 in *Popular Science Monthly*, the second of his series "Illustrations of the Logic of Science." During the prior year, Wright's main essays, including "Speculative Dynamics" and "The Philosophy of Herbert Spencer," appeared in the posthumous volume, *Philosophical Discussions*. Then in 1878, Wright's friends issued his *Letters*. Given the publication dates of these writings by Peirce and Wright—and given Newcomb's acquaintance with Peirce as well as his friendship with and admiration for Wright—it seems more than coincidence that Newcomb's midcareer writings parallel those of his two colleagues, especially Wright. Indeed, it seems more than coincidence that Newcomb first publicly presented a meticulous statement of his linguistic, empirical method late in 1878. As we shall soon see, Newcomb echoed in his speeches and essays the views of Peirce and Wright along with those of Comte, Darwin, and particularly Mill.

CHAPTER V

Midcareer

A "big, lusty, joyous man" is how Frederic Howe, one of Newcomb's American acquaintances, described him in his middle years. A "first rate man—full of *go*" was the reaction of William Thomson (later Lord Kelvin) on first meeting Newcomb in 1876. The Washington astronomer stood a little over five and one-half feet tall, with his weight having increased from about 140 pounds during his twenties and early thirties to about 175 pounds beginning in his forties. (We know these vital statistics because, in his typically tenacious fashion, he charted his growth and health over a three-decade period.) His sister recollected: "The entire physique indicated great power of endurance and capacity for rapid recuperation after long-continued exertion, either mental or physical." A friend similarly recalled that Newcomb was physically "imposing," even "elemental" in appearance. Though his quiet, reserved, and unpretentious manner belied his robust physical presence, strongly held personal interests complemented it. He had a passion for travel, both within North America and abroad. And he loved to read history and literature, purchasing volumes by authors ranging from Milton to Dickens and often memorizing lengthy poems. "His wide and varied reading, combined with accurate memory and unusual interest," an associate recalled, "made his conversation virile and enlightening." Yet another colleague found him to be "ruggedly independent in thought and in speech." This colleague added: "The essential quality of his mind is that of a philosopher, rather than that of a mathematician or an astronomer merely. . . . In his treatment of all questions, it is the philosophical habit of his mind which is the most remarkable and the most valuable."[1]

Newcomb's most distinctive physical feature was his head, which his sister described as "massive and solid, well developed in every way" and a friend depicted as "Olympian" and "superbly poised." "His massive head," another colleague felt, "suggested at once the rugged intellect he possessed." Further possessing, in his sister's words, "a fair, ruddy skin, dark brown hair and beard, and large, full eyes far apart and of such depth of blue that in the shadow they seemed black," he had an imposing mien. And the impact of his appearance, according to another colleague who knew him after he achieved fame, intensified with age:

> As a presiding officer he was a most extraordinary and impressive figure. Not unconscious of the high worth implied by the numerous honors, doctorates, and decorations heaped upon him from all parts of the world, he conducted himself with great dignity, heightened by the massive leonine head with its crown of iron-grey hair and the strong mouth framed by beard and moustache. To see him presiding at a meeting of astronomers was indeed a serious sight, well calculated to inspire awe of the profession in a youthful mind.[2]

Having done farm work as a youth and become "addicted" to gymnastics and rowing while in Cambridge, Newcomb maintained during his middle years a vigorous exercise program centered on walking. His walking regimen took on an extra attraction when at age forty-eight, in 1883, he spent a year abroad and developed an "insane" fascination with hiking and climbing in the Swiss Alps. While physically strong, he suffered from an infirmity traceable to his first years in Washington when he started working at the Naval Observatory with its swampy, mosquito-infested setting; he began to experience "a bilious or malarial attack" once or twice a year. Usually the episodes lasted only a week or two, but one "bilio-malarial fever" that struck in his early thirties persisted for several weeks and left him weak for months. Though the attacks tapered off beginning in the 1870s, they weakened his legs so much that, try as hard as he might, he could no longer walk beyond eight or ten miles.[3]

Paralleling Newcomb's physical setbacks and recovery, the nation's scientific enterprise was still recovering from the setbacks of the Civil War. During the middle years of his career, few organizations fostered studies on physical science. Of the organizations that did, the most respected included the Smithsonian Institution, the Harvard Observatory, the U.S. Coast and Geodetic Survey, and the navy's Nautical Almanac Office. From the mid 1870s through the early 1880s, with his reputation burgeoning, Newcomb received either direct offers or informal pleas to head each of these organizations.

TAKING CHARGE OF THE ALMANAC OFFICE

In one case, Harvard zoologist and geologist Alexander Agassiz (son of the prominent Louis Agassiz) wrote to Newcomb in 1878 expressing support for him as the successor of Smithsonian secretary Joseph Henry, the deathly ill dean of American physical science. "I would frankly say that you are my first choice," Agassiz confided, "and should you feel it possible to accept this position and at the same time go on with your scientific work there are many of our scientific friends who would gladly do all in their power to see you at the head of the Smithsonian." While rejecting Agassiz's offer of support in this "flattering letter," Newcomb revealed in his reply that Joseph Henry himself concurred with Agassiz's choice. "If I had no important unfinished work, nothing would be more agreeable to me than the administration of the Smithsonian. Some months ago Professor Henry expressed to me a desire that I might be his successor but I am not willing to give up the great life work which I am getting fairly under way, that of perfecting the theories of the planetary motions and constructing tables of the planets corresponding in accuracy to the present state of astronomy. The administration of the Smithsonian Institution would be incompatible with the mental leisure necessary to the successful prosecution of this work."[4]

Henry's desire to have Newcomb succeed him reflects the deepening of their friendship after Newcomb moved from Cambridge to Washington in late 1861 to serve at the Naval Observatory. Indeed, it was Henry who introduced Newcomb to the woman who in 1863 would become his wife—Mary C. Hassler, granddaughter of the founder of the U.S. Coast Survey, Swiss-born geodesist Ferdinand R. Hassler. Newcomb and his old mentor had become intimate friends, having frequent chats regarding personal as well as professional matters. For example, Newcomb jotted in his diary for 2 February 1867 that he "called on Prof. Henry this evening, and had quite a talk with him, mainly on the teaching of Physics. He expressed great regret that he had ever come to Washington." Newcomb had no regrets that Henry was in Washington and able to provide moral support—and Smithsonian financial support for research on, for example, the orbits of Neptune and Uranus. By the mid 1870s, Henry held Newcomb in high esteem. In recommending his younger friend for membership in the French Academy of Sciences, he could write "that for scientific attainments, power of original investigation, habits of industry and perseverance and a genuine love of science, I am acquainted with no one in this or any other country who is his superior."[5]

In addition to the feelers to become Smithsonian secretary, an enticing offer also had come three years earlier from Charles W. Eliot, the president of Harvard and an old friend. (During Newcomb's student days at the Lawrence Scientific School, Eliot served as a mathematics tutor.) Eliot had offered Newcomb, still a "professor" at the U.S. Naval Observatory, the directorship of the Harvard Observatory—a position "held for life, with pleasant surroundings and increasing resources." After some exchanges regarding salary, a house, computational assistants, and other conditions of employment, Newcomb declined. He had been losing interest in observational astronomy and had been turning increasingly to theoretical calculations involving the planets, sun, and moon. Also, through his marriage, he had established strong family ties in Washington. Even a pessimistic prognosis for government science voiced by Eliot and reiterated by Newcomb's superior, Secretary of the Navy George M. Robeson, could not induce him to take the position. Nor could the prodding of his former teacher at Harvard's Lawrence Scientific School, Benjamin Peirce, persuade him to accept. Toward the end of the negotiation, Peirce had written to Newcomb, greeting him with the declaration: "Your scientific friends at Harvard are greatly chagrined that you are not at the head of our observatory." He closed the letter with equal force: "Harvard needs you and must have you at whatever cost." Although Newcomb went on to excel in the type of lunar and planetary orbital theory in which Benjamin Peirce had pioneered, he chose not to do it as director of the Harvard Observatory.[6]

Another feeler for a job came in 1881, when a vacancy occurred in the superintendency of the Coast and Geodetic Survey, a position held in earlier years by Peirce (before the organization had added "Geodetic" to its name). Again, Newcomb was unresponsive to the suggestion of James J. Sylvester, a foremost British mathematician then at the Johns Hopkins University, to apply for the post.[7] In fact, the post Newcomb sought primarily during the period of all these offers was the superintendency of the organization for which he had first worked in the late 1850s: the navy's Nautical Almanac Office. He felt suited to this astronomical agency, with its computational rather than observational emphasis.

After serving at the Naval Observatory since 1861 and after much political maneuvering, Newcomb was appointed superintendent of the navy's Almanac Office in 1877. The office was now located in Washington, having been moved from Cambridge in 1866. He believed, as he later stated in his *Reminiscences*, that as head of this Washington agency dealing with applied and basic astronomy, he was "in a position of recognized responsibility" where he "could make plans with the assurance of being able to carry them out." In addition, he felt that as a top scientist in government he could help remedy what he perceived in the United States to be "an absence of touch between the scientific and literary classes on one side, and 'politics' on the other." Writing in 1881 to Harvard's Alexander Agassiz, he had earlier expressed these convictions "with a frankness which I have never before ventured upon, even with my most intimate friends." (Agassiz and Newcomb, both born in 1835, had been fellow students at Harvard in the late 1850s.) In deflecting yet another invitation to join the "best men of the century" at Harvard, Newcomb explained to Agassiz that by persevering with his government-sponsored research program, he retained the possibility of seeing his name "associated with one of the great astronomical works of the century." In addition, while denying that he spoke with "over confidence" or "any pretensions," he entertained the further possibility of succeeding Henry as the scientific community's informal liaison with the government: "Since the death of Professor Henry there has been here no recognized representative of the general interests of Science in connection with the government. Perhaps there may be no need of one, but it seems to me there is." Newcomb's appraisal of the scientific opportunities at the Almanac Office, an agency that he aggressively restructured, would prove to be reasonably accurate. His appraisal, however, of the opportunities in government service to improve communication between scientists and politicians would prove to be overly optimistic. After retiring from the Almanac Office in 1897, at the mandatory naval retirement age of sixty-two, he would still be lamenting the "want of touch between our academic and political classes." Somewhat disillusioned, he would even come to question the sagacity of his decision to remain a government scientist.[8]

THE PLANETS AND THE MOON

What had Newcomb accomplished in his early astronomical research to merit possible head positions at the Smithsonian Institution, the Harvard Observatory, the Coast Survey, and the Almanac Office? Though he spent his first years in Washington principally making exact determinations of stellar positions, he had enough free time at the Naval Observatory to reassert his interest in mathematical astronomy—an interest that dated even to the days before his immersion in the subject at Cambridge. As the decade of the 1860s unfolded, he returned to theoretical studies of the planets and the moon. Following in the footsteps of Benjamin Peirce and other mentors, he felt challenged to formulate abstract mathematical expressions to account for actual planetary and lunar observations. This involved making complex calculations of orbital deviations caused by the gravitational perturbations of interacting celestial bodies and then constructing positional tables that would allow comparisons with observational data. (In fact, as he was well aware, the "empirical" and "theoretical" components of this research enterprise blurred. The empirical observations, on the one hand, had to be "reduced" through a variety of mathematical and statistical techniques. The theoretical equations for the orbital parameters or "elements," on the other hand, incorporated various empirical "constants" such as the planets' masses and the earth-sun distance.) By engaging in this research program, Newcomb was answering a high calling. As Arthur Cayley, English mathematician and president of the Royal Astronomical Society, commented in 1874: "The formation of the tables of a planet may, I think, be considered as the culminating achievement of Astronomy: the need and possibility of the improvement and approximate perfection of the tables advance simultaneously with the progress of practical astronomy, and the accumulation of accurate observations; and the difficulty and labour increase with the degree of perfection aimed at." Seeking a new degree of perfection, Newcomb tackled a particularly attractive pair of planets, Uranus and its recently discovered companion, Neptune (observed in 1846 following the predictions of Urbain J. J. Leverrier and John Adams). By 1868, he had completed provisional studies of the two planets and was ready to begin a five-year investigation that culminated in definitive tables for Uranus—tables that included the perturbational effects of Neptune.[9]

Newcomb also found time at the Naval Observatory to hone the mathematical theory of the moon's motion. Drawing Newcomb to the lunar problem were not only recent theories by Peter Hansen and Charles Delaunay but also the intrinsic complexity of the moon's motion. Newcomb would later come to view his lunar research as the centerpiece of his life's work in mathematical astronomy. By 1869, having completed his reappraisal of star positions, he decided to pick up the strands of some of his earlier lunar investigations by initiating a concerted study of the moon's motion. Concerned that his post in an observational facility precluded him from engaging in this intensive theoretical study—and seeking, more generally, to advance his career—he petitioned for a transfer to the more mathematically inclined Nautical Almanac Office. Though he gained the support of Peirce and Henry, his superiors judged the move unnecessary; they agreed, however, that Newcomb could proceed with the research under the auspices of the Naval Observatory. In the next few years, he published a series of innovative lunar studies in French, German, and British journals.

What he considered his biggest lunar coup came, however, in 1871 as part of a memorable first trip to Europe—the same extended trip that included a conversation with John Stuart Mill and the observation of a solar eclipse in Gibraltar. Traveling on to France, he resurrected old records at the Paris Observatory and ferreted out lunar positional data extending back to 1675. This six weeks of archival digging (made in the thick of the civil hostilities surrounding the Paris Commune), followed by three or four years of calculation and analysis, added seventy-five years of data to the lunar record, dramatically demonstrating the deficiency of accepted lunar tables. About this time, he also experienced a more personal "triumph." Like Henry, Benjamin Peirce had come more and more to respect Newcomb as a scientist. Though slight friction arose in the mid 1860s over the study of Neptune's orbit, Peirce increasingly envisioned him as successor to his life's research on the moon's motions. In the handwritten manuscript titled "Autobiography of My Youth," Newcomb recalled Peirce confiding in 1872 that "he was too old to go on with the work, and had become convinced that I was the one to do it. His immediate object was to turn over to me work on the lunar theory." Newcomb added, with glee probably similar to that expressed by Johannes Kepler when he assumed control of Tycho Brahe's research program: "Thus, my triumph may be said to be complete. From this time forward no obstacle was ever placed in the way of any work I desired to undertake."[10]

Though Newcomb placed high stock in Peirce's gesture, the event that perhaps did more to remove obstacles from Newcomb's scientific path occurred early in 1874. In Britain, the Royal Astronomical Society presented its gold medal to the thirty-nine-year-old American. In the formal citation read at the award ceremony, President Cayley proclaimed that all of Newcomb's astronomical writings exhibit

> a combination, on the one hand, of mathematical skill and power and on the other hand of good hard work—devoted to the furtherance of Astronomical Science. . . . The Memoir on the Lunar Theory contains the successful development of a highly original idea, and cannot but be regarded as a great step in advance in the method of the variation of the elements and in theoretical dynamics generally; the two sets of planetary tables [for Neptune and Uranus] are works of immense labour, embodying results only attainable by the exercise of such labour under the guidance of profound mathematical skill—and which are needs in the present state of Astronomy.

He added, "We have done well in the award of our medal."[11]

To be sure, even before winning the medal for studies of the moon, Neptune, and Uranus, Newcomb was gaining professional recognition, as evidenced by invitations to become a member of elite national and international scientific organizations. But the British award signaled the beginning of a cascade of honors that would persist for the rest of his career. The principal European scientific societies of which he was not yet a member began to elect him to special categories of membership such as "foreign associate," "corresponding member," or "honorary fellow." He also began to garner honorary doctoral degrees, a trend that would eventually involve most of the major universities in Europe and North America. And merely four year after the British award, the Dutch Academy of Science presented him with the Huygens gold medal. Awarded only every other year, this prestigious medal rotated in twenty-year cycles between the different natural sciences, with the award of 1878 going to the one astronomer who over the prior two decades "distinguished himself in an exceptional manner."[12]

Newcomb completed his tenure at the Naval Observatory with two major observational projects. In 1873, his superiors placed him in charge of a new telescope—not just any telescope, but the nation's largest operating refractor. Having helped initiate and guide the effort to obtain this massive telescope, with its twenty-six-inch aperture and thirty-five-foot tube, he found himself awestruck when he became the first astronomer to test the instrument: "I was filled with the consciousness that I was looking at the stars through the most powerful telescope that had ever been pointed at the heavens, and wondered what mysteries might be unfolded." The mystery that he first tried to solve involved a possible second satellite of Neptune; though unsuccessful in his search, he did collect orbital data on the known moons of Neptune and Uranus to use in calculating more exact values of the planets' masses, critical constants in the construction of the planetary tables.[13] He also took the lead in a second major project: mounting an American expedition, composed of eight separate parties, to observe the 1874 transit of Venus. The Americans joined European astronomers, all stationed at different sites in the eastern hemisphere, in applying triangulation techniques to the passage of Venus over the solar disk to better fix the earth-sun distance. Though Newcomb introduced innovative photographic apparatus, "unpropitious" weather largely frustrated the efforts of the American and European parties. This disappointing experience would prompt Newcomb to argue against new expeditions for the coming transit of 1882, though it would be the last until 2004. He felt that the sun's distance could be better determined using calculations involving the velocity of light and the earth's orbital velocity (that is, involving what is technically known as the aberration of light). Whereas he went on to exploit the latter method, he deferred to his colleagues' preferences, consenting to lead an 1882 expedition to the Cape of Good Hope. While the new international transit data enabled him to refine the earth-sun distance for personal use in his planetary tables, he lamented that the American results were never published.[14]

Newcomb proved a capable administrator after taking charge of the Nautical Almanac Office in 1877. "Practically I had complete control of the work of the office," he recalled, "and was thus, metaphorically speaking, able to work with untied hands." Freedom to set the research agenda was important to Newcomb: While still at the Naval Observatory, he had chafed under the leadership of those naval administrators who lacked scientific sensitivity. (Over the next twenty-five years, he would work behind the scenes in support of civilian astronomers' attempts—ultimately unsuccessful—to gain control of the Naval Observatory.[15]) He soon reinvigorated the Almanac Office, securing new quarters in a recently completed government building and assembling a staff of eight or ten mathematical assistants (that is, "computers," as he had been in Cambridge). Impressed by the success of Sir George Airy, the Astronomer Royal, in systemizing activities at Greenwich—and, to a lesser extent, Leverrier in organizing the Paris Observatory—Newcomb adopted a managerial approach characterized by efficiency and economy. He insisted, for example, that promotion be based on merit rather than seniority and that salary be commensurate with time spent on a job. "These economies went on increasing year by year," he explained, "and every dollar that was saved went into the work of making the tables necessary for the future use of the Ephemeris."

While he always carefully justified the office's work in terms of its indispensability to American ships navigating the world's oceans, he was personally and primarily interested in the basic science behind the navigational tables. In this scientific realm, he also proved to be a capable administrator, charting a systematic and exhaustive course of research.

> The programme of work which I mapped out, involved, as one branch of it, a discussion of all the observations of value on the positions of the sun, moon, and planets, and incidentally, on the bright fixed stars, made at the leading observatories of the world since 1750. One might almost say it involved repeating, in a space of ten or fifteen years, an important part of the world's work in astronomy for more than a century past. Of course, this was impossible to carry out in all its completeness. In most cases what I was obliged practically to confine myself to was a correction of the reductions already made and published. Still, the job was one with which I do not think any astronomical one ever before attempted by a single person could compare in extent. . . . The other branches of the work were . . . the computation of the formulae for the perturbation of the various planets by each other.

To ease publication of findings, Newcomb launched in the early 1880s the *Astronomical Papers Prepared for the Use of the American Ephemeris and Nautical Almanac*. This complemented the office's mandated issuance of the *American Ephemeris and Nautical Almanac*.

By 1894, seventeen years after taking over the Almanac Office and thirty-three years since joining the Naval Observatory, he had completed the bulk of the research program. Except for the final step of constructing tables for the planets beyond Mars and a few other loose ends involving the moon's orbit, he had largely succeeded in bringing to a close the reduction of the observations and the determination of the planetary orbits. One colleague later described the effort as being "of herculean and monumental proportions." Twentieth-century commentators would look back, for example, at his analysis of Mercury's orbit, noticing that he had pinpointed the modern value of a slight orbital anomaly (known as precession of the perihelion and first detected by Leverrier). This anomaly, which Newcomb suspected defied conventional Newtonian gravitational explanations, would become intelligible only through Albert Einstein's general theory of relativity. Indeed, Einstein would describe Newcomb's lifework as being "of monumental importance to astronomy."[16] But Newcomb's "preliminary results," which he published early in 1895 as *The Elements of the Four Inner Planets and the Fundamental Constants of Astronomy*, also generated a more immediate response. As we will see, the results helped actuate an international movement to set the world's astronomical ephemerides on a more homogeneous basis—something that Newcomb had been urging for many years.

In light of his technical achievements, his accolades from the international scientific community, and his professional connections through friendships and marriage, it is easy to understand why, in just the years from 1875 to 1881, Newcomb received offers or feelers to head the Smithsonian Institution, the Harvard Observatory, the Coast Survey, and the Almanac Office. We also begin to understand why he gained such a high reputation among the general public: his popular renown grew, in large part, as a consequence of his professional reputation. But his solid scientific credentials do not account fully for his high standing in the eyes of the American public. Admittedly, the public could appreciate Newcomb's more palpable and uncomplicated activities, such as his association with the nation's largest refracting telescope and his expeditions to exotic parts of the world. They could also appreciate, and be impressed by, his gold medals and honorary degrees. However, the staples of Newcomb's research program—performing the laborious perturbational calculations of classical planetary mechanics and constructing the complex tables of an astronomical ephemeris—simply were too murky and arcane for even highly educated Americans to appreciate. Add to this that Newcomb, for all his contributions, never made that one dramatic discovery or devised that single influential theory to which the public could easily attach his name. We must look beyond his personal, technical accomplishments to comprehend more fully his popular appeal.

At least part of Newcomb's allure arose not out of anything particular to him but out of a general mystique common to astronomy. Astronomers in nineteenth-century United States, as in other scientific nations, enjoyed tremendous prestige. It mattered little that their traditional image as heroic, isolated discoverers—a public image perpetuated by the astronomers themselves—had less and less connection to the modern reality of professional managers of workshops housing mechanized instruments and regimented assistants. Journalists celebrated the romance of stargazing. Philanthropists, politicians, and educators encouraged the building of observatories. The same aura that had lured young Newcomb to astronomy served later to attract the American public to Newcomb the astronomer. And Newcomb's peculiar trappings as a navy astronomer enhanced this inherently positive image. He later recalled that the young astronomy "professors" found it pleasant "to wear the brilliant uniform of their rank, enjoy the protection of the Navy Department, and be looked upon, one and all, as able official astronomers." "As things go in Washington," he added, again looking back on his own circumstances, "the man who does his work in a fine public building can gain consideration for it much more readily than if he does it in a hired office."[17] Already a practitioner of a charmed profession, Newcomb appeared doubly charming bedecked in military finery and begirded by federal architecture.

Newcomb did not, however, simply wait for popular recognition to come to him. A born writer with an almost obsessive drive to participate in public forums, he initiated a dialogue with the American public. Beginning in earnest around 1873, he penned an intermittent but ongoing series of astronomical articles, reviews, and often anonymous notes in widely read journals such as the *Nation*, *Scribner's Monthly*, *Harper's Magazine*, the *North American Review*, and *Popular Science Monthly*. Occasionally, he even submitted letters on astronomy to newspapers such as the *New York Tribune*. His most ambitious early popularization appeared in *Harper's Magazine* during late 1874, half a year after winning the Royal Astronomical Society's medal. Running in two consecutive issues and titled "Some Talks of an Astronomer," the thirty-page article offered "a very short and summary survey of the work of astronomers in exploring the heavens." Newcomb took advantage of this public forum to promote his specific branch of astronomy ("The principal problem of the astronomy of the present day is the determination of the motions of all the heavenly bodies") and to advertise its benefits for a sea captain ("On the accuracy of this information he must often risk his ship and all on board"). Around this kernel, Newcomb would build a more comprehensive survey, published four years later as *Popular Astronomy* by Harper Brothers—a thriving New York City publisher that, during this period, issued many of Newcomb's writings on astronomy and political economy. Thirty years later, after the book had gone through many American and British editions as well as translations into German, Norwegian, and Russian, one respected astronomer would comment that "it still remains the best composition on the subject." Another astronomer, pondering all of Newcomb's popular books and articles, would conclude more generally: "His profound knowledge, logical and orderly mind, together with a facile pen, made him the best popular writer on astronomy since Sir John Herschel." Of course, Newcomb's popularizations did more than entertain and inform the general public; they also helped recruit budding scientists into the astronomy profession. One university student who redirected his studies to astronomy after reading Newcomb's *Popular Astronomy* in the mid 1880s was William W. Campbell, later director of the Lick Observatory, president of the University of California, and chronicler of Newcomb's professional achievements.[18]

PROFESSIONAL INVOLVEMENTS

During his middle years, not only governmental and academic organizations desired Newcomb as a leader but also scientific societies. In 1877, he served as president of the American Association for the Advancement of Science (AAAS). In 1879, he succeeded the recently deceased Joseph Henry in two annual terms as president of the Philosophical Society of Washington. Henry had presided over this "society for the advancement of science" since founding it in 1871 with the help of Newcomb and other Washingtonians.[19] Then, in 1883, the National Academy of Sciences elected Newcomb as vice-president, a post he held until 1889. And in 1885 and again in 1886, even though he was skeptical of psychic phenomena, Newcomb served as the first president of the American Society for Psychical Research. William James, an organizer of this unorthodox but fashionable research society, believed that Newcomb gave the group credibility and respectability, or as he expressed it to a colleague: "I think Newcomb, for President, was an uncommon hit."[20]

Universities also courted Newcomb as a teacher, especially in his home area of Washington, D.C. He served on the nonresident staff at the Johns Hopkins University in its first year of operation, 1876, participating in a highly visible program of afternoon public lectures. Specifically, President Daniel C. Gilman agreed to pay him $1000 for a course of twenty lectures "on the Hist. of Mathematical science, or on some other theme." Newcomb settled on the history of astronomy, eventually incorporating the lectures in his best-selling *Popular Astronomy*.[21] From 1884 through 1893, he held the more formal position at Hopkins of professor of mathematics and astronomy—after having declined Gilman's offer to head the mathematics department. Though officially James Sylvester's replacement after the British mathematician accepted a new appointment at Oxford, Newcomb still directed the Almanac Office and thus had to limit his Hopkins position to half time. He somehow arranged enough open hours, however, to succeed Sylvester as editor of the *American Journal of Mathematics*, the pioneering scholarly periodical that he had helped Sylvester launch at Hopkins. He also found time to upgrade the university's astronomy department, urging the trustees to purchase enough instruments to enable students to make their own celestial observations. In addition, he was a lecturer from 1873 to 1884 and then professor of astronomy from 1884 to 1886 at Columbian University, later renamed George Washington University. And although he presented a series of four lectures on political economy at Harvard during the academic year 1879–1880, he reacted with reserve the following year to a feeler from Alexander Agassiz to join "the best men of the century" as professor of mathematics at Harvard. Finally, in 1885, the University of California unsuccessfully tried to enlist him as president of the school.[22]

Having become identified with the state-of-the-art twenty-six-inch telescope at the Naval Observatory, Newcomb found himself in demand as a consultant on subsequent major refractors. Around 1874, perhaps influenced in part by an article about the new telescope that Newcomb wrote for *Scribner's Monthly*, wealthy Californian James Lick decided to build an even bigger telescope. Over the next fourteen years, Lick's agents would rely on Newcomb as their principal advisor, soliciting his counsel on everything from design to staffing of the world's largest refractor, with its thirty-six-inch lens and fifty-seven-foot tube. In fact, he was probably offered the directorship of the new Lick Observatory on California's Mount Hamilton. Similarly, beginning in the late 1870s, Newcomb endeared himself to Otto Struve and other Russian astronomers by advising them on the construction of a thirty-inch refractor for the Pulkovo Observatory. The Russian czar would recognize Newcomb's services in 1889 with the presentation of an inscribed jasper vase, while the Imperial Academy of Sciences in Saint Petersburg would award him their Schubert Prize in 1897.[23]

Even with the demands created by the Almanac Office, scientific societies, universities, and observatories, Newcomb found time to produce a prodigious quantity of technical, popular, and pedagogic books and articles. (After his death, his sister recalled that he "worked not like an avalanche, but like a glacier, slowly, steadily, irresistibly."[24]) Newcomb's bibliography of lifetime writings includes 318 works in astronomy, 35 in mathematics, 42 in economics, and 146 dealing with physics, philosophy, religion, society, and other miscellaneous topics.[25] Almost half of this total of 541 items appeared in the decades of 1870 and 1880. And this half included not only major technical monographs but also many of his most widely reprinted and translated books, such as *Popular Astronomy* (1878), *Elements of Plane and Spherical Trigonometry* (1882, part of the series "Newcomb's Mathematical Course"), and *Principles of Political Economy* (1885). This half also included the bulk of his general essays on science.

Besides publishing at this pace, Newcomb also read relentlessly. Over the years, he assembled a massive personal research library.[26] He also maintained a voluminous correspondence with scientists and other thinkers throughout the world, this still being an era when informal informational networks were crucial to scholarly pursuits. Through midcareer alone, Newcomb regularly exchanged letters with North Americans such as Henry Adams, Alexander Graham Bell, Charles W. Eliot, Daniel C. Gilman, William James, Samuel Langley, and Henry Rowland, to name only a few. European correspondents included scientific luminaries such as Norman Lockyer, James Clerk Maxwell, Otto Struve, and William Thomson.[27] And as eight jammed boxes of invitations and calling cards attest, Newcomb and his wife lived an increasingly active social life, receiving at home and visiting while abroad the topmost scientific, educational, and political leaders. As head of a key scientific office within the government, he even had occasional dealings with the president. And as a distinguished representative of the United States, he gained audiences with most of the reigning monarchs of Europe.[28]

Of his myriad writings and professional involvements, those which brought him most recognition during midcareer dealt, of course, with mathematical astronomy. A close colleague later drolly remarked that "a continual flow of medals, prizes, degrees and honorary memberships came for his reception, till the possibilities were exhausted." The possibilities narrowed dramatically when, in 1890 at age fifty-five, Newcomb received the Copley medal of the Royal Society of London. Great Britain's highest scientific award—indeed, perhaps the world's most coveted scientific prize in this pre-Nobel era—the Copley medal recognized Newcomb's cumulative contribution not just to astronomy (especially lunar theory) but to natural science overall. The medal had gone to only three other North Americans: Benjamin Franklin in 1753, Swiss-born Louis Agassiz in 1861, and James Dwight Dana in 1877. As his midcareer years drew to a close, Newcomb found himself amply rewarded for his exhaustive updating of the data, theories, constants, and tables for the motions and positions of the moon, planets, sun, and principal stars.[29]

Considering that the list of Newcomb's honors could be greatly expanded by including lesser awards and carrying the list through his retirement in 1897 and death in 1909 (at age seventy-four), it becomes easy to understand the epitaph offered by Raymond C. Archibald, one of Newcomb's old colleagues: "No other American scientist has ever achieved such general recognition of eminence." Indeed, as early as 1898, William Alvord, president of the Astronomical Society of the Pacific, in awarding the group's international medal to Newcomb, could unabashedly speak of "the undoubted fact, that he has done more than any other American since Franklin to make American Science respected and honored throughout the entire world." Similarly, Lehigh University's William Harper Davis, in the lead article in *Popular Science Monthly* a few weeks after Newcomb presided over the international scientific congress held with the 1904 Saint Louis world's fair, could label him "perhaps of all Americans the most honored throughout the world among the peers of the realm of science." Finally, one can appreciate why the announcement of Newcomb's death in July, 1909, was featured by newspapers and professional journals around the world and why his elaborate state funeral was attended by President William H. Taft and other national and international dignitaries. Holding the "relative" rank of rear admiral in the Navy, Newcomb was buried with full military honors in Arlington National Cemetery. In the funeral procession from the family church to graveside, the U.S. Marine Band, three companies of marines, and one company of U.S. Navy bluejackets led Newcomb's black caisson, which was draped in an American flag, driven by a detail from an artillery regiment, and followed by a long file of carriages.[30]

CHAPTER VI

American Science, Scientific Method, and Social Progress

From 1866 onward, Newcomb regularly contributed reviews and articles, especially on political economy, to the *North American Review*. By 1874, editor Henry Adams had gained enough confidence in Newcomb's work that during an absence from the journal's office he could instruct his assistant, Henry Cabot Lodge, to publish without delay or further consultation writings submitted by a select group of "responsible" authors including Newcomb. Adams himself, in fact, had just followed this procedure with Newcomb's latest contribution, an article titled "Exact Science in America."[1]

In "Exact Science," this recent recipient of the Royal Astronomical Society's gold medal presented his first comprehensive analysis of the deficiencies of physical science and mathematics in the United States. The article, intended ultimately to advance American science, linked Newcomb to the germinal campaigns mounted in preceding decades by Henry, his fellow Lazzaroni, and other early science boosters. Over the next half dozen years, he would elaborate his analysis, contending that if the institutional framework of science was to be strengthened within this nation dedicated to democratic doctrines, there needed to be fuller public support of basic research. Such support, he felt, would be forthcoming if educated citizens could be convinced of the value of science. Hoping to help persuade them, Newcomb chose to argue that the value of science lay in its method. When applied to political and economic issues, for example, the scientific method would lead to social progress.

Edward L. Youmans, the editor of *Popular Science Monthly*, was impressed by "Exact Science in America." "So important is the subject, and so excellent its presentation," Youmans wrote in the December 1874 issue of the *Monthly*, "that we shall make copious extracts from the article." Youmans proceeded to quote and endorse Newcomb's main conclusions, even though "his results are not flattering to our national vanity." Another editor, Benjamin Silliman, Jr., of the venerable *American Journal of Science and Arts*, also praised Newcomb's "excellent article." Chemist Silliman wrote to his colleague agreeing with the article, even though, like Youmans, he found the analysis to be "very little flattering to our self love." He added that Newcomb had adopted "the course best calculated to correct the evils," in particular, bringing the evils to "public notice."[2]

INSTITUTIONAL DEFICIENCIES AND PUBLIC INDIFFERENCE

What were Newcomb's unflattering findings? He began his article by asserting that in the United States the exact sciences—physics, astronomy, and their close ally, mathematics—were foundering. Contrasting the quantity of research published by Americans and Europeans in these fields, he surmised that the Americans lagged behind their overseas colleagues in level of achievement by a generation. He cautioned that this lag was not because of inadequate facilities or unqualified personnel. Especially in the eastern states, researchers had access to good libraries as well as expensive instruments and apparatus. And the country possessed its share of first-rate practitioners in physics and astronomy. In actuality, Newcomb explained, the lag correlated with deficiencies in research journals, universities, and professional societies. Newcomb devoted the bulk of the essay to documenting these institutional shortcomings and linking them ultimately to the lack of support in the United States for research—basic research, that is, as contrasted to research having overt practical applications.

Unlike exact scientists in Europe, Americans published no journals that were devoted exclusively to reporting original research in the single field of either mathematics, physics, or astronomy. Regarding mathematical journals, Newcomb commented, "We have none and never have had any." He acknowledged that physicists and astronomers did have one domestic outlet, but this was not a journal for specialists but the multidisciplinary *American Journal of Science and*

Arts, or as it was familiarly known, Silliman's journal. Newcomb also drew stark contrasts between Europe and the United States in science education. Whereas in Germany professors were expected and encouraged to initiate original investigations, in the United States most were not. Finally, Newcomb noted the dearth and ineffectiveness of national professional organizations. The only large groups that catered to the exact sciences were the Geographical Society of New York and the AAAS. Regarding these and smaller scientific organizations in the United States, he complained that "with a few exceptions, they exhibit a total lack of cohesive power, vitality, and that undefinable something which may be called weight and importance."[3]

Newcomb linked the institutional problems to a lack of personal incentives offered to researchers. In Germany, England, and France, various combinations of journals, professional societies, and universities served to encourage and reward scientists engaged in basic research; these institutional structures, in turn, had the support of each nation's overall scientific community and educated citizenry. In the United States, neither the scientific community nor the public rewarded researchers with wealth or status. Scientists themselves as well as the public doled out the few existing perks on the basis of "simple age and social position" rather than "talent and industry" — a vestige of a more genteel, less professional era. Although bothered by his fellow scientists' disregard for fundamental research, Newcomb was particularly upset by the public's indifference to the worth of such research. "The main fact with which we have to deal is," Newcomb insisted, "that original scientific research does not by itself command the public consideration which the same talent would if directed in other ways, nor which it would if exercised in the same way by a European." This low regard for basic research translated into weak journals, societies, and universities.

Newcomb concluded his essay with a confident statement of a general solution to the problem. The public simply must be taught to respect and support basic science. He wrote: "The remedy is to educate the intelligent public into an appreciation of the importance of scientific investigation, and of the necessity of bestowing upon those who are successfully engaged in it something in the way of consideration which may partially compensate them for devoting their energies to tasks which, from their very nature, can bring them no pecuniary compensation." This education of the "intelligent public," he suggested, could take place through articles and columns in newspapers and journals. Appealing to nationalistic sentiments, he added that public support when coupled with a realignment of values within the scientific community would result in the coming of age of American physical science. "If we had an equally rigorous system of intellectual natural selection and equal public encouragement for talent of the highest class, America would rapidly take a leading position among the scientific nations of the world."[4]

Intent on establishing that a problem existed with American exact science, Newcomb devoted little space in the 1874 article to his proposed solution—educating the public to appreciate and support basic science. At one point, however, by briefly instructing his readers on the general virtues of scientific methods, he presaged the thrust that he would take in his later writings and speeches when he enlisted more fully in the campaign of public education. Reflecting the influence of colleagues such as Wright and authors such as Mill, he declared that the "first proposition" of modern science is that its "methods and objects . . . are distinguished by their purely practical character, using the word 'practical' in its best sense. Indeed, the most marked characteristic of the science of the present day . . . is its entire rejection of all speculation on propositions which do not admit of being brought to the test of experience." Taking as an example the prediction of an eclipse using the law of gravitation, he further asserted: "This prediction is complete with respect to the phenomena and to everything connected with it which can influence the material interest of mankind, yet it is entirely independent of the question, What causes the moon to gravitate toward the earth and sun?" Scientists solved the problem of planetary motions, Newcomb advised his readers, only by ignoring questions of ultimate cause and "confining the attention to the purely phenomenal aspect of the problem."[5] This "practical" attention to the legitimation of propositions through "the test of experience" as well as this rejection of metaphysical "speculation" are themes that Newcomb would soon elaborate and promote more fully in his campaign to garner national support for exact science.

Apparently, Newcomb's recital of deficiencies touched a raw nerve among American scientists. Besides expressions of agreement from two of the nation's most active science journalists, Youmans and Silliman, he also received an endorsement of his "excellent article" from Philadelphia entomologist John L. LeConte. "You have most admirably expressed several ideas which I have uttered crudely in conversation with friends in this city, & have certainly done a good service to your colleagues in science by calling attention to the duties which the community owes to them."[6]

THE NATION'S NEED FOR SCIENTIFIC METHOD

In August 1875, Henry Adams wrote to Newcomb: "I propose to issue next January a centennial number of the *North American Review*, to contain six articles of forty pages each on the movement of American thought in Religion, Politics, Literature, Law, Science, and Economy.... Will you undertake the subject of progress in Science?" Newcomb agreed, joining with other contributors such as William Graham Sumner (politics) and Daniel C. Gilman (education) to commemorate the nation's first century. During the next few months, through letters and conversations, he sought advice from editor Adams on writing the article. To Newcomb's worry that the centennial article might duplicate his previous article, "Exact Science in America," Adams responded: "I think perhaps you might at once supplement your former and complement the other [centennial] reviews by seeking the causes of that indifference to abstract research which you have observed, and giving us a diagnosis of the mental condition of our country which may offer some light as to the probable tendency of our future."[7]

Adams also offered specific advice. One of his more colorful suggestions was for Newcomb to comment on the scientific roles of certain cities, especially Philadelphia. "Philadelphia is always an unfailing resource. One can prod it with the same amusement with which a pork-packer cuts up a hog. Our New England kine are lean and hungry, but Philadelphia is juicy and streaked with rich veins of fat. Its influence on science has always been an interesting study." Though Newcomb did not heed this bit of advice, Adams liked what Newcomb eventually wrote: "I have just finished reading your essay aloud to my wife. I cannot fairly dismiss it without writing you one line to say how admirable I think it."[8]

When the essay appeared in print, two editors of the *American Journal of Science and Arts*, Benjamin Silliman, Jr., and his brother-in-law, James D. Dana, also dropped lines of praise. Though he quibbled with Newcomb about early contributors to American chemistry, Silliman appreciated his general message; he related that he had read the centennial article with a "sense of bruised patriotism at the poor showing we make in *Abstract Science*!" Writing independently of his brother-in-law, Dana told Newcomb he hoped "that your excellent paper on American Science during the century past might be the means of stirring up our scientific students to greater thoroughness and industry."9

In later years, Newcomb recalled that he had written this 1876 appraisal of American basic science "with a view of influencing the thought of the public." "I was therefore much pleased, soon after the article appeared," he further recalled, "to be honored with a visit from President Gilman, who had been impressed with my views, and wished to discuss the practicability of the Johns Hopkins University, which was now being organized, doing something to promote the higher forms of investigation among us."10 One action that Gilman took to promote such investigation was to hire Newcomb as a part-time lecturer beginning in the university's inaugural year of 1876 and then as a regular (though only half-time) professor beginning in 1884.

Newcomb's centennial article surveyed a broader range of sciences and a wider span of years than his 1874 essay. Titled "Abstract Science in America, 1776–1876," it covered events of the prior century in not only the physical sciences and mathematics but also the life sciences. His basic message regarding the current status of American science, however, remained the same. Although Americans working in the realm of basic or "abstract" science had adequate physical facilities — observatories, laboratories, and the like — the majority of researchers were failing to generate and publish original ideas. Most acute in mathematics, physics, and astronomy, the problem plagued even the biological sciences, including popular areas of investigation such as Darwin's theory of evolution. Once again, Newcomb explained that this deficiency reflected institutional flaws, not failings in individual practitioners. He wrote: "When we inquire into the wealth and power of our scientific organizations, and the extent of their publications, — when, in fact, we consider merely the gross quantity of original published research, — we see our science in the aspect best fitted to make us contemplate the past with humility and the future with despair." As he had in 1874, he attributed the institutional flaws to one cause: the scientists were being inadequately rewarded by society. It was not the case that the American people were totally uninterested in science; rather, their interest simply did not include the promotion of basic research.11

In a democracy, how does one garner a suitably high level of support for basic scientific investigation? According to Newcomb, one must convince the citizenry and their elected representatives of the worth of such research. "In other intellectual nations," Newcomb elaborated, as he reviewed less egalitarian approaches to supporting research, "science has a fostering mother,—in Germany the universities, in France the government, in England the scientific societies; and if science could find one here, it would speedily flourish. The only one it can look to here is the educated public; and if that public would find some way of expressing in a public and official manner its generous appreciation of the labors of American investigators, we should have the best entering wedge for supplying all the wants of our science." But in a nation where there exists a gulf between "the political and business classes of our community on the one side, and the literary and scientific classes on the other," how does one go about convincing the citizenry and elected officials of the worth of basic research? Newcomb's strategy was simple: to point out that basic research can provide the nation with something it desperately needs but now lacks. The missing element was not, as one might have expected, the international prestige that comes through a first-rate research program. Nor was it the improved technology and associated increase in personal and national wealth often promised by researchers and administrators such as Joseph Henry in discussing the application of basic science to practical fields such as agriculture. Rather, it was the *method* used in abstract science—a proven set of procedural rules that could be adapted for use in critically analyzing pressing social problems.[12]

According to Newcomb, the American people and their leaders currently lacked in the "dialectic faculty." That is, while Americans displayed a knack for practical reasoning on a common-sense level, they stumbled when required to logically analyze the "first principles" of a complex subject. This "national one-sidedness of the judging faculty" not only inhibited the practice of science in the United States, but also hindered the enactment of rational political and social policies. In the realm of political economics, the nation unnecessarily suffered from flawed policies regarding, for example, protective tariffs, the interest rates charged by banks, and the issuance of currency without adequate gold reserves. The adoption of the scientific method could correct this slackness in the judging faculty of the American people and ensure clear thinking in critical areas like political economy.[13]

Newcomb closed the centennial article with a forceful restatement of his argument. "No want from which our nation suffers," he informed his readers, "is more urgent than that of a wider diffusion of the ideas and modes of thought of the exact sciences, and nothing is more fallacious than to look upon the results of such thought as purely ornamental." Maintaining that the reward of basic research lay not in its discoveries or applications but in its methodology, he specifically recommended the views of John Stuart Mill and called for science educators to focus in the classroom on methodology:

> What is required to insure us against disaster is not mere technical research, but the instruction of our intelligent and influential public in such a discipline as that of Mill's logic, to be illustrated by the methods and results of scientific research. The present great movement in favor of scientific education will be productive of one excellent result, if it serves to direct the minds of the rising generation toward the methods of science, and the ways in which those methods must be applied to the study of societary laws rather than to the technicalities of science, or to its practical applications to the ordinary operations of industry.

Viewed from this perspective of the mastery of method rather than specialized content, he concluded, "science presents itself as a system of national liberal education, to be maintained for the same reasons that we maintain the liberal education of the individual." Mill, incidentally, had promoted a similar pedagogic position.[14]

In this manner, then, through a somewhat roundabout chain of reasoning, Newcomb linked an assessment of the institutional factors that were inhibiting basic research in the United States with a call for the American people to adopt the scientific method. Recalling our threefold classification of the rhetorical uses of method claims— internal, disciplinary, and public uses—we see that he was operating on the public level. He was using pronouncements on method and its civic applications to present a positive image of science to the broader society in the hope of garnering increased support for his fellow research scientists. Of course, he had a high personal stake in the campaign to extend science's cognitive authority and boost public backing: as a government scientist beholden to elected officials and military bureaucrats (the supposed representatives and agents of American citizens), he realized, perhaps more than academic or private scientists, the importance of gaining public favor. Sustaining such favor loomed doubly important because his particular branch of government science required extraordinarily large expenditures for precision apparatus, sizable support staffs, complex publications, and occasional world travel.[15] Through a major speech that he would make in 1880, he would put his plan into practice as he further carried the message of the social utility of method to the public.

SOCIAL PROGRESS AND THE SCIENTIFIC USE OF LANGUAGE

In his diary for 6 April 1870, Newcomb wrote: "In response to an invitation from Prof. Henry I went to the Smithsonian this evening to attend a conference of the Washington members of the National Academy of Sc. respecting the formation of a local society, or the greater localization of the Academy. It was unanimously determined to do one or the other." Henry and his colleagues eventually decided to organize an independent, local group, "having for its object the free exchange of views on scientific subjects, and the promotion of scientific inquiry among its members." The charter members of the new "Philosophical Society of Washington" held their first meeting at the Smithsonian Institution early in 1871, at which time Joseph Henry was elected president and Newcomb joined a small circle of distinguished scientists on the General Committee. In his "Anniversary Address" at the close of the society's initial year, Henry expressed his pleasure that Washington now had its own flourishing scientific organization. Such an organization had been needed, in Henry's opinion, because "in no other city in the Union are there so many men, in proportion to the population, connected with scientific pursuits, or so many faculties for scientific investigation." In fact, by 1876 Washington led all other American cities, except greater Boston-Cambridge, in total number of scientists who were distinguished enough to be fellows of the AAAS. Though the Civil War had temporarily constrained Washington's scientific enterprise as the Union capital gave itself over to marshaling troops, armaments, and provisions, the governmental growth that accompanied the war eventually enlivened the local research community.[16]

Henry continued as the unanimous choice for president of the society until his death in 1878. As his successor, the membership elected Newcomb. He served as presiding officer for two terms, from November of 1878 until November of 1880. For his official speech as retiring president, he departed from his usual topic of astronomy. The minutes of the meeting for 4 December 1880 record that forty-eight members heard him present a "weighty, instructive, and interesting address" titled "The Relation of Scientific Method to Social Progress."[17] The address was soon widely disseminated, being published as a separate pamphlet, reprinted in two journals including the *Smithsonian Miscel-*

laneous Collections, and then reissued in 1906 in a widely sold book containing Newcomb's major essays.[18]

In this address, Newcomb carried out the strategy he had articulated in the *North American Review*. His plan had been to attest to the societal benefits of scientific method in the hope of gaining public support for basic science—support which, in turn, would redound to the strengthening of the institutional framework of American science. For Newcomb, this was anything but a cynical misuse of claims about method to enhance his own profession; he believed sincerely in method's social utility. Without drawing attention to the underlying strategy itself, he now simply proceeded to call for the application of scientific method to current social problems. Speaking in the nation's capital to an influential group of Washingtonians, he minced no words about the desperate need for method. "I make bold to say," he explained, "that the greatest want of the day, from a purely practical point of view, is the more general introduction of the scientific method and the scientific spirit into the discussion of those political and social problems which we encounter on our road to a higher plane of public well being." For examples of particularly acute problems, he once more turned to the realm of political economics, singling out the controversy over import tariffs. This controversy, like others involving financial matters, illustrated the inability of the concerned parties to achieve a consensus on the facts of the case, let alone on an appropriate national policy. With the introduction of scientific method, however, dissension would give way to the establishment of a "common basis" of discussion. Scientific method, being above human bias and political prejudice, would allow the quarreling parties to evaluate the tariff issue objectively and arrive at a consensus on policy. The outcome would be "the increase of the national wealth and prosperity."

Implicit in Newcomb's argument were the assumptions that there was one general method, that it was applicable to fields outside natural science, and that it worked. In particular, Newcomb commented: "Every one knows that, within the last two centuries, a method of studying the course of nature has been introduced which has been so successful in enabling us to trace the sequence of cause and effect as almost to revolutionize society. The very fact that scientific method has been so successful here leads to the belief that it might be equally successful in other departments of inquiry." The challenge, of course, lay in persuading persons involved in the other departments to adopt the method. Again, as in the *North American Review* articles, education provided the solution. It would be a mistake, Newcomb explained, to foist on the public merely additional technical instruction in the natural sciences or even in a particular system of political economy—after all, "which of several conflicting systems shall we teach?" Instead, the nation's college students should be taught "the scientific spirit" and "the scientific discipline." By scientific spirit he meant the desire to seek knowledge for its own sake rather than for practical benefit. By scientific discipline he meant the system of rules, procedures, and criteria that guide the behavior of practicing scientists. That is, by discipline he meant method. In an 1884 article in *Science* titled "What Is a Liberal Education?" he repeated this call, arguing that training in method should be the foundation of a true liberal education. If there is to be social, political, and economic progress, college students must be exposed not merely to the technical content of science but to its method.[19] This proposal meshed well with traditional educational thought that emphasized the training of mental "faculties" (such as logical reasoning) rather than the mastery of particular subject matter (especially natural science).[20] The proposal also meshed well with Newcomb's more basic desire to broaden public support for science by expanding its intellectual authority; inclusion of scientific method in the formal curriculum offered a specific way of legitimating scientific knowledge and sanctioning it as authoritative.[21]

What was this set of methodological rules—this discipline—that should be taught to the American public? Newcomb answered in his 1880 address with a succinct statement of the basic linguistic, empirical precept that he had been nurturing since his student years in Cambridge, since reading Mill's *System of Logic* and conversing with Wright.

> The scientific discipline to which I ask mainly to call your attention consists in training the scholar to the scientific use of language. Although whole volumes may be written on the logic of science there is one general feature of its method which is of fundamental significance. It is that every term which it uses and every proposition which it enunciates has a precise meaning which can be made evident by proper definitions.... If I should say that when a statement is made in the language of science the speaker knows what he means, and the hearer either knows it or can be made to know it by proper definitions, and that this community of understanding is frequently not reached in other departments of thought, I might be understood as casting a slur on whole departments of inquiry. Without intending any such slur, I may still say that language and statements are worthy of the

name scientific as they approach this standard; and, moreover, that a great deal is said and written which does not fulfill the requirement. The fact that words lose their meaning when removed from the connections in which that meaning has been acquired and put to higher uses, is one which, I think, is rarely recognized.[22]

Newcomb's basic methodological rule was that concepts must be grounded in clear definitions; adherence to this rule ensured that members of a community could communicate intelligibly with one another. For Newcomb, such unambiguous communication distinguished science. A question, however, remained. What exactly constituted "proper definitions" of terms and propositions?

Newcomb clarified what he meant by proper definitions through an analysis of the circumstances under which "language can really convey ideas." Specifically, he considered how we might communicate with an intelligent person who was completely unfamiliar with any languages or words that we know. After tracing the steps used in teaching the person our language, Newcomb concluded: "Every term which we make known to him must depend ultimately upon terms the meaning of which he has learned from their connections with special objects of sense."[23] Newcomb elaborated this point through a series of caveats and illustrations.

First, he acknowledged that there exist only indirect sensory ties for "abstract terms" and "words expressive of mental states." That is, he considered language to involve a series of steps: beginning with direct correspondences between words and "sensible objects"; moving next to terms signifying relations between objects; moving on to terms that apply inductively to complete classes of objects; and finally arriving at abstract terms and the other complexities of a complete language. "If we transgress the rule of founding each meaning upon meanings below it, and having the whole ultimately resting upon a sensuous foundation," he warned, "we at once branch off into sound without sense." He added that the unconstrained verbal emphasis of the current system of education in the United States encouraged persons to "transgress the rule"; consequently, to remain faithful to the scientific use of language, a person needed "severe mental discipline."[24]

Second, he guarded against the charge of being an extreme sensationist of John Stuart Mill's bent or skeptical empiricist of David Hume's tenor. That is, he distanced himself from both those who denied any role for the mind in the creation of ideas and those who doubted man's ability to know material objects. He did this by broaching epistemic issues and distinguishing, as he had in an earlier unpublished writing, between the origin and the empirical warrant of knowledge:

> Of course the mind, as well as the external object, may be a factor in determining the ideas which the words are intended to express; but this does not in any manner invalidate the conditions which we impose. Whatever theory we may adopt of the relative part played by the knowing subject, and the external object in the acquirement of knowledge, it remains none the less true that no knowledge of the meaning of a word can be acquired except through the senses, and that the meaning is, therefore, limited by the senses.[25]

Newcomb's third caveat entailed the difficulty of achieving in actual practice rigorous and unassailable definitions. A science attains the ideal of unambiguous expression, he acknowledged, only when using symbolic or mathematical languages. "To secure the same desirable quality in all other scientific language it is necessary to give it, so far as possible, the same simplicity of signification which attaches to mathematical symbols. This is not easy, because we are obliged to use words of ordinary language, and it is impossible to divest them of whatever they may connote to ordinary hearers."[26] Having added these caveats, Newcomb moved on to examples of proper definitions.

THE LANGUAGES OF PHYSICS, BUSINESS, AND PHILOSOPHY

To illustrate the scientific use of language, Newcomb turned to the terms and propositions of physicists, businessmen, and philosophers. He focused first on the physicists' concept of force, a concept that he had scrutinized during his Cambridge days when sorting the views of Mill and Whewell. Physicists, he reminded his audience, had become embroiled in a controversy around 1700 over the meaning of the phrase "force of a moving body." The scientists had been uncertain whether "force," when expressed mathematically, was proportional to velocity or velocity squared. Newcomb pointed to this dispute to show that disagreements sometimes arise, even among scientists, when the opposing parties do not underpin their seemingly contrary terminology with precise definitions. Here, the disputants had failed to specify their meanings of "force" in terms of "the measure of force." Newcomb confidently assured his listeners, however, that this dispute was "almost unique in the history of science during the past two centuries, and that the scientific men themselves were able to see the fallacy involved, and thus to bring the matter to a conclusion." While throughout this 1880 speech Newcomb elaborated and individuated his particular view of scientific method, he frequently echoed the thoughts of Chauncey Wright and John Stuart Mill. In his account of and lesson drawn from the "force" controversy, Newcomb paraphrased a passage by Wright in his "Speculative Dynamics" (1875, reprinted 1877). Wright's words, in turn, were reminiscent of those by Mill in his *System of Logic*.[27]

Although he commented on physicists' use of language, Newcomb took his main illustrations from an odd combination of professionals—businessmen and philosophers. The combination, however, served Newcomb well in that it built on stereotypes and took advantage of Americans' presumed respect for practical businessmen but disdain for impractical philosophers. To add credibility to his thesis that scientific method was relevant to social progress, Newcomb associated scientists with businessmen by claiming that businessmen were naturally attuned to scientific method. He used philosophers as his foils. Specifically, in the down-to-earth business world where clear communication was essential, persons by necessity grounded all their terms in precise definitions. Newcomb argued that scientists followed the identical method. The language of science corresponds to the language of business "in that each and every term that is employed has a meaning as well defined as the subject of discussion can admit of." Each of the two languages succeeds by "confining its meaning to phenomena." Along this same line, he endorsed the definition of science as "organized common sense" given by William K. Clifford, the recently deceased English mathematician and philosopher. Again, it was Mill who had opened this line of argument: While discussing the "homogeneity" of scientific method in *System of Logic*, he had identified the "logic of the sciences" with the "logic of practical business and common life."[28]

In stark contrast to businessmen stood philosophers. "There is nothing in the history of philosophical inquiry more curious," Newcomb declared, "than the frequency of interminable disputes on subjects where no agreement can be reached because the opposing parties do not use words in the same sense." For a primary example of "the danger of using words without meaning," Newcomb returned to his favorite philosophical topic from earlier years—free will. He cited his prior correspondence with "one of the most acute thinkers of the country"—Chauncey Wright, an ally—to reinforce the point that philosophers lacked clear definitions of the word "freedom" and, thus, were locked in futile debates. As in his exchange with Wright, Newcomb offered his own elaboration of Mill's compatibilist conception: whereas human actions can be "free" in the sense of not coerced, they are still "determined" in the sense of subject to the law of causality. Specifically, Newcomb defined "free" in terms of a *relation* between "some active agent or power, and the presence or absence of another constraining agent." Next, he noted that, while human acts can be free in the sense of unconstrained, everyday experience shows that they are "as much the subject of external causal influences as are the phenomena of nature." "All that the opponents of freedom, as a class, have ever claimed," he added, "is the assertion of a causal connection between the acts of the will, and influences independent of the will." Accordingly, when the term "free" is clearly defined, nothing is left for the philosophers to debate.[29]

In presenting his alternative to the philosophers' "unscientific use of language" concerning free will, Newcomb spoke of causal influences and connections. Later in his address, he returned to this issue and further berated the philosophers for misunderstanding the word "cause." Mill had similarly linked misapprehensions concerning free will with misapprehensions about the doctrine of causation. The philosophers, according to Newcomb, were critical of the way scientists employed the word "cause," finding their use to be ambiguous. Newcomb responded that the word "cause" functioned in science as a purely neutral, descriptive term similar to its role in common, everyday life. He objected to the philosophers' accusation that "the idea of power" was connoted in the scientists' usage of the word. Using a favorite ploy based on his linguistic, empirical method, he sarcastically asked: "But what meaning is here attached to the word power, and how shall we first reduce it to a sensible form, and then apply its meaning to the operations of nature?" Once again, this discussion of the word "cause" closely reflected in both language and sentiment a similar discussion in Wright's most famous essay, having been publicly praised by Darwin, "Evolution of Self-Consciousness" (1873, reprinted 1877). Wright's discussion, in turn, paralleled a passage from Mill's *System of Logic*.[30]

The language of philosophy furnished Newcomb with more than a dramatic contrast to the language of business—and, hence, the language of science. It also offered another opportunity to call for the extension of scientific method to nonscientific realms of thought. With noticeable conviction, he stated: "I cannot but feel that the disputes to which I have alluded prove the necessity of bringing scientific precision of language into every demand of thought." Similarly, in the final sentences of the address he returned to his main theme and, raising the prospect of "unceasing progress," called once more for the application to economics and politics of the method used in business and science. Not just the advancement of philosophy but the welfare of the nation depended on the adoption of scientific method.

CHAPTER VII
Political Economics
Old versus the New School

Newcomb did more than merely extol, as in his 1880 address to the Philosophical Society of Washington, the desirability of using scientific method to attack current political and economic problems. He himself had developed into a political economist of some repute, publishing numerous technical and popular expositions on finance, trade, taxation, currency, and labor. These were lively topics in the Reconstruction period following the Civil War—a period characterized by large national debt, high import tariffs, uncertainty about the value of paper money, rising discontent among workers, and a severe depression beginning in 1873. It was also a period in which Americans perceived the bosses in politics and business to be self-indulgent if not outright corrupt, hence Mark Twain's label "the Gilded Age." (Twain and Newcomb were the same age. In fact, as in Newcomb's case, the 1835 and 1910 visits of Halley's comet coincided with Twain's birth and death, but even more exactly.) Finally, it was a time in which practitioners of the social sciences, while edging toward professionalization, attempted to balance calls within their ranks for partisan social reform and dispassionate investigation.[1] Through a steady stream of books, articles, notes, reviews, and speeches, Newcomb sought to provide a dispassionate analysis of the political and economic issues of his day, thus demonstrating rather than merely describing the social utility of scientific method.

During these years of national turmoil, many sought Newcomb's council on political economics. Editor Henry Adams, for example, turned to the scientist in late 1873 to get "some good sense written in the North American [Review] on our present financial difficulty." Edwin L. Godkin, editor of the *Nation*, also relied on Newcomb, especially through the 1870s, for a steady stream of anonymous reviews and editorials. On a more popular level, in 1875, *Harper's Weekly* ran Newcomb's nine lessons on "The ABC of Finance" and then expanded the series into a book. Likewise, the Harper publishing house, which had also printed Newcomb's works on astronomy, issued a book version of his "A Plain Man's Talk on the Labor Question," which Newcomb originally presented in the *Independent* over a five-month period in 1886. Apparently, these popularizations filled a niche. Nationally prominent educator William T. Harris found Newcomb's treatment of the labor question "admirable." "You have succeeded wonderfully," he wrote to Newcomb, "in adapting your deep thoughts to the popular mind." Universities also sought Newcomb's insights regarding political economics. In 1879, Harvard president Charles W. Eliot invited him "to give three or four lectures at Cambridge this coming autumn on any subject in political economy"; he eventually lectured on taxation. And while a professor of mathematics and astronomy at Johns Hopkins, beginning in 1884, he participated regularly in the doctoral examinations of graduate students in economics. He also taught an undergraduate course during the 1887–1888 academic year on American business and financial institutions.[2]

Moreover, Newcomb found a following among professional colleagues. In 1884, he began a long sinecure as president of the newly founded Political Economy Club of America. Meeting periodically to discuss pertinent issues, and representing a variety of ideological persuasions, the club was composed of about thirty of the nation's foremost educators, financiers, journalists, and statesmen who were active in political economics. Primarily from the northeast, they included Henry C. Adams of Cornell, Charles F. Dunbar of Harvard, Richard T. Ely of Johns Hopkins, Edmund J. James of the University of Pennsylvania, Arthur L. Perry of Williams College, William Graham Sumner of Yale, and Francis A. Walker of the Massachusetts Institute of Technology. Members outside academic circles included Charles Francis Adams, Edward Atkinson, Edwin L. Godkin, John Jay Knox, and David A. Wells. When Newcomb offered to step down as president of the club in late 1887, treasurer and secretary J. Laurence Laughlin of

Harvard wrote that there was unanimous agreement that the versatile astronomer and economist continue in office: "We are too glad to have a double star at the head of our constellation to give up the chance so easily. Your reign is undisputed."[3]

In all of these economic forums, public and professional, Newcomb's stated goal was to give the discipline a more logical, mathematical, and scientific formulation. That is, he was trying to demarcate the disciplinary boundaries of the "science" of political economy. This effort involved both the realignment of practitioners within and the elimination of dabblers outside the "science" — disciplinary constrictions that ultimately would expand the public role of what Newcomb sanguinely took to be scientists with proper credentials. In retrospect, we also realize that Newcomb's unstated and perhaps unconscious goal was to use method to promote a particular political-economic agenda.

Newcomb was not the only American political economist advancing this type of hidden agenda. Neither was he alone in attempting to demarcate the profession of economics through methodological criteria. Late nineteenth-century social scientists increasingly viewed their intellectual and moral authority as deriving from adherence to shared scientific methods.[4] In fact, throughout the nineteenth century, methodological arguments in economics assumed rhetorical forms as practitioners struggled to distinguish between what they variously construed to be legitimate and illegitimate "science."[5] In this effort at demarcating and legitimating their "science," late nineteenth-century neoclassical economists specifically drew on metaphors and images from mid-century mathematical physics, thus associating themselves with this venerable natural science.[6] Of course, the rhetorical coupling of economics to physics does not mean that these economists were necessarily being duplicitous or self-deceiving or were hindering the pursuit of economics.[7] It merely means that persuasion and advocacy were central to method-based discourse in late nineteenth-century economics. Certainly, Newcomb was ideally positioned to transpose the methods and metaphors of the physical sciences to economics.

NEWCOMB AS POLITICAL ECONOMIST

Newcomb had broached the general theme of bringing scientific rigor to political economics as early as the closing year of the Civil War. At his own expense, he published a book illustrating how "financial science" could clarify the government's wartime monetary policies and the residual impact of those policies. In delineating this science, he followed Mill's lead and stressed its objective character, its attentiveness to phenomena but indifference to the rightness or wrongness of human conduct:

> Political economy, as an abstract science, considers every thing as wealth which men desire, and which they can obtain by labor, and in no other way. It does not discuss the ethical question whether men *ought* to desire it; the bare fact that men *do* desire it, and are willing to labor in order to enjoy it, is all that concerns it. The value of any article of wealth is measured by the least amount of labor adequate to its possession by the individual who desires it. If a laborer is willing to work all day for a quart of whiskey to get drunk upon, political economy does not question his wisdom; it argues that the quart of whiskey must afford him more enjoyment than any thing else he could obtain at the same price, else he would have bought something else.

Not wishing to leave the impression that he favored such misguided behavior, he explained that "statesmanship" rather than political economics was responsible for regulating the ethical course of the nation. Once the citizenry agreed on a proper course, "political economy steps in as a concrete science, and shows how the good may be encouraged, and the evil discouraged." By thus contributing to the national welfare through its objective analyses, political economy did, in Newcomb's opinion, serve ethical ends.[8]

Newcomb elaborated the same general themes in 1872 in what has since become one of his most notable economic statements: a review of W. Stanley Jevons's recent monograph on political economy. A British economist, Jevons worked out in mathematical detail the marginal-utility theory of consumer behavior—the theory that the utility or satisfaction that a consumer derives from a commodity ultimately decreases as the quantity of the commodity increases. With his principle of diminishing marginal utility, Jevons triggered the mathematization of economics in the English-speaking world; methodically and steadfastly throughout the middle decades of the nineteenth century, he pressed for the incorporation of mathematical methods into economics.[9] And this campaign entailed links between economics and physics. In particular, the principle of marginal utility—a cornerstone of neoclassical economics—derived from mid nineteenth-century physics, specifically the mathematical techniques and formalisms of energy analysis.[10]

Newcomb not only conditionally endorsed Jevons's theory of marginal utility—thereby becoming one of the first Americans to appreciate and highlight a perspective that later came to be viewed as a major addition to economic thought—but also wholeheartedly approved Jevons's aim of providing political economy with an exact mathematical expression, thereby becoming one of the first Americans to call for the mathematization of economic theory. His endorsement is not surprising in that his commitment to the quantitative research programs of mid-century physics predisposed him to Jevons's physics-based principle. While Newcomb supported Jevons's mathematical goal, he disapproved of some of the economic variables the Englishman chose for quantitative study. In rebuking Jevons, Newcomb insisted that the "philosophical mathematician" would never attempt to measure human feelings, like doubt and belief or pleasure and pain. Visible acts and phenomena were the only meaningful objects of quantitative analysis. "We may make the acts of man undertaken with a view of gaining pleasure and avoiding pain the subject of a calculus," he wrote in 1872, "but this can hardly be considered as measuring pleasure and pain themselves."[11]

In a longer essay-review—an 1875 work titled "The Method and Province of Political Economy"—Newcomb suggested that subject matter alone did not offer a means of demarcating political economy. He made this point while reviewing a text by John E. Cairnes, a British political economist in Mill's classical mold whom Newcomb judged to be "one of the ablest and clearest of recent writers on the logic of political economy." Newcomb felt that, to distinguish the "abstract science" of political economy from other forms of inquiry concerning the production, distribution, and consumption of wealth, one needed to identify the particular method used in the inquiry. What differentiated the science of political economy from other types of inquiry was the use of scientific method, especially the precise use of language. Once again, Newcomb expanded on the importance of mathematical language—an importance that Cairnes, in his otherwise exemplary book, failed to recognize: "Mathematical analysis is simply the application to logical deduction of a language more unambiguous, more precise, and, for this particular purpose, more powerful than ordinary language. That a vague and indefinite language can for any purpose of thought be better than a precise one, no one will maintain." He also rephrased his earlier advice concerning which economic variables were amenable to quantitative examination:

It is not degrees of mental feeling which it is necessary to express in numbers, but only the phenomena to which these feelings give rise. . . . It would be utterly hopeless to attempt expressing hunger and thirst in numbers. But this fact does not make it impossible to say precisely how many barrels of flour the inhabitants of a city have consumed in a given period, nor how many they are likely to consume in time to come.

Once more, he added that this method of impartially analyzing economic phenomena did not enable one to make ethical judgments. However, the political economist aided the moral philosopher by providing him with scientific appraisals of the outcomes of various human actions—appraisals that enabled the moralist to make judgments regarding good and bad. "It is only by keeping within its proper sphere," Newcomb advised, "that economy can efficiently help morals."[12]

Newcomb's Millian view on freedom of the will—the compatibility of freedom and determinism—also crept into this 1875 discussion of the method and province of political economy. On the one hand, he stressed that political economists must always be aware of the role of "human volition," or the ability of people to make economic choices. On the other hand, he cautioned that volition could not affect basic economic laws. "In so strongly insisting on human volition as a link in every economical chain of causes, we must not be understood as countenancing the vulgar notion that the prices of certain things—public securities, for instance—are arbitrarily fixed every morning by the brokers over their coffee. The will of any individual is as powerless in altering economical laws as it would be in altering the course of nature." As we saw earlier, Newcomb acknowledged that individuals have liberty of choice to the extent that their actions are not coerced; nevertheless, he maintained that acts of the will are not independent of but actually subject to natural causes, whether physical or economic.[13]

As the 1870s drew to a close and the nation sought to resolve whether paper money, a legacy of the Civil War, should be redeemable in gold, Newcomb brought his scientific approach to bear on the issue of the fluctuating value of the dollar in the American economy. He sensed that most persons failed "to apprehend clearly that the word 'dollar' is only a name, and is not in itself a standard of value at all." Increases and decreases in the purchasing power of the dollar "entirely elude all ordinary investigation, and are made known only by a collation of facts which can not be effected without long and painstaking research." By insisting on a "precise determination" of the concept of "value," he showed that the monetary unit of the dollar was not stable in value as was commonly thought, but actually depreciated and appreciated over time. Newcomb felt that "one of the greatest social *desiderate* of our day" was the elimination of fluctuations in prices through the establishment of a dollar with uniform value. Toward achieving this goal, one shared by other nineteenth-century economists working with their own national currencies, he suggested that the dollar be explicitly defined as a definite quantity of something; that is, he proposed the establishment of a standard of comparison. Following his mathematical predilection, he suggested that the standard be the statistical average of prices of a group of representative commodities sold in the public markets. As the average fluctuated, the government would make corresponding changes in the amount of gold for which each paper dollar could be redeemed if desired. In subsequent years, he amplified his views on this system of "variable coinage" and a "tabular standard of value," factoring in other variables such as wages paid to laborers and improvements in manufacturing efficiency.[14]

By 1885, Newcomb had reflected and written much on the theme of economics as a rigorous science whose practitioners must be mindful of language and meaning. His thinking culminated that year in a massive textbook titled *Principles of Political Economy*—the same title adopted by many eminent economists of the day, including Mill, who had assigned it to his widely read 1848 text. In this tome, which incorporated various of his earlier writings, Newcomb sought to subsume the subject of political economics "in a scientific form as an established body of principles." That is, distinguishing himself from judgmental and moralistic social commentators who attempted to mold behavior, Newcomb sought to study the "mechanism" of the "social organism" just as an impartial physician studies the workings of the human body. In line with this goal, he devoted all of "Book I" of *Principles* to a preliminary explication of the "Logical Basis and Method of Economic Science." As we will see, he not only systematically reviewed the formal characteristics of scientific method but also warned against its limitations and misinterpretations when applied to economics. Scrutiny of definitions and clarification of terms constituted integral aspects of this methodological program. Consequently, he proceeded in the opening twenty pages of "Book II" to clarify the clouded terminology of economics. "An exact nomenclature," he premised, "is one of the first requirements of an exact science." The meaning of "value," for example, entails not merely a qualitative specification of its characteristics but also a description of "how it shall be measured." Though the main chapters of *Principles* reflected the influence of classical political economists, especially Mill, Newcomb again developed Jevons's mathematical notion of marginal utility. He also elaborated his own views on pegging the value of the dollar to the average price of commodities and worked out a mathematical equation for the circulation or exchange of money in society.[15] Coming five years after his address "The Relation of Scientific Method to Social Progress," *Principles* represented Newcomb's most ambitious attempt to achieve his goal of creating among political economists a community of understanding through the scientific use of language.

Harper and Brothers, publishers of *Principles*, reissued the book in 1887, 1890, and 1895, printing a total of some twenty-five hundred copies. The book, in other words, had fair circulation. Apparently, it also had devotees. Irving Fisher, a neoclassical, mathematical economist at Yale who had trained under physicist J. Willard Gibbs and whose career spanned the period from about 1890 to 1930, built on Newcomb's ideas, even dedicating one of his books to Newcomb's memory. (Fisher, strongly committed to the mathematics and metaphors of physics, went on to help spearhead the econometrics movement in the United States.) Later, in England, John Maynard Keynes commended *Principles* as being "one of those original works which a fresh scientific mind, not perverted by having read too much of the orthodox stuff, is able to produce from time to time in a half-formed subject like Economics."[16] More recent historians of economics have also given *Principles* high marks. One judged that the book, although little known outside the United States, had "a number of outstanding features." "One of the noteworthy sections of Newcomb's book," he explained, "is the opening one on the 'Logical Basis and Method of Economic Science.' Whether or no the opening chapter of a textbook of principles is the best place for a disquisition on scientific method as applied to economics, Newcomb's treatment in its clarity, precision, and balance, must still be among the best that has been given." In a more general vein, a historian of American economics concluded: "It was unfortunate for economics that Newcomb's primary interest was in astronomy. His talents were such that he might easily have been the outstanding contributor to economics in his time."[17]

PARTISAN LIBERAL POLICIES

While later economists and historians found much of value in Newcomb's *Principles of Political Economy*, how did contemporary practitioners react? One reaction eclipsed all others. Edmund J. James (1855–1925), professor of public finance and administration in the Wharton School of Finance and Economy at the University of Pennsylvania, mauled the author and the book in a savage review in *Science*. James began by asserting that Newcomb, by training and profession, was completely unqualified to discuss the specialized field of political economics. He sarcastically observed that "we have a great and successful astronomer and physicist wandering over into the economic field and undertaking to set things to rights." Moreover, he proceeded to chastise Newcomb for knowing "next to nothing" of "the recent literature of the science either in England or on the continent." This ignorance resulted in Newcomb's presenting principles and drawing conclusions that were not only antiquated but false. In one of the few positive statements in the review, James granted that Newcomb provided a service when "he calls attention, in his chapter on economic method, to the necessity of more exact definition and careful reasoning," but he quickly added that this service was rendered useless in light of Newcomb's actual treatment of economic problems in the remainder of the book.[18]

How do we account for James being so intense and unequivocal in his critique? And how do we explain the disparity between the reactions of James and later commentators? After all, it seems that Newcomb in writing *Principles* was engaged in an innocuous scholarly pursuit. Ostensibly, he was using scientific method to improve the utility of political economy by demarcating the science of political economy from nonscientific versions. In actuality, along with most political economists of this period, he was also using method for a political purpose—to legitimate his own economic and social ideology. And it so happened that James was one of the leading exponents of a contrary ideology.

In particular, in the 1880s, a rift that had been developing among the nation's political economists opened fully, reflecting a split already present in Europe. Two schools took shape in the United States, each commonly perceived to associate a particular theory with both a particular method and a particular set of national policies. Seen from James's "new school" perspective, Newcomb was squarely within the "old school" (even though, in retrospect, we realize that Newcomb was breaking new ground in his emphasis on mathematics and endorsement of Jevons's marginal-utility theory). According to the popular view, members of the old school of political economy coupled (1) classical British theory in the mold of Smith, Mill, and David Ricardo; (2) hypothetico-deductive methodology that relied on abstract, immutable laws; and (3) laissez-faire, individualistic policies that encouraged, among other things, free trade between nations. Members of the new school, generally younger and trained in Europe, coupled (1) more recent historical and statistical theories as advanced especially in Germany; (2) a Baconian, inductive methodology that emphasized the importance of beginning with historical and statistical facts that were particular to a time and place; and (3) policies that emphasized government intervention and the establishment of, for example, protective tariffs. New schoolers often linked their endorsement of state intervention with an explicitly Christian commitment to domestic social reform.[19]

With the publication of *Principles* and James's subsequent review, Newcomb did not inadvertently stumble into someone else's fight. Actually, a few years earlier, he had helped provoke the dispute between the new and old schools. Though he had not thrown down the original gauntlet, he was the American economist who had picked it up. Specifically, Newcomb had accepted a challenge to old-school economics issued by self-proclaimed new schooler Richard T. Ely (1854–1943). A German-trained economist teaching at the Johns Hopkins University, Ely was gaining a national reputation as a crusader for America's working class and as a political reformer with socialist and Christian leanings. Early in 1884, the university published a talk that Ely had recently delivered in which he not only labeled and distinguished the two economic schools, but denigrated the old school. When Newcomb read the speech—noticing that he could "claim the paternity" of some of the ideas under attack—he took it upon himself to quell the upstart Ely, nineteen years his junior. Writing to President Gilman in May, he asked for a university forum in which he could respond, confiding that he felt "stirred up." In a follow-up letter, he identified Ely's new school with a disregard for scientific method and cuttingly noted Ely's affiliation with Hopkins: "It looks a little incongruous to see so sweeping and wholesale [an] attack upon the introduction of any rational or scientific method in economics come from a university whose other specialties have tended in the opposite direction." The upshot of Newcomb's ire was a paper on the two schools, a rebuttal that he quickly drafted and sent to Gilman for comments. A few months later when he himself assumed a regular, half-time position at Hopkins, Newcomb continued to press his case, giving a public talk in defense of political economy as a mathematical rather than, as Ely would have it, a purely descriptive science. Finally, in November, his paper "The Two Schools of Political Economy" appeared in the *Princeton Review*.[20]

Following Ely's provocative statement and Newcomb's retort, the controversy spread, embracing economists such as Edmund James. One recent commentator on the "Battle of the Schools" explains: "The Ely-Newcomb controversy quickly degenerated into an academic free-for-all of classic proportions." Another commentator, while granting that the debate was strident and the issues were "momentous," cautions that the controversy lasted only a short time, that journalists perhaps blew it out of proportion, and that primarily only two men, Ely and Newcomb, sustained it. Both of these analysts agree that few Americans in the mid 1880s actually fit the sharp categories of the two schools; by then, for example, almost no one except Yale sociologist and economist William Graham Sumner endorsed a strict laissez-faire outlook.[21] Nevertheless, in the mid 1880s, businessmen and economists with laissez-faire inclinations tended to align themselves with Newcomb while German-trained economists, liberal clergymen, and social reformers drifted toward Ely.

Frustrated by the past dominance of the old, or English, school and perceiving the Political Economy Club to be a bastion of conservative thinkers, members of the new, German, historical, or statistical school—as it was variously called—established in 1885 their own organization, the American Economic Association (AEA). Ely, the main organizer, desired not only to reform American political economy but also to achieve a level of professionalization within the ranks of academic practitioners, especially. After a few years, the association's overt political agenda gave way to the professional concerns. Initially, nevertheless, Ely went so far as to write into the organization's original statement of principles a preference for active intervention by the state and the use of historical and statistical rather than hypothetico-deductive methods. He consciously attempted to use method to enforce a partisan policy and thus to exclude from the association members of the old guard, a group that he deridingly called "the Sumner, Newcomb crowd." Though the AEA leadership immediately dropped Ely's blatantly exclusionary language and after two years even dropped a more lenient but still restrictive statement, traditional economists remained piqued by what appeared to be stringent preconditions for membership. Newcomb later complained that Ely had intended the association "to be a sort of church, requiring for admission to its full communion a renunciation of ancient errors, and adhesion to the supposed new creed." Naturally, Newcomb and his confederates found it in their self-interest to oppose a pattern of professionalization that would exclude them. And oppose it they did.[22]

Soon after James's new-school review of Newcomb's *Principles*, Sumner had written to Newcomb grumbling about the review's exclusionary tone. "I did not like your book," he admitted forthrightly, "but I am in for a fight against the position that these fellows can simply toss a book aside because it ignores the Dutch drivel of the last ten years." One of Newcomb's younger mathematical colleagues at Johns Hopkins, Fabian Franklin, similarly had written a letter to the editor of *Science* stating that it was "presumptuous" for a member of the new school like James "to regard a general adherence to the methods of Mill and Cairnes as evidence of ignorance or incompetence." Franklin went on to defend the quality of Newcomb's credentials and the soundness of his analysis. He also pointed out instances where James, to enhance his critique, had misquoted Newcomb. Newcomb himself, bridling his feelings somewhat, had written an impassive letter to *Science* calling attention to James's flagrant alterations of his words and challenging him to deal with the substantive arguments of *Principles*, not with just his credentials and theoretical preferences.[23] And whereas James had attacked Newcomb for ignoring new-school approaches, an anonymous reviewer in the *Nation* instead commended Newcomb for bridging the divergent methodologies of the two schools. The reviewer, who went on to praise Newcomb for accenting "the actual things to be observed" in economics but to fault him for allowing these same "actual affairs" to cloud his analysis, summarized Newcomb's conciliatory view: "The author holds that the strictly *a priori* method and the historical method are both alike defective; that the true method lies in a union of the two."[24] The fact remains, however, that Newcomb, James, and Ely, along with members of their schools, were using method to promote disparate political economic agendas.

Newcomb's political program implicitly pervades most of his writings on social issues. It stands out most explicitly in the lead article that he wrote for the January 1870 issue of the *North American Review* and incorporated fifteen years later as a chapter in *Principles*. Indeed, his ideology is encapsulated in the article's title, "The Let-Alone Principle," a title that harks back to the "hands off" policy of Adam Smith. Reflecting the British and American tradition of political and economic liberalism fostered by Smith and later Mill, Newcomb believed that the government should not interfere with the freedom of each individual to follow his own economic self-interest. Specifically, he endorsed two alternative expressions of the let-alone principle:

> In the first case, the principle declares that society has no right to prevent any individual who is capable of taking care of himself from seeking his own good in the way he deems best, so long as he does not infringe on the rights of his fellow-men. In the second case, the principle forms the basis of a certain theory of governmental policy, according to which that political system is most conducive to the public good in which the rightful liberty of the individual is least abridged.

This did not mean opposition to all government intervention. Irving Fisher recalled that while Newcomb endorsed the let-alone policy he distinguished it from the "keep out" policy. That is, "he believed in the economic activity of individuals, but did not advocate the exclusion of government from economic activity."[25]

Newcomb isolated what he felt was a particularly crucial example of the let-alone principle, the natural right of individuals to enter into contracts broadly defined. In Newcomb's opinion, the "most onerous forms of legislative interference" arose through the government's meddling with contractual agreements of all types and at all levels. Consequently, he opposed government interference as manifested in protective tariffs that restricted open trade; usury laws that capped interest rates for loans; regulations that set prices for commodities; and labor laws that dictated salary or restricted the number of hours a person could work. He also objected to legal-tender laws that allowed the government to breach its contractual obligation of backing up paper money in full value with gold. In fact, a main reason for writing his book on financial policies during the Civil War was to criticize the government's decision to issue dollars that could not be redeemed for gold coins or "specie."[26]

Newcomb's opposition to paper currency marked his entry into the partisan world of Reconstruction politics. In the words of one historian who scrutinized this era, Newcomb initially was a member of "a gifted corps of financial freelances and journalists" who supported the postwar campaign of academic economists, Protestant clergy, and others hoping to restore hard money. His book on wartime finance brought him into contact, for example, with two of the country's most powerful financiers, Hugh McCulloch and George S. Coe. Coe was an influential New York banker; McCulloch was the secretary of the treasury in President Andrew Johnson's postwar administration. Early in 1866, Coe wrote to Newcomb: "I have read your little work with *very* great pleasure, and have commended it to my professional brethren as containing 'sound words.' " Later in the year, Coe sent Newcomb a letter of introduction to Secretary McCulloch, to whom Newcomb subsequently wrote. In responding, McCulloch commented that he already knew of Newcomb's book, through a recommendation from Joseph Henry, but had not yet read it. "I will avail myself of the earliest opportunity to read not only the chapters to which you refer especially," he added, "but the whole book." Newcomb became better acquainted with McCulloch through the Washington Scientific Club, an informal group of scientists and others. The group's meeting sites in the late 1860s included the secretary of the treasury's home. Though McCulloch was somewhat older than Newcomb, the two men and their wives eventually became friends, attending dinners and social events at each other's homes over the next twenty-five years (including the time when McCulloch resumed his cabinet post in the administration of President Chester Arthur). Newcomb later commented that McCulloch, as secretary of the treasury, was "my beau idéal of an administrator." McCulloch reciprocated these sentiments as they pertained to Newcomb as an astronomer, but expressed a paternalistic concern that his younger colleague, by becoming entangled in subjects such as political economy, was "scattering his fire" too widely. Newcomb, it turned out, harbored this same concern. "I have never been able," he acknowledged in his autobiography, "to confine my attention to astronomy with that exclusiveness which is commonly considered necessary to the highest success in any profession."[27]

Newcomb developed close relationships with other influential political figures as well, especially within the Republican party. These included Senator Charles Sumner and Congressman and later President James A. Garfield, two men he met accidentally through rooming and boarding in Washington. In his diary for 7 July 1866, the year after publishing his book on the government's financial policies, he wrote: "Called on Hon. James A. Garfield, and had quite a talk with him on protection and finance. Was unexpectedly pleased with his views. Left him my draft of a bill for returning to specie payments." Consisting of five handwritten pages with two additional pages of explanation, the bill would have enabled Congress to "restore the currency by recommencing specie payments." Though the bill's fate is unknown, Garfield was active in the House of Representatives' Committee on Banking and Currency, and in 1874 Congress authorized resumption of specie payments beginning five years later. During this decade, Newcomb remained close to the congressman. The tenor of their ongoing friendship is captured in an 1872 note in which Garfield asks Newcomb: "Will you come and take a family dinner with us at 5 o'clock this (Wednesday) afternoon?" In charge of House appropriations, Garfield apparently extended this dinner invitation to discuss American astronomers' request for congressional funding of expeditions to observe the 1874 transit of Venus; Congress soon granted the astronomers $50,000, followed by an additional $125,000, generous amounts in an era of temperate government support for science. When Garfield was fatally wounded in 1881 during his first year in the White House, Newcomb joined Alexander Graham Bell and others in attempting to locate by electromagnetic means a musket ball remaining in the president's body, as well as to cool the dying president's room.[28]

Throughout the 1860s and 1870s, Newcomb used scientific method to justify his liberal stance on free trade, a gold standard, and the like. That is, he implied that his arguments favoring these positions were credible because he had formulated them through proper method. Conversely, he used the criterion of scientific method to debunk proponents of protective tariffs, legal tender laws, and other forms of government regulation of the economy. Newcomb would have his readers and listeners believe that adherence to his version of scientific method was the essential credential of a political economist. For example, in skirmishes that portended the later battle between the two schools, he repeatedly criticized the writings of Henry Carey, a prominent American economist who backed protective tariffs and opposed classical theory, for an "absence of logical method and scientific ideas." In an 1866 review of Carey's *Principles of Social Science* and again in 1875, Newcomb specifically found him guilty of using what Mill in *System of Logic* had called the "chemical method" of social inquiry. In contrast to this method that treated the actions of separate persons as a collective phenomena, Mill had advocated methodological individualism—treating complex social phenomena as manifestations of individual human actions. Newcomb disparaged Carey's method: "It views mankind as the chemist views a compound whose properties are to be learned by trial alone. Of individual men, their motives and their springs of action, it knows no more than the chemist knows of the molecular forces which produce the changes he observes." In a draft of this 1866 review, Newcomb explicitly criticized Carey for studying "the moral world by a method similar to that of Bacon for the physical world." According to Newcomb, because Carey mistakenly believes that social laws "are to be determined by observation alone, he seeks to generalize from historical examples." Newcomb deemed this Baconian inductive method to be overly restrictive. He favored Mill's deductive edifice of social science as built on causal laws governing individual human behavior.[29]

In preferring Mill's methodological individualism over Carey's chemical method, Newcomb joined Mill in merging political liberalism and scientific method. That is, Newcomb's liberal belief in the primacy of the individual found expression in his belief that social phenomena must be understood in terms of the individual. This literal blurring of political ideology and method appears, for example, when Newcomb summarized what he took to be the ridiculous consequences of Carey's chemical method: "The natural result is to look upon man the individual as having no more power over his own destiny than the particles of a chemical mixture. Of an animal capable of adapting himself to circumstances, applying means to ends, alive to his own interests, sharp at a bargain, disposed to take time by the forelock, the author seems but in one or two instances to have any conception. His 'Man' is a mere puppet."[30] Consistent with his Millian view that, although human acts can be "free," they are still "caused," Newcomb did not mean to imply that the individual could alter basic economic laws. His point was merely that political leaders and social scientists, in their policies and methods, needed to acknowledge the primacy of the individual and his or her volitions.

DEFENDING THE OLD SCHOOL

In taking Carey to task, Newcomb drew on method as a weapon of attack. This offensive posture gave way to a defensive one when, in the mid 1880s, Newcomb found himself confronted by vocal members of the new school. A master of method, he wound up in the curious position of having to defend his—and, more generally, the old school's—reliance on a hypothetico-deductive approach. And the stakes were high: the authority to speak in the United States as "scientific" and, hence, professional political economists—experts deserving recognition and support.[31]

In 1884, responding to Ely's initial challenge, Newcomb sought to set matters straight with his article in the *Princeton Review*, "The Two

Schools of Political Economy." Parading his scholarly credentials, he signed this defense of the old school's methodology, "Prof. Simon Newcomb, LL.D., F.R.S.," the former referring to various honorary doctorates that he had received, most recently from Harvard, and the latter highlighting his foreign membership in the Royal Society of London. Newcomb's goal was to answer the criticism that the old political economy was built on the untrustworthy methodological and epistemic foundation of hypothesis. His tactic was not to deny the general claim but to argue that all science—even the purportedly empirical science of the new school—is based on hypothesis. "You will notice," he had commented to Gilman who was previewing a draft, "that my paper is founded on the fact that I have never been able to see any essential difference between the objections raised against political economy from the new school point of view and the general objections of the public against the value of theoretical science."[32]

In the article, Newcomb first specified the charge against the political economy of the old school. According to critics, particularly Ely, the old system is "a deductive science founded on a-priori hypotheses respecting human nature, which are too wide of the actual facts of the world to form a sound basis for any practical conclusion. It assumes to subject all economic phenomena to a few formal laws, and fails to consider how these laws are modified or even reversed in practice. . . . The result of thus substituting ideal for actual conditions is a body of doctrine which, however logically it may be reasoned out, does not agree with the state of things which actually exists around us." In responding to this charge, Newcomb aimed to disarm the critics by granting their general argument, but then pointing out the argument's applicability to all of science. "Formidable as this indictment looks," he explained, "we can easily show that it applies with equal force to every branch of pure science, when we consider the science in its relation to practical applications. It is in fact a most valuable illustration of a truth which every logical student should know, but which hardly any one always bears in mind—that all scientific propositions are in their very nature hypothetical." Critics of the old school simply have forgotten "the limitations which are placed upon human knowledge in every department of inquiry, and the necessary imperfections of all scientific statement." To illustrate the disparity between practical results involving actual phenomena and the imaginary, ideal, and hypothetical propositions of the sciences, Newcomb briefly reviewed how instructors in schools and colleges taught such apparently concrete subjects as arithmetic, algebra, and physics.[33] This was familiar ground for Newcomb: as early as the Cambridge draft of his physics textbook, he had explained why the lack of perfect agreement between hypothesis and evidence does not necessarily invalidate the hypothesis.

Would it be possible to construct a science of political economy that eliminated the gulf between the science's hypothetical propositions and actual practice involving real phenomena? Newcomb replied with an emphatic no, asserting that "the imperfections alluded to are inseparable from all exact knowledge." These imperfections, however, do not diminish the significance of basic science. "Paradoxical tho it may appear, the fact that the phenomena of nature cannot be reduced to simple formal laws does not render less necessary the consideration and study of such laws." People simply needed to remember that, because of the overwhelming complexity and even opaqueness of nature and human society, all expressions of natural or social laws involve stipulations of simplified conditions or circumstances under which the laws hold. To the extent that these simplified circumstances do not exist in the actual natural or social world, the laws are ideal or hypothetical. Whether political scientist or physicist, the scientist is responsible for judging what restrictions or modifications are necessary in applying a hypothesis in particular circumstances. Calling attention to the subjectivity of scientific inquiry, Newcomb emphasized that "no science that ever existed professes to give formal rules by which conclusions can be worked out without any exercise of judgment on the part of the individual."[34]

Newcomb concluded his arguments concerning the methodological and epistemic limitations of science by returning explicitly to the two schools of political economy. The old system, built around hypotheses concerning human nature, bears "the same relation to the transactions of the commercial world that theoretical physics bears to the working of machinery." Critics of the old school apparently were overlooking this parallel:

> The objections to the deductive features in this school can arise only from a misapprehension. Its deductions being only hypothetically true, are not to be applied in practice unless the actual case is shown to apply to the hypothesis. But it does not follow that the method is useless because it needs modification when applied to particular cases, because this is true of all science.

In other words, the old political economy with its hypothetico-deductive method is still viable when properly applied and appropriately adjusted to particular circumstances. What about the new school? "The one fundamental principle of this school is," Newcomb recalled, "that instead of beginning with certain hypothetical principles of human nature it professes to start from the great facts of history and statistics." A review of the literature demonstrates, however, that "the new school has not really put any new system into practice." When the new-school economist "tells us that he has found out a better way of developing the subject,—a method by which the incompleteness inherent in all scientific systems is avoided,—he takes a position which he lamentably fails in making good."[35] In other words, the new-school revisionists have not lived up to their promise of creating an empirically secure, Baconian science.

Within weeks after reading this article on the two schools, Harvard's James Laughlin wrote to Newcomb in support of his position on method. Laughlin, a younger member of the old school with a particular attraction to the classical views of Mill and Cairnes, was founder as well as secretary and treasurer of the Political Economy Club. Newcomb, recall, served as first and continuing president of the club. Probably like many readers of Newcomb's article, Laughlin attached extra credence to Newcomb's pronouncements on scientific method because of his impeccable credentials in astronomy, physics, and mathematics: "I saw your last paper on the 'Two Schools of P. E.' in the Princeton Rev. Your statements on method have peculiar strength because you speak as a scientific man, from experience in purely scientific work. I don't see where the 'new method' men have a peg to hang a new system on." A year later, when the club's future was threatened by the recent formation at a gathering in Saratoga Springs, New York, of Ely's self-assertive American Economic Association, Laughlin again wrote to Newcomb expressing his hope that Newcomb would offer his counsel on the issue of scientific method at the club's next meeting. "It is needless to say that I hope you will sacrifice something to get to our meeting at New Haven, Oct. 10. The new organization at Saratoga will no doubt be discussed in our Club, & our objects may be defined. Be sure & come. We need your 'scientific method.'" To be sure, Laughlin and Newcomb had sought to keep overt partisan politics out of the club and to maintain a balanced membership, admitting even Ely and James. Nevertheless, politics were beginning to intrude, adding to the popular view of the group as a stronghold for proponents of free trade.³⁶

Newcomb renewed his campaign for hypothetico-deductive method in his *Principles of Political Economy*, which appeared in 1885, the year following the *Princeton Review* article. Recall that he devoted "Book I" of the volume to the "Logical Basis and Method of Economic Science." By giving this priority to method, Newcomb was reflecting, on the one hand, his lifelong commitment to the centrality of method in scientific inquiry as well as his decade-old commitment to convincing the American public of the value of basic science. Thus, he stressed that scientific method should be extended from the natural sciences to the problems of political economy and that, when applied appropriately, it could lead to "understanding" rather than mere factual "knowledge." On the other hand, he was reflecting an immediate imperative to answer further his new-school critics. Newcomb opened the discussion with a straightforward summary of the scientific method, a less didactic and constricted summary than the one that led off his physics textbook more than two decades earlier. Though he declares, as had become his wont, that scientific method parallels common sense in providing principles to guide reasoning, he quickly moves on to a technical discussion of the hypothetico-deductive method. He explains that laws of nature are merely conditional propositions expressing only hypothetical relationships between causes and effects. As a general consequence, "all scientific conclusions are to be regarded, not as particular truths, but as things which are or would be true under certain assumed conditions." He also explains that while all scientific propositions are grounded in "a study of the facts of experience," it is still "generally impossible to infer a law from mere observation." Rather, because of the overwhelming complexity of causes behind an actual event, scientists must go beyond simple empirical analyses and invoke "abstractions" in which only certain basic causes are assumed operative and all others are ignored. Just as the physicist must deal with idealizations such as uniform gravitational fields and frictionless movements, so too must the political economist engage in abstraction and "begin with a hypothetical man." At least this is the case in "pure economics," it being a "pure science" like thermodynamics rather than an "applied science" like steam-engineering. While economic science shares features with the physical sciences and while the basic principles of scientific method "are common to all science," political economy diverges in certain respects from fields such as physics and, therefore,

Newcomb comments, the scientific method must be applied differently. The basic dissimilarity is, of course, that humans are more complex than entities such as molecules in that they have individual wills, motives, and desires.[37]

Newcomb reserves the final chapter of "Book I," "Fallacious Views of Economic Method," for specific responses to new-school critics, though he never identifies the group or its leaders by name. Renewing the thrust of his *Princeton Review* article, he states that misapprehensions concerning the application of the hypothetico-deductive method to political economy rise from one root: the mistaken notion that the resulting theoretical propositions are "absolute truths which can be applied without regard to time, place, or circumstance." Those interested in political economy manifest this fallacy in two ways, by committing either the "Doctrinaire's Error" or the "Popular Error." The former error plagues scientifically naive thinkers who actually believe that the theoretical propositions are uniformly and directly applicable to all situations regardless of particular circumstances. Newcomb dismisses such doctrinaire views, pointing out that the science of political economy "is not like a map in which is laid down every stone and pitfall in some mammoth cave, but rather like a lantern in the hands of an explorer by the aid of which he can discover all the stones and pitfalls for himself." As he mentioned in earlier pages, the emphasis should be on the process of understanding through method, not on the accumulation of factual knowledge.[38]

The second or "Popular" error is the error of the new schoolers. They built a straw man out of the doctrinaire thinker and proceeded to destroy him. That is, they asserted that traditional political economists professed a belief in a consummate body of economic doctrine that enabled investigators to arrive at truth in every circumstance directly through deductive reasoning; on realizing that the traditional science did not fulfill this promise, they insisted that it be abandoned. Once again, Newcomb attempted to dissolve the new schoolers' argument by not only agreeing that traditional political science failed to achieve comprehensive applicability but also pointing out that every abstract science suffers under the same limitation. Indeed, the student attempting to master political economy need not be discouraged once he realizes that

> the imperfections which we have just been describing are only those which are common to all human knowledge. No knowledge of the future affairs of mankind is perfect, because we cannot possibly tell what causes may

come into play to disappoint our expectations. But notwithstanding these imperfections, we can form more or less probable judgments of the action of causes and effects in the world generally which are of the greatest value. The imperfections of political economy are less than those of meteorology.

Newcomb ends "Book I" with one last fallacy concerning economic method. With new-school moralists and social reformers apparently in mind, he again pleads for political economists to differentiate between dispassionate, descriptive statements and partisan, value-laden statements. Only the prior language merits a place in the science of political economy.[39]

While Newcomb's *Princeton Review* article and his textbook were reassuring to old-school economists such as Laughlin, the writings did not convince new-school practitioners of the errors of their way. Edmund James's scathing review of Newcomb's *Principles* appeared in November 1885. Neither did Newcomb's writings sway or silence Ely; he "severely reviewed" the textbook in a talk at Hopkins. In addition, he encouraged Albert Shaw, one of his former students and editor of the *Dial*, "A Monthly Journal of Current Literature," to pen a harsh review. Detailing the shortcomings of the textbook in a letter to Shaw, Ely branded Newcomb "an ignoramus as regards the investigations of scholars during the past generation." When he wrote his review, Shaw actually adopted more prudent language and offered a mixed judgment of the textbook. He began, like James, by questioning Newcomb's credentials, the same credentials that Laughlin found so credible. Shaw commented that while it might be appropriate for the "eminent" astronomer to involve himself in theological issues—as Newcomb had recently done in a widely publicized exchange with a group of prominent Christians (see chap. 8)—it seemed questionable having an astronomical specialist pronounce on economic issues. "When on one occasion he turned from his tables of Uranus and his nautical almanacs to annihilate the theological doctrine of 'final cause,' the transition was in some sense natural. Most great astronomers have gone into theology more or less. But it is not a little noteworthy that there should have come forth from the naval observatory at Washington the most ambitious treatise on the 'Principles of Political Economy' that any American has written in recent years." The "vitiating" consequence of having this astronomer venture into economics was that Newcomb treated what he called the "economic organism" in strict analogy to the solar system, mistaking it solely for a mechanical system subject to inexorable natural law.[40] Evidently, Newcomb's exhortations on the limitations of all science had not registered with Shaw.

CONFRONTING RICHARD ELY AND EDMUND JAMES

In response to the further controversy generated by James's review and the creation of the American Economic Association, the editor of *Science* proposed a written forum to air the issues. He sought to give members of the new school "an opportunity of propounding the fundamental principles which they think should rule at the present time." On first learning of the forum and apparently unaware of the preliminary skirmishes, William Graham Sumner wrote to Newcomb asking for background to the dispute and inquiring: "Is there to be a grand battle, two or three on a side, between the 'schools'?" Of those writing on the side of the new school, there turned out to be three principal spokesmen: Edwin Seligman of Columbia, James, and Ely. Their series of three articles was followed by a discussion by Newcomb, who happened to be a member of *Science*'s board of directors. Later, the original essayists submitted replies to Newcomb, who in turn wrote his own rejoinder. Soon published in a separate volume, the *Science* economic forum was, to quote one historian, "the high point of conflict between the old and new schools."[41]

In his main discussion in the forum, Newcomb returned to the issue of the scientific use of language, which he had developed in his 1880 address "The Relation of Scientific Method to Social Progress." He faulted the three authors for not clearly stating their positions on what supposedly was the distinguishing policy feature of their school—the encouragement of government intervention in industry and trade. Struck by the ambiguity of the three new-school articles, he reminded the authors that "the familiar terms 'government intervention' and 'state interference' are themselves so vague, that in discussing them we must exactly define the sense we attach to them." He continued: "There can be no reasonable discussion over such vague propositions as, 'the state ought to interfere,' or 'the state ought not to interfere,' " because everyone can imagine cases where these generalities apply. In a more specific reply to Seligman, a moderate representative of the new school, he repeated the argument that he had given in the *Princeton Review* defending classical political economy. Acknowledging that the abstract propositions of the classical system do not always seem to apply to present society, he nevertheless maintained that the propositions are adequate to the task when interpreted and applied appropriately. Newcomb also replied specifically to Ely, whom he labeled as having "socialistic" leanings. Coming back to the issue of language in science, he faulted this would-be social reformer for advocating that economic scientists talk about "what ought to be" rather than restricting themselves to "things as they are." Though political economists were entitled to their own views about the ethical implications of social policies, they needed to exclude such considerations from their scientific discussions—a point that Newcomb had been pressing since his study of Civil War financial policies. "Times without number," Newcomb warned, "I have seen educated men refuse to accept a statement of fact, not on the ground that it was not a fact, but that it was not *necessarily so*, or *might* be different, or *ought* to be different." Such talk, Newcomb contended, leads to confusion.[42]

As for the methodological differences between the two schools, Newcomb adopted a more conciliatory tone a few weeks later in another, shorter article in *Science*. In answering the question posed in the article's title, "Can Economists Agree upon the Basis of Their Teachings?" he suggested that just as physics instructors taught both "the experimental method" and "mathematical deduction," so too should the economics instructor teach the methods of both schools. Comparing the bickering loyalists in the two schools to medieval philosophers, he held the physicists up as models of tolerance regarding alternative approaches. Thus, he had a mitigating answer to the specific question, "How shall we proceed to acquire the necessary knowledge of society,—by purely deductive processes from general principles or by the study of the facts as developed by history and statistics?" His reply: "We can attain no result except by a judicious combination of both processes." Moreover, rather than mastering particular principles and facts, the student needs to acquire a general, methodological understanding of the proper scientific posture toward principles and facts. In an aside, however, Newcomb betrayed his bias when he mentioned that students who are taught abstract principles have an advantage over students taught facts in that they alone have a structure to guide their study.[43]

Though Newcomb portrayed himself merely to be a peacemaker seeking accommodation between the two schools, this public stance held a tactical benefit. Unable to dislodge the competing school from the science of political economy, he at least was claiming equal professional standing for his own school. Compatibility of the two schools entailed the sharing of cognitive authority—and, thus, the protection of the old schoolers' voice in economic affairs. To assent in principle to the scientific legitimacy of the new school, however, differed from accepting in practice the agendas of particular new schoolers. Availing himself of the cover provided by an unsigned review in the *Nation*, Newcomb withheld the hand of reconciliation to his old adversary, Ely. Specifically, a few months after his conciliatory article in *Science*, Newcomb anonymously assailed Ely and his new book on the American labor movement. The book struck Newcomb as merely moralistic apologetics for labor unions, particularly the Knights of Labor. Adopting an unrestrained and acrimonious tone, he charged that Ely "shows a lack of logical acumen and a narrowness of view which, in a university teacher, are most remarkable." "His worst defect," Newcomb continued, "is an intensity of bias, and a bitterness toward all classes of society except one, to which it would be hard to find a parallel elsewhere than in the ravings of an Anarchist or the dreams of a Socialist." Overall, the book "is marked by a general puerility of tone and treatment, a scrappiness of narrative, and an absence of everything like strength of touch, mental grasp, or logical unity—faults which deprive it of all real interest." Saving his most devastating blow for his final sentence, Newcomb surmised: "Dr. Ely seems to us to be seriously out of place in a university chair."[44]

Ely's allies rallied to his aid as threats to his Hopkins position came from the *Nation* and other quarters. Ely himself again marshaled support from his prior student, Albert Shaw. Probably realizing that Newcomb contrived the *Nation*'s attack, Shaw countered with an essay-review in the *Dial* in which he cannily juxtaposed Ely's book and a new book by Newcomb that also happened to cover the labor question. Shaw credited Ely with having "an enviable and well-earned reputation" and certified his economic analysis to be "trustworthy," "thorough and mature," "well-considered," and "creditable from the scientific standpoint." As in his earlier review of Newcomb's *Principles*, he dismissed Newcomb as an astronomer biased by a mechanistic view of society based on an inappropriate mechanical analogy to the solar system. Moreover, as an astronomer, Newcomb was isolated from actual problems of the working class, such as those created by the self-serving proprietors of railroads. Readers simply should not trust the astronomer's "casual observation of social phenomena from the altitudes of his observatory tower." As to be expected, old schoolers held an opposite view. Sumner responded, for example, with enthusiasm to Newcomb's new book, *A Plain Man's Talk on the Labor Question*. He wrote to Newcomb that he had read the work with "great interest & pleasure" and found it to be, as Newcomb had intended, "a very successful popular attempt."[45]

Newcomb visited the camp of the enemy in early 1888 when he presented a lecture at Edmund James's home institution, the University of Pennsylvania. Actually, by that date, the sharp lines of battle of the two schools had already begun to blur. The American Economic Association had loosened its criteria for membership and old-school economists were joining, giving the group an eclectic spirit; professional rapport was displacing factionalism. Concurrently, the Political Economy Club, though it continued to meet, was fading in importance, perhaps a reflection of declining support for pure laissez-faire theory.[46] Newcomb, as his private draft of the lecture shows, maintained a conciliatory tone when he addressed his Pennsylvania audience on the "labor question." Specifically, he affirmed the message of his *Science* article from two years earlier. "Were you to ask me which of the schools of political economy I believe in, the deductive or the historic and statistical school," he opined early in the talk, "I should say that I believed equally in both, and saw not the slightest occasion for antagonism between them so far as principles were concerned." Continuing in the same vein, he granted that he was sympathetic to laissez-faire doctrines but acknowledged, as even Mill had done, that occasions arise when government participation is appropriate.

Moving through the core of his lecture, an appraisal of recent thought on the labor problem, he offered his view on the most common defect of the thought. The defect was not in the doctrines of the old or new school but in something more basic, the unscientific use of language. He complained that "instead of dealing with the actual practical material things on which our interests really depend we concern ourselves too much with mere words and phrases which at the very best only represent adumbrations of real things." Specifically, instead of restricting themselves to discussions of "ploughs, fields, fences, cattle, factories, cloth, railways, locomotives," and other "really practical things," both popular commentators and political economists inappropriately discussed "capital," "labor," "capitalism," and other equally abstract terms without providing proper definitions. Confusion of thought resulted. Later in the talk, Newcomb generalized further about the basic methodological defect of current political economics:

> If I am right the greatest vice of our political economic thoughts is not so much false doctrine as unsound methods of thinking. A doctrine which is good and true of one state of things may be entirely at fault when applied to another. You cannot therefore assign any high degree of merit or demerit to any doctrine considered merely in itself. But loose and disjointed thinking, vague assertions which have no definite meaning, the substitution of prejudices for opinions, and of wishes for facts are vices which so long as they exist must prevent our reaching any sound or valuable conclusions. . . . On the other hand if you adopt and practice sound methods of thought, analyzing all your propositions in order to see where they carry you and what they mean; opening your eyes to all the facts of the case, taking the widest possible views of things by tracing them back from all their beginnings to their ends finding out where it is true and where it begins to be false then whatever circumstances may arise you will be prepared to meet them in a rational spirit.

The "greatest foe" of political economists, Newcomb surmised, is the person "who surrounds himself with an impassable swamp in the shape of meaningless words and phrases, mistifications and distortions and every other species of logomachy." Though he perhaps had Ely as well as James and his review of the *Principles* in mind when he spoke these particular words at the University of Pennsylvania, Newcomb seemed truly to picture himself, as he expressed it later in the talk, not a partisan but an impartial analyst.[47]

While willing to cede a place for new schoolers among economic scientists, Newcomb continued to resist imputations against old schoolers. Thus he objected to an address delivered to the American Economic Association by Francis A. Walker, head of the Massachusetts Institute of Technology and president of the AEA during its first seven years. In an anonymous commentary in the *Nation* during 1891, Newcomb rebuked Walker, normally a moderate regarding economic theory, for too readily siding with the public in favor of the labor movement and against established economic science. He particularly protested Walker's renewal of the charge, one that Newcomb had repeatedly rebuffed, that the theoretical principles of traditional economists were unserviceable because of their "arbitrary and unreal character." As the president of a scientific and technical university, Walker should realize that even the physical sciences relied on fundamental principles that were decidedly "unreal," that is, removed from actual, practical applications. Just as an engineering student must master the abstract science of thermodynamics to understand steam engines, so too an economics student must master Ricardian theory to understand the business of railroads.[48]

Newcomb reinforced this general point when, using a curious journalistic artifice, he responded to his own anonymous commentary on Walker's address with a signed letter to the editor of the *Nation*. As in earlier writings, he disarmed Walker by not only agreeing that abstract economic principles had practical limitations, but also declaring that these very limitations are what need to be emphasized to the American public. People need to comprehend that a disparity between abstract principles and practical applications pervades every branch of theoretical science. Newcomb also censured Walker for perpetuating in his presidential address a fatuous two-school distinction.

> I am sorry to see so sound a thinker talk about the "two schools" of political economy. After much careful examination, I have concluded that the qualification required to constitute a new-school man is half a page or so of slighting remark about the economy of Ricardo, Mill, and the English school. Had the President of the Economic Association been a man of less sturdy independence, one might have suspected that the remarks to which the *Nation* takes exception were suggested by previous failure on his part to thus qualify himself, and were the repetition of the creed necessary to his admission to full membership in the new economic church.[49]

By 1893, Newcomb sensed that the divergence between the two schools was an exhausted topic. "In some of its aspects this divergence has become so trite a subject," he wrote in the *Quarterly Journal of Economics* published at Harvard, "that it might seem doubtful whether anything new and useful could now be said about it." To the extent that opponents of the old school had come to realize that "modifying circumstances" must be taken into account when applying abstract principles to human affairs, the battle had become an "intelligent one" between well-meaning adversaries. The real problem facing political economists was not their internal dispute, but the public's ignorance of the established conclusions of their discipline. In other words, Newcomb had returned once again to the message of his 1880 address to the Philosophical Society of Washington, "The Relation of Scientific Method to Social Progress." As he now expressed it, professional economists simply were not making lasting impressions on the thinking public; most Americans still adhered to simplistic ideas that were centuries old. Newcomb argued "that in the every-day applications of purely economic theory our public thought, our legislation, and even our popular economic nomenclature are what they would have been if Smith, Ricardo, and Mill had never lived, and if such a term as political economy had never been known." Whether belonging to the old or new school, political economists needed to educate the public about their science.⁵⁰

While Newcomb had reconciled himself to the new school of political economy and returned to more fundamental issues confronting all economists, he remained unable to overcome personal animosities. Toward James, he remained chary. When Francis Walker invited Newcomb to his house for a meeting of the Political Economy Club in late 1888—eleven months after Newcomb had spoken at James's University of Pennsylvania—Walker felt it prudent to assure him: "Dr. James is *not* to be here; so there will be no danger of a riot." Toward Ely, Newcomb remained irreconcilable. Availing himself of a book review in the *Journal of Political Economy*—a periodical newly established by Laughlin, his close colleague, now at the University of Chicago—Newcomb in 1894 aimed one last shot at his old adversary. Ely, disenchanted with Johns Hopkins, had taken a position at the University of Wisconsin in 1892; just a few months before Newcomb's review, he had barely weathered a widely publicized university trial, instigated in part by the *Nation*, for teaching and practicing subversive economic principles.⁵¹ Newcomb commenced his review by chiding the battle-scarred social reformer for having ever attempted to mold the American Economic Association into a discriminatory economic "church." As for two new textbooks by Ely, the reason for writing the review, Newcomb found them tendentious. "Principles, not doctrines," Newcomb inveighed, "should be the motto of the teacher." And though he conceded that Ely's writing contained some insights, he reported that he was "profoundly impressed on almost every page with the absence of those powers of analysis and logical reasoning which form the fundamental requirements in a teacher of a scientific subject."⁵²

This 1894 review would be Newcomb's last public comment of any substance on not only Ely but also economic science. His next mention of Ely would come after the turn of the century when, writing in his autobiography, he would merely reminisce about his old adversary, respectfully saluting him, especially as a teacher at Johns Hopkins. And in 1905 at a gathering of the AAAS, he would simply offer retrospective comments on economics as an exact science. Apparently, having retired from the Nautical Almanac Office early in 1897, this major voice in American economic affairs gave himself over to other concerns, mainly the study of planets. In his autobiography, he remarked on his disengagement from economics and on his earlier contribution: "Being sometimes looked upon as an economist, I deem it not improper to disclaim any part in the economic research of to-day. What I have done has been prompted by the conviction that the greatest social want of the age is the introduction of sound thinking on economic subjects among the masses, not only of our own, but of every other country."[53]

In sum, scientific method served Newcomb's interests in a threefold manner. When displayed and touted as being useful to the study of political economy, as we saw earlier, method provided the vehicle through which he hoped to convince the public to provide fuller support for basic science in the United States. Secondly, it furnished him with a powerful means of demarcating the mathematical science of political economy, both inwardly among the various "scientific" factions and outwardly among lay participants. This was a quest for professional legitimation and consequent public influence that was part of a broader rhetorical tradition in economics, one dependent on the metaphors of physics. Finally, as just discussed, scientific method supplied Newcomb with a subtle agency for sanctioning and defending his own liberal political and economic position.

CHAPTER VIII

Religion
A Clash with Gray, Porter, and McCosh

In August of 1878, two years before his Washington speech on scientific method and social progress, Newcomb gave a major address in Saint Louis as retiring president of the American Association for the Advancement of Science (AAAS). On the day preceding his address, he had written to his wife: "The meetings here, will I fear, be the smallest and dullest ever held, because the intense heat has driven every one off, and the yellow fever scare has kept many from coming." Two days later, in a follow-up note, he had added: "The speeches of the vice-presidents were made on Wed. evening to the most 'beggardly array of empty benches' with which such a meeting was ever honoured, so far as I can remember. They did a little better for me last night, and I think I got through very well indeed."[1] Though Newcomb's speech had an unpropitious debut, it soon reached a large audience and induced a spirited reaction. This was due not only to the attention normally afforded a presidential address issuing from the estimable AAAS but also to the controversial nature of Newcomb's topic: the relationship between natural science and Christian religion. Four American periodicals, including *Popular Science Monthly*, and one British journal reprinted Newcomb's text. The AAAS, besides publishing the speech in its *Proceedings*, also reissued it as a separate pamphlet.[2]

The relationship between science and religion had been at issue in the United States for many years as Christians, especially, grappled with troublesome theories such as those concerning the development of the solar system and the evolution of plants and animals. The relationship became an especially sensitive subject in the United States following, among other events, the lecture tours of two British scientists: Thomas Huxley, the evolutionist and agnostic who visited in 1876, and John Tyndall, the physicist and religious skeptic who visited in 1872–73. A recent analyst comments that Huxley, Tyndall, and other "great Victorian scientific publicists" from the mid 1840s through the late 1870s "employed the theories of evolution, atomism, and the conservation of energy as instruments to challenge the cultural dominance of the clergy, to attack religion and metaphysics in scientific thought, and to forge a genuinely self-conscious professional scientific community based on science pursued according to strictly naturalistic premises." Newcomb's speech, in fact, seemed reminiscent of Tyndall's presidential address on science and religion delivered in Belfast in 1874 to the British Association for the Advancement of Science.[3] Like Tyndall, Newcomb called for the separation of scientific reasoning and theological arguments. This was an untenable position in the eyes of those educated Christians who still accepted the central claims of natural theology as advanced especially by Englishman William Paley. Though their numbers had dwindled since earlier in the century, such persons believed that science augments religion and that God can be revealed through the study of his design in nature—especially the design of human organs, such as the eye or ear.[4] Indeed, in the very midst of the controversies generated in Great Britain and America by Darwin's theory of evolution, many Christians reconciled their religious faith with even evolutionary thought; these persons ranged from those who meshed orthodox theistic theology with Darwin's actual theory to those who meshed a more liberal theology with Lamarckian and other variations of evolutionary theory. Some cynics claimed, however, that these Christians were merely temporizers seeking to preserve their religion.[5]

That Newcomb voiced a separatist position is not surprising, given his desire to encourage full and open scientific inquiry as envisioned by his mentors and thinkers such as Comte, Mill, and Darwin. But his motives were not merely intellectual, the staking out of science's cognitive domain; they were also institutional. Along with other late nineteenth-century Anglo-American "positivists" seeking to separate science from religion, he wanted not only to counter outmoded ideas but also to eliminate their advocates from the scientific community.[6] Newcomb, with his high aspirations for "American science," felt this institutional imperative probably even more than his relatively established British brethren in that the community of American scientists was still in an emergent stage of professionalization. A sharp distinction between scientific and theological discourse informed the vision of modern science that he sought to convey to the public as he worked to gain popular support, with the ultimate intention of strengthening the institutional underpinnings of American science. But he needed to draw the boundary with finesse: in excluding the old guard of natural theologians from the ranks of professional scientists, he had to be careful not to denigrate (at least in public) Christianity itself. Adroitly, he portrayed scientific knowledge and religious belief as distinctive but complementary—as alternative approaches to understanding. This enabled him to claim a province for science in American culture without, he hoped, alienating the Christians who typically dominated the culture.[7]

THE SAINT LOUIS SPEECH

Titled alternatively "The Course of Nature" or "Simplicity and Universality of the Laws of Nature," Newcomb's address contained a detailed comparison of scientific and religious modes of thought. After some introductory niceties but before engaging the main topic, he set forth a group of guiding methodological premises. He began with a restatement of the general empirical "proposition" that he had presented formally for the first time in his 1874 essay on the deficiencies of American science.

> The key-note of my discourse is found in a proposition which is fundamental in the history of modern science, and without a clear understanding of which everything I say may be entirely misunderstood. This proposition is, that science concerns itself only with phenomena and the relations which connect them, and does not take account of any questions which do not in some way admit of being brought to the test of observation.

He went on, as in his 1874 and later essays, to compare the modern scientist to a successful businessman and to explain, "Scientific investigation is, in a certain sense, purely practical in both its methods and aims."[8]

Newcomb then became more specific. In contrast to his blanket assertions of the mid 1870s that "propositions" and "questions" be amenable to the test of "experience" or "observation," he now sharpened his analysis and insisted that individual terms be definable through sensory experiences.

> To speak with a little more precision, we may say, that as science only deals with phenomenon [sic] and the laws which connect them, so all the terms which it uses have exact literal meanings, and refer only to things which admit of being perceived by the senses, or, at least, of being conceived as thus perceptible.[9]

It is noteworthy that such a meticulous methodological statement appeared in Newcomb's public comments for the first time in August of 1878. (Recall that the equally exacting methodological statements in his speech to the Philosophical Society of Washington on "The Relation of Scientific Method to Social Progress" initially appeared in 1880. His commentary on the methods of the economic schools fell in the mid 1880s.) Chauncey Wright's various essays having pragmatic leanings had been republished the previous year, and six months before this Saint Louis speech Charles Peirce's only early pragmatic essay had appeared. Newcomb's statement parallels Wright's insistence that individual terms have "precise, unambiguous definitions" based on "measures of the phenomena" and "sensible properties." It also parallels Peirce's pragmatic maxim published in January of 1878 as well as his accompanying discussion of the meaning of individual conceptions based on "sensible perception" and "sensible effects of things." And just as Peirce carefully indicated that a conception need not refer to actual effects but only to those "which might conceivably have practical bearings," Newcomb qualified his demand that terms "refer only to things which admit of being perceived by the senses" by adding that the things admit, "at least, of being conceived as thus perceptible."[10]

Newcomb continued this preliminary statement with an observation also reminiscent of Wright's and Peirce's recurrent laments regarding the misuse and blurring of both scientific and nonscientific terminology by metaphysically minded philosophers and theologians. "This purely literal meaning of all scientific language," he concluded, "is in strong contrast to the metaphorical and poetical forms of expression into which we are apt to fall in discourse upon abstract subjects generally, where our ideas cannot be at once referred to sensuous impressions." In another published version of the speech, Newcomb presented this point more graphically:

> The use of plain language appears to be an actual source of difficulty with some in trying to understand the philosophy of science. Long habit in the

use of figurative language in which ideas not readily comprehensible are symbolized by common terms leads one to look for hidden meanings in all philosophic discourse, and to see difficulties in terms which, to a scientific thinker, are as plain and matter of fact as an order for breakfast to an hotel-waiter.[11]

Having established a methodological keynote for his address, Newcomb was now ready to state his case regarding the relationship between science and Christian religion. In particular, he drew on his preceding characterization of the linguistic, empirical method of science for a means of distinguishing scientific from religious thought — two forms of thought sharing at times a common focus, nature. Put most simply, "no question is a scientific one which does not in some way admit of being tested by experience." In Newcomb's opinion, persons obstruct the progress of science by ignoring this criterion and injecting religious issues into scientific discussions. To be sure, religious questions might be worth thinking about in their own right, but to disregard the "well defined limit" between scientific and nonscientific questions is to invite "confusion of thought." "The current desires that science shall consider man as something more than an animal," he explained, "are as unreasonable as if we wanted to make algebra a help to moral philosophy."[12]

Newcomb also contrasted the theological and scientific notions of "truth." For the Christian theologian, certain doctrines are true in the ultimate, metaphysical sense of the term. "His idea of truth," Newcomb reports, "is symbolized in the pure marble statue which must be protected from contact with profane hands, and whose value arises from its beauty of form and the excellence of the ideas which it embodies." In contrast, for the scientific investigator, the idea of truth "is symbolized by the iron-clad turret, which cannot be accepted until it has proved its invulnerability." Continuing the artistic and military similes, Newcomb maintains that, unlike the theologian's statue that is protected from violence, the scientist's iron-clad turret is intentionally exposed to attack.

> Its weak points are sought out by eyes intent on discovering them, and are exposed to the fire of every logical weapon which can be brought to bear upon them. A scientific theory may thus be completely demolished; it may prove so far from perfect that its author is glad to withdraw it for repairs or reconstruction; or it may be hammered into an entirely new shape. But however completely it may stand the fire, it maintains its position as a scientific theory only by being always in the field ready to challenge every new comer, and to meet the fire of every fact which seems to

militate against it. A countless host of theories have thus been demolished and forgotten with the advance of knowledge, but those which remain, having stood the fire of generations, can show us a guarantee of their truthfulness which would not be possible under any other plan of dealing with them.

In speaking of "invulnerability" and "a guarantee of their truthfulness," Newcomb does not mean to imply that scientific propositions, theories, or laws are true in any ultimate, metaphysical sense. Nor does he mean to suggest that scientific truth is established through revelation or authority. Rather, truth is "decided by the human judgment." That is, he hastens to add, the truth of a proposition is established "not by anything in itself, but by a more or less long and painful examination of the evidence for and against it." When a scientist says that a "proposition is worthy of being received as true, he means, not that it bears any recognized seal of truth, but that the evidence in favor of it entirely preponderates over all that can be brought to bear against it." Along similar lines, he later refers to Newtonian gravitation and comments that, in a certain sense, "laws of nature are simply *general* facts, distinguished from special facts by their dependence upon certain antecedent conditions." As such, there is "no profound philosophy involved in their action or expression any more than there is in such statements as that all unsupported bodies fall toward the centre of the earth."[13]

Throughout the remainder of the speech, in general accord with his opening remarks on language, meaning, and truth, Newcomb advocated a complete separation of scientific and religious thought; he continued, however, to acknowledge the possible validity of each within its own domain. The sole domain of scientific inquiry, with its "mechanical" mode of explanation, should be the visible world of phenomena. The sole domain of religious inquiry, with its "teleological" explanations, should be the unseen universe of final cause and ultimate purpose. Regarding this distinction, he explained that persons who favored mechanical explanations believed that causes acted in nature without any regard to observable consequences. In other words, as the alternative titles to his address implied, the course of nature is uniform and its laws are universal, operating blindly throughout cosmic history and across all cosmic realms. Furthermore, according to the mechanical thinker, the laws of nature "do not possess that character of inscrutability which belongs to the decrees of Providence; but are capable, so far as their sensible manifestations are concerned, of being completely grasped by the human intellect, and expressed in scientific language."

Unfortunately, in Newcomb's view, many persons of a teleological persuasion still disregarded these distinctions and overstepped the boundary between science and religion. Such persons believed not only that God is intimately involved in the processes of nature but also—and this was the bothersome point—that his involvement is discernible through scientific investigation:

> The teleological explanation of nature, presupposes that her operations are akin to human actions insomuch as they are under the control of, and directed by one or more intelligent beings having certain ends in view; that the events are so directed as to compass these ends; and, finally, that the relation of the events to the ends, admits of being discovered by observation and study. This last condition is a very important one, because, without it, the teleological explanation of the cause of nature would not be a scientific one.

Newcomb went on to suggest that adherents of the teleological school were entitled to discourse as much as they pleased about God's ends in nature, but they were not justified in claiming that the ends were discoverable through scientific investigation.[14]

Newcomb urged that, contrary to past patterns, scientists should stop speculating on or investigating God's plan in nature. Conversely, theologians should refrain from pronouncing on the material course of nature and its laws. Although he did not go so far as to endorse Darwin's theory of evolution, he implied that the theory with its positivistic tone was helping to eliminate teleological speculation from scientific explanation. Darwinian studies, he explained, "are designed to show that those wonderful adaptations which we see in the structure of living animals, and which in former times were attributed to design, are really the result of natural laws, acting with the same disregard to consequences which we see in the falling rock." Accordingly, he added, "the eye was not made in order to see, nor the ear in order to hear, nor are the numberless adaptations of animated beings to the conditions which surround them in any way the product of design."[15] Like Wright, Newcomb advocated the metaphysical neutrality of scientific inquiry: Christian teleology should have no hold on natural science.[16]

This divisive view of scientific and religious modes of thought, though it entailed the complementarity of the modes, met with immediate criticism. During an era in the United States when various Christian theologians, philosophers, and even scientists—including defenders of Darwin's theory of evolution—still believed not only in "scrutable design" in nature but also in divine providence and the power of prayer to effect change, Newcomb's restrictive perspective was an affront. Some of these Christians objected to his contention that mechanical laws alone sufficed to explain all natural phenomena; this appeared in the cases of certain phenomena to be an encroachment of science into realms that were further or better understood in terms of God's design and providence. Other Christians who sought the accommodation of scientific interpretation and Christian belief objected to Newcomb's unequivocal separation of the two; this seemed to be an unnecessary surrender to the rising agnosticism and skepticism of scientists like Huxley and Tyndall. Newcomb soon felt it necessary to respond to the various, published criticisms, especially the forceful attacks of three eminent Christians: Asa Gray, Noah Porter, and James McCosh.[17]

ASA GRAY, NOAH PORTER, AND JAMES MCCOSH

The most sustained reaction to Newcomb's speech appeared in the *Independent*—a respected Christian weekly that had been founded by antislavery Congregationalists but which now reported on a wide range of topics for a general readership of about fifteen thousand.[18] The *Independent* not only reprinted the text of the speech but also during late 1878 and early 1879 published numerous articles, letters, and editorials critical of the speech. Two series of particularly incisive attacks were submitted by two anonymous "country readers." Both authors feigned innocence regarding science, philosophy, and theology. Knowledgeable subscribers to the *Independent* probably realized that the first "country reader" was Harvard's Asa Gray (1810–1888), America's most respected botanist and foremost defender of Darwin's theory of evolution.[19] "Another Country Reader" eventually revealed himself to be Noah Porter (1811–1892), one of the nation's leading clergyman, a Congregationalist, and the president of Yale College.

An orthodox theist of Calvinist persuasion, Gray consulted with kindred spirit George Frederick Wright before responding to Newcomb. Having clarified his strategy, he invoked Joseph Henry's name to argue that Christian and scientific thought are compatible and that the design of the "supreme intelligence" lies open to scientific investigation. "I find it simply impossible," he wrote, "to doubt that there are 'scrutable designs,' or what Prof. Henry terms definite ends, *in some parts* of Nature—notably in the animal and vegetable kingdoms."[20] In countering this view in a series of responses carried by the *Independent*, Newcomb critically dissected Gray's language. When Gray misinterpreted Newcomb to have maintained that "so far as Science shows or *can show*, a *will* has nothing to do with the course of events," Newcomb responded: " 'Can show!' No, sir! Not one sentence of my discourse, so far as I can remember, is devoted to limiting what Science *can* show. I only ask what she *does* show, as the result of three centuries of observation." And scientific study shows, Newcomb concluded, no concrete evidence of a divine being modifying the course of nature. In a similar vein, Newcomb later complained that he did not know what Gray meant by "influence of a will"; before conversing further with Gray, Newcomb insisted on having the concept applied to a "concrete case." In general, Newcomb objected to "the fog of ambiguity and sentimentalism" that enveloped Gray's type of theist perspective.[21]

Noah Porter followed Gray in the *Independent*, publishing critiques of Newcomb's views as they pertained to Christianity. Porter suggested that it was atheistic as well as unscientific to deny teleological ends and providential effects in the world of natural phenomena. He argued that an exhaustive understanding of a natural phenomenon could be obtained only when a "teleological explanation" was added to the "mechanical." Like Gray, he believed that design could be discerned in nature and that there could be "a scientific theory of Providence and Prayer."[22]

Newcomb's stand irked not only a leading scientist and a noted clergyman but also a well-known philosopher. The *Princeton Review* published a disapproving essay by James McCosh (1811–1894), the president of the College of New Jersey (later Princeton). McCosh, who a few years earlier had similarly rebutted John Tyndall's Belfast address, was a principal proponent of the Scottish, intuitionist philosophy. In line with his intuitionist doctrine and the doctrines of introspective psychologists and others who disdained positivist reductive analysis, he singled out Newcomb's insistence that scientific terms be definable through sensory experiences. He quoted Newcomb's key methodological statements, including the assertion that science "refers only to things which admit of being perceived by the senses, or at least of being conceivable as thus perceptible." McCosh faulted this "narrow view of science" for excluding the a priori truths of mathematics. In addition, he faulted it for ignoring the invisible sphere of "mental action, such as perception, memory, imagination, reason, emotion, conscience, will." "In this way," McCosh complained about Newcomb, "he shuts out from the domain of science mental action, which I hold admits of scientific treatment quite as much as material action, though it is not visible, and is not made known by the telescope, the microscope, or any like appliances."[23]

More generally, McCosh questioned Newcomb's credibility regarding theological issues. He described Newcomb as being "a gentleman who is eminent as an astronomer, but who has gone out of his way to take up a topic with which he is not specially acquainted, and who has uttered words of which he does not see the meaning, nor the consequences, nor the use which will be made of them." (Recall that Edmund James and other new schoolers would level a similar charge against Newcomb six years later when the astronomer published his textbook on political economy.) In other words, Newcomb was one of a group of upstart scientists including Tyndall and Huxley who unfortunately "have not enjoyed the advantage of comprehensive instruction in philosophy and in the science of the human mind, such as is required in our higher universities." Thinkers who were properly schooled realized that not merely mechanical but also teleological principles were at work in nature. "I discover not only force which hurries on like a railway train," McCosh elaborated, "but rails to restrain it and intelligence guiding it. I find not only mechanism, but machines constructed for ends. The mechanical doctrine, if carried out exclusively, would strip nature of all that endears it to us—of all its sunshine, of all its beauty and beneficence, and leave nothing to call forth our admiration, our gratitude, our love." A theological liberal, McCosh insisted that we can discern final cause even in the evolutionary "development" of living creatures, not just in phenomena of the physical universe. Moreover, we can achieve a full understanding of nature only by acknowledging the compatibility of mechanical and final cause.[24]

Newcomb responded by initiating a formal dialogue with Porter and McCosh. This occurred through a written symposium titled "Law and Design in Nature" published in the *North American Review* during May 1879. He opened the interchange with a terse restatement of a "postulate" that had colored his writings since his Cambridge years and had provided the keynote for his 1878 AAAS address—a postulate which, in fact, had its origin in the writings of Wright and, before him, Mill (specifically, Mill's "axiom" of the uniformity of the course of nature and "law" of universal causation).[25] Newcomb phrased this "fundamental postulate of the scientific philosophy" with care: "The whole course of Nature, considered as a succession of phenomena, is conditioned solely by antecedent causes, in the action of which no regard to consequences is either traceable by human investigation, or necessary to foresee the phenomena." This postulate, which found specific expression in the theory of evolution, meant the elimination from scientific discussion of "all those abstract conceptions which are frequently associated with phenomena, but which do not serve to assist in defining phenomena." These conceptions included "the opposing ones of potentiality and necessity" as well as "those of the invisible forces or causes which may lie behind the visible course of Nature." Newcomb again emphasized, as he had originally in his 1878 address, that prior misunderstandings between scientists and theologians had arisen "partly from a failure to distinguish between phenomena as such and the abstract ideas with which they may be associated." Accordingly, in the present symposium, he encouraged the use of "concrete examples" over an "abstract definition" in trying to give "meaning to the term, laws of Nature."[26]

Newcomb's positional statement appeared along with the reactions of four religious commentators, including Porter and McCosh. These two men reiterated their earlier statements that design is discernible in nature and therefore relevant to scientists. No right-minded person, according to Porter, "would deny that design or purpose is as clearly traceable in many of the arrangements and phenomena of Nature as the causes or laws that are ascertained by experiment or induction." McCosh concurred: "He who asserts that there is no regard to consequences traceable in mundane action is setting aside that argument for the Divine existence which the Scriptures sanction (Ps. xix. 1, Rom. i. 20), and which the great body of mankind have acknowledged to be valid." For an example, McCosh reconsidered a phenomenon discussed originally by Newcomb, a fire in a busy theater; even with such a tragic and seemingly pointless event, McCosh declared, it is sometimes possible to discover "an intended connection" between a "sin and its punishment." Newcomb, at this point in his personal copy of McCosh's published remarks, jotted a sarcastic reply: "Now if he will give the date and place of a certain fire, and name and residence of person punished."[27]

A month later, the *North American Review* carried a "rejoinder" wherein Newcomb publicly answered the commentators' criticisms, especially those pertaining to evolution. Contrasting the natural theology of his adversaries to his own scientific outlook, he reemphasized that all talk about design should be relegated to discussions of spiritual matters and not be extended to the realm of actual scientific practice. "It is one thing," he wrote, "to say that there is design in Nature, or that all things were designed to be as they are, but an entirely different thing to say that we know these designs, and are able to explain and predict the course of Nature by means of them." Newcomb contended: "The scientific philosophy entirely excludes design as affording that explanation of Nature which it desires, that is, such an explanation as will enable men to foresee the course of Nature."[28]

Newcomb closed this rejoinder with a reappraisal of his own fundamental "scientific postulate" regarding the inappropriateness of metaphysical considerations in scientific studies of the "course of Nature." Specifically, in line with his linguistic, empirical methodology, he reexpressed the postulate in language more directly tied to sensory experience:

> All definitions of the phenomena of Nature, in general and abstract terms, such as we have used in formulating the postulate, are subject to this inconvenience: that we apprehend the meaning of the terms used only by their unconscious reference to special objects. As our ideas of a man, an animal, a metal, or a color are derived only by having special objects presented to us to which we have learned to apply these names, so the ideas which we attach to the most general philosophic terms are derived in the same manner. We may, therefore, avoid a possible failure to understand correctly the idea presented, by dispensing with general definition of the course of Nature, and considering the postulate as expressive of the doctrine that Nature always has been what we now see it, and is in all its realms as we see it around us every day.

In other words, straightforward observation shows that nature is controlled by fixed and uniform laws. Newcomb added that this doctrine of the "unity of nature" might be labeled "monism" in that it rejects his Christian critics' inclusion of both natural and supernatural causes in the operation of nature; however, because monism also traditionally denotes the additional doctrine of the unity of mind and body, he preferred to label his view the "scientific philosophy."[29]

NEWCOMB'S PERSONAL RELIGIOUS SKEPTICISM

In all of his follow-up commentaries, as in his AAAS speech, Newcomb avoided direct references to his personal religious beliefs. Even when his "religious friends" urged him to disclose his personal views, he replied: "I respectfully but firmly decline to express any opinion upon any theological question whatever in connection with this discussion."[30] He consistently portrayed himself as a disinterested scientist who was seeking objective clarification of the relationship between scientific and Christian modes of thought and, thus, the elimination of obstacles to knowledge. At least to his satisfaction, the clarification came through using scientific method as a criterion to differentiate scientific and religious outlooks. While maintaining a tone of impartiality, he could argue that traditional natural theology failed to meet the methodological test and therefore should be divorced from modern science. This argument served, of course, the ulterior purpose of delineating a professional community of scientists in an era when Americans such as Newcomb were attempting to shore up the institutional moorings of their nation's science. Whether the exponents of natural theology were considered to be practitioners within the scientific community or theologians and philosophers outside the ranks, they needed to be purged from the profession. Yet, Newcomb was not indicting Christian religion. Though he urged the exclusion of natural theologians, he scrupulously acknowledged the complementary character of religious and scientific knowledge, a prudent tact for a scientist not wanting to jeopardize his profession's claim to fuller public support and cognitive authority.

Newcomb, however, had a personal motive for challenging orthodox religious opinion. Just as his personal political ideology lay behind his seemingly selfless desire to bring scientific method to the aid of political economy, so too did personal religious convictions underpin his professedly impartial attempt to ensure the progress of science by eliminating religious obstruction. In private, he believed that Christianity was an untenable, dying religion. But in an era when members of polite "Christian society" maintained airs of piety, he was unwilling to voice the opinion openly — an unwillingness compounded by the damage that such frankness might do to the cause of science in the United States. Though it still would turn out to be controversial, the Saint Louis speech, with its academic flavor, offered a seemingly acceptable alternative to full personal disclosure. By analyzing whether natural theology met a methodological criterion, Newcomb could maintain the guise of scholarly neutrality while still criticizing aspects of Christian theology. Furthermore, by invoking the methodological criterion and concentrating on the process rather than the product of scientific inquiry, he could sidestep the troublesome fact that Asa Gray and others were advancing evolutionary science though working within a Christian intellectual context.

Newcomb, as we have seen, had a long-standing interest in religion. Indeed, his interest bordered on being a fixation—a fixation perhaps precipitated in childhood by unresolved tensions between his mother's piety and his father's rationalism. Curiously, we know about his religious life largely because of the controversy that followed his AAAS address; in an effort to sort out his ideas on Christianity in light of the dispute, he began in the late summer of 1879 "writing out my religious views and difficulties at various times of my life." "I have been led by the discussions of the last year," he comments, "to examine into the whole subject more carefully, and think it over more thoroughly than before."[31] In the consequent drafts of his religious autobiography, Newcomb detailed his religious development, beginning with the active belief and terror associated with the orthodox Calvinism of his childhood. As we saw earlier, this period of emotional intensity gave way during his teenage years to a period of indifference in which he began to disassociate sacred and profane history. Later while at Harvard, through church attendance, discussions, and reading, he experienced a more liberal and benign form of Christianity; though he was sympathetic to this form, he remained unable to give himself over to the faith.

When he moved to Washington, Newcomb continued to wrestle with religious issues. As we have seen, he found himself drawn to the objective spirit though not the doctrines of Darwin, Huxley, and other exponents of "modern thought." In Washington, he also continued to attend church. (Participation in church, especially the Presbyterian church, had social as well as spiritual advantages; on an autumn Sunday in 1861, he mentioned in his diary that not only Joseph Henry was at the service that morning but also President Lincoln.)[32] His marriage in 1863 to Mary Hassler, a member of an influential scientific family in Washington, further immersed Newcomb in the ecclesiastical life. Specifically, it involved him in "an orthodox family, maintaining the tenets of the Presbyterian church" and in "a fashionable city congregation." Whereas he attended church, he resisted formally joining: "no influence could overcome my early shrinking from the subject of personal religion, to say nothing of the doubts as what I could consciously say I believed, and my taste for an independent attitude." Thrown back into a Calvinistic church that exhorted nonbelievers to seek conversion, Newcomb recoiled: "I found myself involuntarily and almost unconsciously listening to my pastor in a spirit of adverse criticism, well calculated to prepare the mind for a more aggressive attitude toward his doctrines." Nor did his attitude soften as he and Mary began to raise three children: Anita (born 1864), Emily Kate (born 1869), and Anna Josepha (born 1871). Even the convention of reading the Bible to his family fed Newcomb's religious skepticism. The regular readings afforded him the opportunity to evaluate the scriptural arguments of Christian commentators; he found their "evidences of Christianity" to be wanting. After completing the 1879 draft of his "religious views and difficulties"—the draft motivated by the controversy stirred up by his speech—he drew a pessimistic general conclusion: "The result is that instead of being surprised at the general decay of faith in what we have always considered *orthodox* religious doctrines, I am surprised that there is any left in some of the tenets among careful readers and students."[33]

As mentioned, it would not only be professionally counterproductive but also unseemly for Newcomb to voice his skepticism openly. Late Victorian propriety dictated discretion regarding religious issues, especially by a well-known government scientist and family man. Indeed, when the *Sunday School Times* responded to the Saint Louis speech by accusing Newcomb of arrogantly presuming to comprehend the limitations on God's role in nature, Newcomb rebutted the charge in a letter to the editor written "at the request of members of my family, who have been pained by the attack."[34] But something more than professional prudence and social propriety lay behind Newcomb's reluctance to voice his religious opinions publicly: he harbored, perhaps partly as a result of the tension between his parents, a deep personal aversion to any open revelation of these opinions. "I could never be induced, could never persuade myself," he wrote in the draft of his religious autobiography, "to speak a word on the subject of my personal religious feelings to another, not even to my mother." So strong was this "life-long and unconquerable aversion," that he could not even confide his feeling to his wife. In a confidential note jotted to either himself or an intimate associate in 1879, he justified not telling his wife about his skepticism to protect her feelings: "I have sometimes felt that a want of sympathy was growing up between me and my wife on account of my want of *sympathy*, irrespective of their *truth*, with those doctrines which she regarded as essential. Perhaps it would have been better had I explained each difficulty to her as it arose, but I knew

Simon Newcomb in 1857 at age twenty-two, while a student at Harvard's Lawrence Scientific School, and a computer at the Nautical Almanac Office in Cambridge. (Photo courtesy of the Manuscript Division, Library of Congress.)

Pages from Newcomb's diary during his Cambridge years. The entry for April 14th, 1859, reads: "A disagreeable day no game of ball. Talked at the castle with Runkle and Wright. The latter had Hamiltons lectures, just published. Dr. Gould has returned to Cambridge. Got a letter from Cousin Simon."

The entry for April 15th reads: "Went to Watertown and took tea with Mrs. Whiting this evening." Newcomb had been seeing Mrs. Whiting's daughter. (Photo courtesy of the Manuscript Division, Library of Congress.)

Simon and Mary Newcomb shortly after their marriage in 1863. Joseph Henry introduced Simon to Mary, the granddaughter of prominent geodesist Ferdinand Hassler. (Photo reproduced from *McClure's Magazine*, Oct. 1910.)

"Professor" Newcomb at the Naval Observatory's telescope. Newcomb, ca. 1873, in the lower-left corner, is at the eyepiece of the new twenty-six-inch telescope. (Photo courtesy of the Library, Naval Observatory.)

Astronomers at the Naval Observatory preparing for the 1874 transit of Venus. Newcomb is standing at the left-center in the foreground. (Photo courtesy of the Library, Naval Observatory.)

Newcomb in the 1870s. He stood about five-feet-six-inches tall and weighed, during this period, about 175 pounds. (Photo courtesy of the Smithsonian Institution Archives.)

The superintendent of the Nautical Almanac Office and president of the Washington Philosophical Society. Newcomb in 1879 at age forty-four. (Photo courtesy of the Manuscript Division, Library of Congress.)

A draft, in Newcomb's hand, of his "Religious Autobiography," ca. 1879–1880. This unpublished manuscript begins: "I was born in the country before the leaven of 'liberalism' had been felt far outside the great cities, and bred in a church which neither glazed over nor softened down the beliefs of the New England Puritans from whom it derived its strength." (Photo courtesy of the Manuscript Division, Library of Congress.)

The elder statesman of American science, Newcomb in 1903 at age sixty-eight, recipient of all the major late-nineteenth-century scientific awards and honors, including the Copley Medal of the Royal Society of London. (Photo courtesy of the Manuscript Division, Library of Congress.)

she was averse to such questioning, and feared to disturb her." His aversion also contributed, of course, to his reluctance to join churches in Cambridge and Washington; he wanted to avoid having to make any "public profession of religion." The Saint Louis speech, as we saw, posed less problem in that it maintained an air of scholarly detachment. In giving a speech that involved only "a nonpersonal doctrine," he did not feel his normal "reserve."[35]

But another avenue existed for Newcomb to publicly express his personal religious skepticism without fear of rebuke or embarrassment, and without anxiety about imperiling science's image. He took advantage of an alternative mode of expression popular during this period, one that he would use to advantage in combating the new school of political economy: the anonymous article. During the very month that Newcomb delivered his AAAS address, the *North American Review* carried his "An Advertisement for a New Religion" under the signature of "An Evolutionist." Then, half a year after his main exchange with Gray, Porter, and McCosh—indeed, as "one result of the discussion"—the *Review* published Newcomb's unsigned essay "The Religion of To-day." In both articles, Newcomb bluntly expressed his low esteem for Christian religion. (Although he apparently never did, Newcomb also intended to publish under a nom de plume the revealing "Religious Autobiography," which he drafted during this period.)[36]

Whereas "An Advertisement for a New Religion" contained many overstatements made with tongue in cheek, it harbored a serious message. All traditional religions, particularly Christianity, are "sick, dying, or dead." To substantiate this point, Newcomb rattled off the damning diagnoses of leading philosophers and scientists, including the three thinkers that he had first studied in Cambridge.

> Comte has demonstrated that we cannot discover either first or final causes—the two dark caves from which all religions have issued, like wild beasts, and into which they retreat when pursued. Mr. J. S. Mill has admitted that, on the principle (which, however, has no evidence in its favor) of causation being universal, there may be some presumption in favor of the existence of a God; but then he proves that this God cannot be an omnipotent God, otherwise he would prevent the evil. Darwin has plucked from man's brow his claim that he was specially created by God and in God's image, and has demonstrated his derivation from the ascidian through the catarrhine monkey.

While adamant that Christianity was failing, he acknowledged that man has deep-seated religious instincts. Consequently, Newcomb saw a need for a new "religion." Unlike the old religions, the new faith should not be based on ideas such as an anthropomorphic God or the personal immortality of the soul. In line with Comte, who had advocated a rejuvenated "Religion of Humanity," and with Mill, who had refined Comte's proposal, Newcomb speculated that the new religion would be humanistic and joyous in form and would be compatible with "the latest natural knowledge." But he was unsure of the exact shape of the new faith, which would only evolve slowly. Thus he desired to "advertise" among "our scientific doctors all over the world" to help with the birth of the new religion.[37]

Whereas his next anonymous article, "The Religion of To-day," had a more somber tone, it contained the same message. Again, Newcomb asserted that there existed a "movement toward skepticism" regarding Christianity. Confronted by modern thought, the church itself was drifting in its doctrines. "Perhaps the most striking example," Newcomb pointed out, "of the readiness of theology to temporize with the irreligious thought of the day, and to explain away doctrines it once held dear, is seen in its attitude toward the now fashionable theory of evolution. No other modern theory is so directly opposed to the doctrine which lies at the basis of our orthodox system of theology." Christians were misguided who, through theistic interpretations of the theory, attempted to reconcile it with church doctrine and thus salvage the threatened doctrine.[38]

With Christianity and other traditional religions failing, the way was becoming clear for a "new religion." In a revealing passage that probably summarized Newcomb's personal religious credo, he reported what a hypothetical follower of the new faith might affirm:

> I have no belief in a personal Deity, in a moral government of the universe, in Christ as more than a philosopher, or in a future state of rewards and punishments. But I was born with a sense of duty to my fellow man. I was imbued in infancy with the view that, as a member of society, it was my duty to subordinate my own happiness to that of others. My sense of right and wrong was thus developed at a very early age, and by the constant endeavor to do what was right my conscience acquired a constant increasing development, and asserted more and more its power over my actions. I am not virtuous from any hope of reward or fear of punishment, but only because I feel that virtue is my highest duty, both to myself and to humanity. This feeling has developed to such an extent that the good of my fellow men is now my ruling motive, and vice the object of my most extreme detestation.

Reflecting the spirit of open scientific inquiry that he valued so highly, he added: "Such a faith fears no false teaching, sets no limit on the freedom of human thought, and views with perfect calm the subversion of any and every form of doctrinal belief, confident that the ultimate result will tend to the elevation of the human soul and the unceasing progress of spiritual development."[39]

Thus, when Newcomb sought in his Saint Louis speech to enhance scientific progress by disengaging natural science from natural theology, he had more than a scholarly or academic interest in the enterprise. He was not merely attempting through the criterion of scientific method to remove an intellectual obstacle to scientific inquiry, but also to bolster the public image of American scientists in a period of institutional inadequacies. In particular, he was seeking to distance himself and other accredited American scientists from promoters of natural theology, whether the promoters be practitioners inside the research community or theologians and philosophers outside. And he was fashioning this boundary with care: eager to cultivate public support, he underscored that religion and science, when confined to their proper spheres, contributed to culture in a complementary manner. But this effort at demarcation, which Gray, Porter, and McCosh resisted, represents only Newcomb's public position. Privately, he was also challenging a religion that, in his opinion, was impeding the progress of humanity.

CHAPTER IX

Physics and Mathematics

Public Understanding and Educational Reform

In realms outside natural science, Newcomb confidently imposed the rules and norms of his rendering of scientific method. Appraising current practice in political economics with the stated aim of transforming the discipline into a mathematical science, he sought both to demarcate scientific from nonscientific studies and to delineate the particular views of scientific factions. Whereas in theology, with the voiced goal of eliminating intellectual impediments to scientific research, he sought to distinguish teleological from mechanical interpretations of nature whether held by practicing scientists or theologians. And while he may have thought that he was disinterestedly engaged in both these pursuits, Newcomb interjected into them his personal ideology—political liberalism and religious skepticism. Recalling our threefold grouping of the rhetorical uses of method, we realize that Newcomb's activities in both political economics and theology fall mainly on the third level. He was operating at the level of the public politics of science, attempting through the rhetoric of method to project a particular image of science to lay people and to elevate the cognitive and cultural status of American scientists—all with the further intention of garnering wider public support for basic research. Specifically, he was portraying science as a socially useful enterprise that enhanced the public's control over the nation's political economy; he also was portraying it as a professional enterprise sharply distinguishable from nonscientific realms of discourse, realms populated by unschooled commentators on political economy and by natural theologians. In addition, he was drawing on the rhetoric of method to advance within a public arena his personal political and religious beliefs.

Newcomb was operating at the second level—the level of the institutional and disciplinary organization and politics of science—to the extent that he was trying to sway opinion on which groups of practitioners deserved the scientific community's allegiance. Thus, he attempted to differentiate two groupings of political economists (the new and the old school) and two groupings of natural scientists (followers of natural theology such as Asa Gray and adherents of Newcomb's "scientific philosophy"). Of course, as mentioned earlier, these initiatives on the disciplinary level can be construed as overlapping the public level; in seeking to disassociate supposedly legitimate scientists from practitioners who adhered to the new school of political economy or to natural theology, Newcomb seemed at times to lump these suspect practitioners into the category of nonscientists. In other words, his various efforts at demarcation are as much outer-directed and aimed at setting the social boundaries of science as they are inner-directed and aimed at delineating disciplinary groupings.

But did Newcomb ever apply his methodological strictures to fields closer to home? Did he bring his rhetoric into play in his frequent excursions into physics and mathematics, let alone in his main research area of astronomy? The answer is a qualified yes. On the one hand, his published technical papers on astronomy, mathematics, and physics are essentially devoid of explicit commentary on method (excepting, of course, method in the sense of observatory procedures and computational techniques).[1] In these formal research papers, he simply gets on with the job at hand, not taking time for lofty philosophical digressions on the nature of scientific inquiry. On the other hand, in various, more general analyses, especially those pertaining to physics, he draws on his methodological skills. These general commentaries are the closest that he comes to making methodological appraisals of contemporary physical science.

Recall that Newcomb had a long-standing interest in physics. During his Cambridge years, he even sought a professorship in physics and drafted portions of a physics textbook for college students. Through midcareer, he continued to delve into the subject, with his most visible contribution coming around 1880 when he successfully redetermined the speed of light; he made this measurement in Washington while in close communication with the young Albert Michelson, later famous for his attempts to detect the luminiferous ether.[2] In his various writings about physics, Newcomb always appeared confident of the conceptual foundation of the science, believing that physical phenomena could be understood through the laws of classical mechanics as applied to an all-pervasive ether and a submicroscopic realm of atoms and molecules. Indeed, in an exchange with philosopher and jurist John Stallo, Newcomb presented a highly emotional defense of the foundations. The defense, however, does not provide an example of Newcomb operating at the first or internal level of our schema of method claims; instead, it provides another example of his operating on the public level. He was demarcating the external boundary of physics and protecting its popular image.

RESPONSE TO JOHN STALLO'S CRITIQUE OF PHYSICS

In 1882, John B. Stallo published *The Concepts and Theories of Modern Physics*. Stallo (1823–1900), a German-born lawyer from Ohio with earlier experience in science and philosophy, presented in this book a critique of nineteenth-century physics. With some justification, he accused physical scientists of harboring a metaphysical commitment to their "atomo-mechanical" research program. That is, he charged that they lacked a solid experiential foundation for their view that physical phenomena were ultimately understandable in terms of underlying atoms that followed the laws of classical mechanics. Harvard philosopher Josiah Royce, about twenty years later, remembered that with the appearance of Stallo's book the "sense of scientific orthodoxy was shocked amongst many of our American readers and teachers of science." Similarly, Henry Adams recalled in his *Education* that "for twenty years past, Stallo had been deliberately ignored under the usual conspiracy of silence inevitable to all thought which demands new thought-machinery."[3] One of the Americans shocked by Stallo's book and partially responsible for initiating the conspiracy of silence was Newcomb. He lambasted the newly published book in his 1882 review "Speculative Science." He wrote the piece for the *International Review*, a short-lived but high-minded journal featuring a diversity of articles and book reviews by leading Europeans and, especially, Americans.[4]

Newcomb's attack on Stallo was, to say the least, somewhat misdirected. It was misdirected because the two men had similar general outlooks on science. Although Stallo did not emphasize sensate definitions as Newcomb did in his linguistic, empirical methodology, he did have strong antimetaphysical and empirical leanings; both Ernst Mach and Percy Bridgman would appreciate this in later years.[5] Why then did Newcomb attack a potential ally? Why did he label Stallo another nonscientific "pretender"? And why did he sarcastically state that the only purpose served by Stallo's book was to show "the possible aberrations of an evidently learned and able author"?[6] To answer these questions, we need to recall the tone of scientistic self-righteousness and missionary spirit that usually accompanied his methodological critiques. Envisioning himself above the fray, he typically purported to be bringing modern scientific method to bear on the confused terminology, and hence confused ideas, of theologians, philosophers, political economists, public servants, and other uninitiated thinkers.

Newcomb's general sense that scientists were free from the linguistic fetters of nonscientists correlated with his specific belief in the nineteenth-century foundations of physical and biological science. Said differently, he expressly criticized the outlooks of nonscientists while simultaneously and implicitly defending the basic doctrines of his colleagues. As we saw in the previous chapter, he endorsed the Darwinian theory of evolution. Furthermore, trusting that the foundations of contemporary physics were inherently above dispute, he agreed with the majority of his American and European colleagues—scientists such as Joseph Henry—that physical phenomena were comprehensible in terms of underlying atoms that followed the laws of classical mechanics. As Newcomb stated in 1878, "The idea now entertained by those who see farthest in this direction is that all the physical properties of matter depend upon and may be reduced to certain attractive and repulsive forces acting among the ultimate atoms of which matter is composed." "It may also be supposed," he added, "that all the operations of the vital organism, both in men and animals, depend, in the same way, upon molecular forces among the atoms which make up the organism." Newcomb spoke with confidence on atomo-mechanical science. In prior years, he had studied in depth the kinetic theory of gases, especially as articulated by James Clerk Maxwell, with whom he had corresponded. Also, he had become well enough versed in atomo-mechanical science to engage another of its architects, William Thomson, in a conversation that caused the British physicist to rethink his hydrodynamic analysis of the earth's rigidity.[7]

Though he accepted this atomo-mechanical research program, Newcomb did stress its provisional and hypothetical character—in accord with his general scientific philosophy and, again, in stride with colleagues such as Henry. That is, as we saw in his statements on political economy and religion, he distanced himself from epistemological and ontological claims regarding the ability of scientists to fathom ultimate physical reality and attain absolute truth. As he expressed it in his AAAS presidential speech, the truth of a proposition depends ultimately on human judgment. To say that a proposition is scientifically true simply means "that the evidence in favor of it entirely preponderates over all that can be brought to bear against it."[8] In reviewing Stallo's book, Newcomb similarly would object to Stallo's characterization of mathematicians who study four-dimensional geometry as "disciples" of a particular metaphysical "faith" and "doctrine." Newcomb would insist that "the foundation of the whole reasoning is avowedly hypothetical."[9]

In this manner, then, Newcomb coupled a distaste for the supposedly metaphysical and doctrinaire speculations of nonscientists with a commitment to a patently hypothetical, atomo-mechanical research program. An awareness of this connection allows us to understand why Newcomb maligned Stallo's book. In brief, Stallo triggered the rebuke by directing his antimetaphysical critique at the very foundations of the prevailing physics. Fearing that Stallo's message might mislead the wider laity, Newcomb concluded that "it becomes almost a public duty in the interests of truth, to point out its defects." Perhaps even more worrisome and annoying to Newcomb, the heresy was coming from an outsider who lacked proper professional credentials. "How far is it possible," Newcomb asked scornfully, "for one not actually engaged in scientific work to write correctly and instructively upon the general principles of scientific method?"[10] To Newcomb, Stallo was just another nonscientific, philosophical speculator naively and ineffectually challenging the dependable theories of contemporary physics. Immersed in the atomo-mechanical conceptual environment of his day, Newcomb was inherently unable to grasp Stallo's iconoclastic perspective. Compounding this nearsightedness was his sensitivity to public criticisms of science at a time when he was working to cultivate popular sentiment; for the scientific community to gain the benefits of fuller institutional support, the community needed the backing of the American polity. He reacted by invoking his linguistic, empirical methodology to discredit Stallo's critique. In hindsight, we realize once again that this was not a disinterested use of method. The rhetoric of method provided Newcomb with the weapon to dispatch an enemy who had the gall not only to intrude into the scientists' camp but also to threaten science's flag. Newcomb was fighting another polemical battle on the front of public science—erecting a barricade past which nonscientists should not advance and shielding the image of a proven science.

Regardless of his allegiance to atomo-mechanical physics and its practitioners, Newcomb presented in his review of Stallo's book one of his most discriminating statements on the scientific use of language. In contrast with his earlier methodological utterances that called generally for definitions of concepts in terms of sensory experiences—a blanket criterion that allowed for passive observation as well as active experimentation—the 1882 statement explicitly called for definitions in terms of actual physical operations, specifically, concrete measurements.[11] Indeed, the statement anticipates Percy Bridgman's often-quoted introductory summary of "the operational character of concepts" in his 1927 book *The Logic of Modern Physics*. Bridgman, using the concept of length as an example, explained: "In general, we mean by any concept nothing more than a set of operations; *the concept is synonymous with the corresponding set of operations.*"[12] Newcomb also turned to the concept of length in his 1882 statement as he identified instances of Stallo's "total misconception of the ideas and methods of modern science."

> The word *mass*, for instance, as commonly used in physics, is an abstract noun like *length*; but he [Stallo] uses it as a concrete term, and in nearly the same sense as we commonly use the word matter. He speaks of the conservation of both mass and motion in a way which shows entire unconsciousness of the fact that this expression has no meaning at all. To give it a meaning we must first define the method in which mass and motion are to be measured, and then, in so many ways as we choose to make this measurement, just so many meanings may the expression have. A bar of metal, for instance, may be measured by its length, its breadth, its solid contents, or its weight. A pile of such bars may be measured by putting them end to end or piling in various ways, and measuring the length of the pile in as many ways as we choose. So, in measuring the motion of a system of bodies, we may adopt almost an infinity of different ways which will give different results. The very first necessity of any exact scientific proposition is a definition, without ambiguity, of a precise method in which every quantitive [sic] measure brought in shall be understood. The conclusions are then valid, assuming that particular method of measurement, but they are not valid on any other method.[13]

Newcomb proceeded to use his operational dictum as a basis for a detailed appraisal of Stallo's views on physics. Stallo, throughout his book, had stressed the flaws of atomo-mechanical theory. In his early chapters, he had isolated four propositions, laden with ontological assumptions, that supposedly constituted the conceptual foundation of current atomo-mechanical theory. Newcomb sought to demonstrate that the four propositions lacked clear meaning, thus revealing that Stallo's critique was directed against propositions that no actual physicist could hold. The implication was that the propositions were fictions fabricated by Stallo. To accomplish this debunking, Newcomb invoked his operational dictum.

For example, Stallo purported that physicists adhered to the proposition that "the elementary units of mass are absolutely hard and inelastic." Newcomb responded by arguing that the concepts within this proposition lack clear meaning in that they are not operationally definable, and, therefore, that a practicing physicist would simply find the overall proposition to be untenable.

> We cannot predicate such qualities as hardness or elasticity of the ultimate atoms of matter. All our conceptions of hardness and elasticity are derived from the qualities of sensible masses which may be supposed to arise from the arrangement of their atoms and from the properties of such atoms. We call a body hard when we cannot compress it, and elastic when it rebounds on being struck. But an isolated atom could not indicate either the presence or absence of any such property, and we might almost as well talk about the color of virtue as the hardness of an atom.

Similarly, to the proposition that "the elementary units of mass are absolutely inert, and therefore passive," Newcomb responded:

> We can see no correct meaning to this proposition. Such words as "active" and "passive" have no application in this case, and they serve no purpose except to produce confusion in the mind of the reader. Scientific investigation is concerned only with the effects produced by matter under different conditions, or perhaps we might speak more accurately by saying that it is concerned only with the effects which follow when matter is placed under certain conditions.[14]

Newcomb's reasoning regarding these propositions is reminiscent of that used by Charles Peirce in his 1878 article "How to Make Our Ideas Clear." Newcomb's comments on "hardness," especially, call to mind Peirce's first example of his pragmatic maxim wherein he explained, "The whole conception of this quality [of being *hard*], as of every other, lies in its conceived effects."[15] Newcomb's reasoning is

Physics and Mathematics 153

also reminiscent of the Millian argument that he himself used, while under Wright's sway in Cambridge, to fault Comte's "partial and inconsonant skepticism." "You experience sensations," Newcomb insisted, in responding to Comte's followers, "and sensations only."

> You experience hardness, softness, heat, cold, colors, sensations of sound, and taste and smell, but you do *not* experience planets moving or bodies falling or seeds germinating or plants growing or animals walking or men moving. But these sensations of feeling and hearing and seeing you desire to account for, and you do it by erecting this vast fabric you call a universe, in which the qualities you experience reside, and which is thus regarded as the cause of your sensations.

More generally, Newcomb's entire review harks back to Wright's 1875 book review "Speculative Dynamics," containing the clear statement on meaning mentioned earlier. That is, Newcomb's criticism of Stallo's analysis of physics suggests Wright's rejection of an earlier nonscientist's philosophical interpretation of physics. Newcomb even assigned his review a parallel title, "Speculative Science."[16]

Incidentally, a few years later, Newcomb found a use in political economy for his operational dictum as illustrated by the concept of length. He drew on this line of argument when in his *Principles of Political Economy* and later articles he restated his views on setting the value of the dollar to a "tabular standard," the average price of commodities. He first commented that the concept of "value as a quality admitting of measurement offers peculiar difficulties to the student, owing to its intangible character." While the concept of "value" seems similar to that of "length" (each is measured in discrete units, dollars and feet), only the latter has meaning in terms of an unchanging, material standard directly evident to the senses. Focusing on the variation in the dollar's value, he warned, "we must always remember that calling a thing, whether metal or paper, *one dollar*, or *one pound*, or *one franc*, no more gives it a fixed value than calling a stick *one foot* makes it a foot long." In keeping with his earlier injunction to Stallo about defining concepts through quantitative measurement, Newcomb went on to recommend that the government take the average price of a group of commodities as the tabular standard of value.[17] The call for rigorous operational definitions served Newcomb equally well in his campaigns in physics and political economy.

Just as political economists of the new school and Christians of a natural theological persuasion did not allow Newcomb's charges to go unanswered, neither did Stallo let pass Newcomb's hostile review. In a lengthy reply in the opening pages of *Popular Science Monthly*, Stallo pummeled Newcomb, matching if not exceeding Newcomb's original level of polemic and sarcasm. Undaunted that Newcomb was "a prominent scientist, at the head of a scientific bureau in Washington," the lawyer and one-time judge from Cincinnati subjected his critic's strictures to a scathing "counter-critical examination." In the process, Stallo added to an already confused dialogue by concluding that Newcomb had misconstrued his book's thesis and had attacked the book for its defense of a naively metaphysical interpretation of atomo-mechanical theory. To set the record straight, Stallo explained that he had actually written the *Concepts and Theories of Modern Physics* from the perspective of a "comparative linguist" seeking to expose the insidious metaphysical core of physics and the tendency among physicists to reify abstract concepts. Stallo's position was "that the mechanical theory with all its implications is founded on a total disregard or misapprehension of the true relation of thoughts to things or of concepts to physical realities; that, so far from being a departure from and standing in antagonism to metaphysical speculation, the propositions which lie at its base are simply exemplifications of the fallacies that vitiate all metaphysical or ontological reasoning properly so called."[18]

Claiming to have advanced this position throughout his book, he was dumbfounded that Newcomb could presume to admonish him for reifying the concept of "mass," a concept that Stallo had used with care.

> According to him [Newcomb], this use of the word *mass* is evidence of my ignorance and intellectual confusion, as well as of my "total misconception of the ideas and methods of modern science." He informs me that the word *mass* is "an abstract noun like *length*," whereas I use it "as a concrete term, and in *nearly* the same sense as we commonly use the word matter." And thereupon he delivers himself of a dissertation (which resembles nothing so much as a sermon of "Fray Gerundio" to his "familiars") on the necessity of using scientific terms only in accordance with their exact definitions, of ascertaining the meanings of the words *mass* and *motion* by a reference to the methods whereby they are measured, and so on. All this is certainly strange news to an author who has devoted several chapters of his book to the task of showing that the great fundamental vice of the mechanical theory is the confusion of concepts with things, and particularly of the connotations of the concept *mass* with the complement of the properties of *matter*—who, in a word, is guilty of the great offense of expressing, in the precise terms of the science of logic, what Professor Newcomb is staggering at with a phrase borrowed from some elementary treatise on grammar!

After justifying his treatment of topics such as transcendental geometry and the kinetic theory of gases, Stallo ended with a final jab perhaps intended to irk Newcomb as an economist and supposed expert in monetary matters. He drew an analogy between current atomo-mechanical theories and coins made of base metals; persons forget that such coins are useful merely as tokens and begin to delude themselves that the coins have intrinsic value. Until physicists arrive at theories made of genuine gold—and, at the same time, refrain from minting spurious theories that are treated as gold—they should carry their "facts about in baskets or bags, and resort to the ancient clumsy method of barter." That is, physicists should resign themselves to a more cautious, empirical approach.[19]

CLARIFYING SCIENTIFIC TERMINOLOGY

Whether Newcomb was chastened by Stallo's vitriolic rebuttal is uncertain. Six years later, nevertheless, he chose his words carefully in his address to the Philosophical Society of Washington, "On the Fundamental Concepts of Physics," the same topic that Stallo had explored in his provocative book. He had not changed the basic message that he had outlined while reviewing Stallo's book: it is inappropriate to attribute to the basic units of matter, whatever matter might be, the fundamental properties of "extension and impenetrability" and the associated notion of absolute position. Though all questions about matter must ultimately be "settled by experiment and observation," he preferred to regard matter not as something capable of "absolute contact" and "passive resistance" but as something "enveloped by a thin sphere of repulsive force which actually prevents any contact with its actual substance." And though the hypothesis of a relationship between matter and the luminiferous ether had yet to receive an unequivocal test (he was speaking a year after Albert Michelson and Edward Morley's null test for the drift of the earth through the ether), he was willing to speculate that the ether was entangled with a fourth dimension of space. He was less sanguine about invoking an ethereal medium to explain the action of gravity and other forces; such hypotheses, he maintained, inevitably involved the substitution of a complex physical process for an already simple one—gravitational attraction.[20]

Making cautious analyses such as this, Newcomb certainly felt that he held himself to the same standard of operational clarity that he insisted Stallo meet. Indeed, along with talks such as that on "Fundamental Concepts" to the Philosophical Society of Washington, he published occasional papers aimed at clarifying ambiguous technical concepts. These papers were not addressed to a lay audience, as had been his review of Stallo's book, but to his scientific peers. The papers offer a minor instance of Newcomb putting his linguistic, empirical method into practice on our first or internal level of technical debate, where knowledge claims are framed, negotiated, and evaluated. It is a minor instance not only because he used the method merely implicitly but also because he was tackling issues of secondary or residual concern to practicing scientists. In *Science* in 1883, for example, he presented a concise, concrete, definitional explication, "The Units of Mass and Force," arguing that a prominent French physicist had misinterpreted the centimeter-gram-second (cgs) system of fundamental mechanical units. He similarly emphasized precise, explicit terminology in an 1889 article in the London *Philosophical Magazine*, titled "On the Definitions of the Terms 'Energy' and 'Work.' " (This article was possibly a published version of his talk to the Philosophical Society of Washington.) And in 1893, as he moved beyond midcareer, he proposed to the international readership of *Nature* a new and unambiguous "Nomenclature of Radiant Energy." In particular, Newcomb felt a need for a single, comprehensive term that would embrace different types of radiant energy; this was because researchers now realized that traditional terms such as "radiant heat" and "light" merely correspond to particular wavelengths of a broad electromagnetic spectrum. He explained that it is "unscientific" to use "the word 'light' for ethereal waves having a length between certain definite limits, while there is no corresponding word for other waves." For a single term that encompasses all electromagnetic waves, he suggested "radiance," defining it by how it was to be measured.[21]

Besides initiating these analytic articles on definitions of technical concepts, Newcomb responded favorably to requests by publishers of popular dictionaries and encyclopedias to provide definitions of key scientific terms. That is, he continued to carry his campaign for clear terminology from his professional colleagues to the public. In the process, certainly, he supplemented his government income. As early as 1878 but more so in the 1890s as he neared retirement, he contributed to volumes such as *Johnson's Universal Cyclopaedia* and Funk and Wagnalls' *Standard Dictionary*.[22] Charles Peirce also contributed to reference books during this period. Recall that in 1889, in fact, Newcomb had publicly criticized Peirce's dictionary definitions of basic terms in astronomy and experimental physics. "The definitions in question are, in many cases," Newcomb declared, "insufficient, inaccurate, and confused in a degree which is really remarkable."[23] Newcomb's ensuing squabbles with Peirce similarly centered on terminology in astronomy and mathematics and again implicitly reflected his linguistic, empirical method. For example, Newcomb objected to Peirce's handling of the mathematical concept of infinity and his insistence along with other mathematicians that two infinitely large numbers can have different magnitudes in the sense that they can be expressed as a ratio. In a personal letter to Peirce, Newcomb wrote:

> I have always held that infinity, considered in itself, could not be treated as a mathematical quantity, and that it is pure nonsense to talk about one infinity being greater or less than another. The ground for this view I think I mentioned in our correspondence a year ago; the very meaning of the word infinity is something without bounds. But we can compare two magnitudes only by comparing their bounds. Therefore I say the reasoning in question is baseless. What more can I say?

A recent Peirce scholar, in discussing the Peirce-Newcomb correspondence, labels Newcomb's literal-minded attitude toward infinity as "extreme conservatism." In actuality, as another researcher has demonstrated by tracing Newcomb's innovative views on non-Euclidean geometries, Newcomb was willing to entertain the speculations and abstract formalisms of contemporary mathematicians; he insisted, however, on differentiating between those concepts that could be related to actual experience and those that, at least for the moment, could not.[24] In defending speculations on four-dimensional space during his 1897 presidential address to the American Mathematical Society, he cautioned: "The wise man is one who admits an infinity of possibilities outside the range of his experience, but who in considering actualities is not decoyed by the temptation to strain the facts of experience in order to make them accord with glittering possibilities." This stance reflects not only that Newcomb was an empiricist but also that he was a practicing physical scientist first and a mathematician second. Or, in the words of a close colleague, it would be a mistake to label Newcomb "a mathematician in the more recent sense of the term." "He was rather a mathematical physicist whose work lay chiefly in the field of dynamical astronomy and ... success in his chosen field depended mainly on a masterful knowledge of both mathematics and physics."[25]

THE TEACHING OF INTRODUCTORY MATHEMATICS

During the final decades of the nineteenth century, American educators felt a need for better coordination between secondary schools and colleges; both types of institutions were experiencing accelerated growth. Accordingly, in 1887, the National Educational Association (NEA) took up the problem of "uniformity" in high school curricula and in college admission standards and, in 1892, convened a committee of ten respected educators to examine the problem fully. The following year, the Bureau of Education published and distributed around the nation thirty thousand copies of the report of this "Committee of Ten." With such wide dissemination and with Harvard's dynamic Charles Eliot as chairman of the committee, the report was ensured an attentive audience. Indeed, its impact was enormous. According to one historian of education, "From 1894 to 1905 almost every treatment of matters educational was referred to, compared with, or distinguished from the report of the Committee of Ten." Appraising the same report, another historian of education insists, "For fifteen years after its publication it served as gospel for the curriculum writers of the burgeoning high schools."[26]

The report actually consisted of two sections: the general findings or recommendations of the basic Committee of Ten, and the statements of nine subcommittees or "Conferences" representing individual disciplinary areas. Newcomb chaired the Conference on Mathematics and, thus, was responsible for making recommendations on one of the most integral components of the secondary-school curriculum, the third of "the three R's." His recommendations, while not containing any explicit mention of scientific method, were an extension of his linguistic, empirical orientation toward science. The orientation underpinned his fundamental belief that students more readily learn mathematical concepts through concrete experiences—not, as had been the practice, through abstract discourse. Newcomb's excursion into mathematics education thus illustrates how his "scientific philosophy" colored—and, as it turns out, was earlier colored by—his pedagogic philosophy. His image of the rules governing learning meshed with his image of the methodological rules governing scientific inquiry. And like the scientific image, the pedagogic image entailed an agenda with specific policy implications.

Newcomb was not a newcomer to secondary-school mathematics instruction. Recall that his father had been a teacher in various schools in rural Nova Scotia and Prince Edward Island. A disciple of British social reformer William Cobbett, the elder Newcomb "held in contempt" rote learning, especially in arithmetic and grammar. Recall also that Simon's first job on moving to Maryland was teaching in a country school at Massey's Cross Roads and then in a village school in Sudlersville; he spent his third and final year in Maryland tutoring in the family of a planter in Prince Georges County.[27] Not content merely to instruct his own students, he attempted in 1854 at age nineteen to articulate his pedagogic philosophy, particularly as it pertained to mathematics. Almost four decades before chairing the NEA's Conference on Mathematics, he drafted a pedagogic statement, "Principles & Maxims," as part of his booklength "Essay on Happiness."[28]

Drawing heavily on examples from mathematics and building on Cobbett's and his father's injunction against rote learning, Newcomb presented five educational maxims. The teacher should always inform the students of the object or goal of their studies, use clear and accessible language in presenting ideas, encourage the students to think and discover things for themselves, convey ideas through concrete illustrations, and use oral rather than written instruction whenever possible. The full expression of the fourth maxim, as we saw when recounting Newcomb's early life, prefigured the linguistic, empirical emphasis of Newcomb's later scientific method. It also prefigured the central element of his later policy regarding mathematics instruction, including that reflected in the report of the NEA Conference on Mathematics.

Young Newcomb provided a terse statement of the fourth maxim: "Every rule and every principle a knowledge of which is to be acquired by the pupil should, if possible, be *illustrated* by some suitable method." "I regard this as one of the most important means of education," he continued, "yet in our common schools it is almost entirely neglected." Newcomb attributed the efficacy of this type of teaching—teaching through concrete illustrations—to the fundamental role that sensory experience plays in learning. "The element of all our knowledge is in the first place obtained through the medium of the senses," he asserted, "and knowledge which we ourselves derive through this medium is far more durably impressed upon our minds than if we received it from another person." In other words, Newcomb at this early date held a theory of learning that was rooted in a decidedly empirical epistemology.

In elaborating on the significance of sensory experience, young Newcomb turned to the example of how one might most effectively teach science—a subject that he compared to a vast and complex temple. He quickly dismissed two common options for providing students with knowledge of the temple: rote memorization of its details and general exposure to a "familiar account" of its main features. Instead, he urged that the instructor actually take the student "to *view* the temple with his own eyes." Only by personally experiencing the temple of science would the pupil acquire ideas that not only were "full and clear" but also "graven on his memory so deeply that time could never efface them." Newcomb ended his explication of the fourth maxim by commenting that students under the ages of twelve or fourteen simply were incapable of learning through abstractions. "The minds of the young," he added, "have been so organized by the Creator that they are wholly employed in acquiring a knowledge of sensible objects. As they grow older they commence comparing those ideas, and then, and not till then can the faculty of abstraction be exercised to advantage." Newcomb could speak with authority on adolescents not only because he was currently teaching them but also because, at age nineteen, he had been one himself only a few years earlier. However, others could, and were, speaking with even more authority: about the time of this essay, Newcomb was beginning to interact with Henry who, in a presidential speech in Washington to the American Association for the Advancement of Education, similarly insisted that pupils be taught the concrete before the abstract.[29]

Though Newcomb never published his "Essay on Happiness" with its section on educating children and adolescents, he put his pedagogic principles into practice from 1881 to 1887 when he wrote seven textbooks in a series that the publisher, Henry Holt and Company, designated "Newcomb's Mathematical Course." For Newcomb, this was a period of heightened activity in mathematics: he became professor of mathematics and astronomy at Johns Hopkins in 1884 and editor of the *American Journal of Mathematics* in 1885. He launched the Holt mathematics series somewhat inadvertently when he wrote an algebra textbook for his oldest daughter, Anita. In the preface to the published version of this first volume, he advised that he was following a basic principle of learning: "that an idea cannot be fully grasped by the youthful mind unless it is presented under a concrete form." Sequels covered topics ranging from geometry and trigonometry through differential and integral calculus. Although Newcomb worried that he was being distracted from astronomy, all of the textbooks sold well and went through many editions, thus supplementing what he considered an inadequate government salary ($3,500 in 1881). Looking beyond the monetary return, however, Newcomb felt that the books contributed not merely to mathematical literacy but to the advancement of all sciences. Including even fields such as political economy and biology, he held that "as science progresses[,] a systematic mathematical training becomes more and more necessary to all who are concerned either with its prosecution in the abstract or with its application to the arts of life."[30]

Not until 1892, however, did Newcomb step back, reflect, and publish a formal analysis of instructing students in mathematics. In October, a month before his appointment to the NEA Conference, the first part of his essay "The Teaching of Mathematics" appeared in the *Educational Review*; the second part, which considered the advanced subject of calculus, came out a year later. Edited by Columbia College's Nicholas Murray Butler and published by Henry Holt and Company, the *Review* printed the writings of America's leading educators, including innovators such as John Dewey. In the first installment of his essay, Newcomb discussed the teaching of elementary subjects such as addition, multiplication, fractions, and geometry. He based his entire analysis on one central premise—what in his 1854 "Essay on Happiness" had been his fourth maxim—the necessity of teaching through concrete illustrations that were tied to sensory experiences. Hand in hand with his growing commitment during the 1870s and 1880s to a linguistic, empirical conception of scientific method, he had elevated the fourth maxim to the pinnacle of his pedagogic philosophy. Just as practicing physicists and political economists should abide by the methodological rule of grounding their concepts in sensory experience, so too should teachers see to it that students ground their mathematical figures, symbols, and operations in the world of experience. He wrote:

> The keynote of what I have to say about elementary teaching, especially that of arithmetic, is that we should devote more attention to embodying mathematical ideas in a concrete form. If we look carefully into the difficulties which the beginner meets with, we find that they can be summed up in the single statement that he has no clear conception of the real significance of the subject which he is working upon. His figures and algebraic symbols do not represent to his mind anything which he can see or feel; the operations upon them are not even representative of any operations with which he is otherwise familiar. So long as this state of things continues, all

his work consists of mere formal processes, which have nothing to correspond to them in the world of sense.[31]

As in 1854, he proceeded to justify this emphasis on concrete experiences with the epistemic argument that "all mathematical conceptions are, in the first place, acquired through the senses." "By no possibility," he continued, "could the conception of a straight line be conveyed to a person who has never seen anything to suggest it, nor can the idea of relations in space be acquired except by first observing them through the senses of touch and sight." For this reason, he opposed the classroom use of abstract concepts and imaginary problems.[32] The epistemic base of this pedagogic stance represents a departure from Newcomb's more general insistence on bypassing the debate over whether the origin of knowledge lay in mind or experience; in other writings dating back to his Cambridge years, he had emphasized concentrating on the empirical warrant of knowledge whatever its source. Now, when addressing the origin of elementary mathematical knowledge within a pedagogic context, Newcomb aligned himself with Mill.

In a handwritten draft seemingly of this *Educational Review* article, Newcomb expanded his epistemic argument by explicitly recalling the controversy between Whewell and Mill over the role of experience in the origin of geometrical concepts. Remarking that the development of non-Euclidean geometry had caused recent mathematicians to rethink the positions of both Whewell and Mill regarding axioms involving the concept of a straight line, he nevertheless favored Mill's position: "Geometrical conceptions are the result of experience; not perhaps in the same sense that the general truths of physics and chemistry rest upon experience, but, at least, in the sense of being conceptions which could not begin in the mind unless suggested and enforced by experience." This being true, it was lamentable that many teachers still believed that mathematical ideas were completely intuitive in origin; this false notion led the teachers to believe that their task involved teaching students only "to express and combine these intuitive ideas in a rigorous logical form." Whereas teachers should eventually guide their pupils to an understanding of mathematics as a purely logical system, they should not start their instruction on this level. Rather they should begin on the level of sensory experience.[33]

In the remainder of the published version of his *Review* article, Newcomb detailed how an instructor might present to "the farmer's boy" this "arithmetic and geometry of sensible objects." For example, he explained how the student could learn addition, subtraction, multiplication, division, and fractions by drawing and manipulating actual chalk lines on his slate. Fourteen years later, in a retrospective address on "Methods of Teaching Arithmetic," Newcomb labeled this approach "visible arithmetic." In this address, delivered in 1906 to the Department of Superintendence of the NEA and subsequently published in the *Educational Review*, he also recalled his motivation in writing his original *Review* article. Although educators had been promoting the teaching of mathematics in a concrete form, their plans had not been put into practice, at least in the United States; consequently, Newcomb had attempted in 1892 to set forth and show the advantages of a system of visible arithmetic. Speaking more than a decade after the effort, Newcomb also assessed its impact. "I am not aware," he began in a modest tone, "that this utterance excited any attention at the time, nor do I know whether it was a factor in the recent tendency of arithmetical teaching in the direction which it advocated." "However this may be," he continued, more assertively, "the whole trend of recent experience and discussion among practical teachers is in the direction of the ideas advocated in the paper referred to; and a system practically identical with the one there found is now embodied in several recent arithmetics, even to the extent of proposing problems and exercises almost identical with those suggested in the paper."[34]

The trend toward visible arithmetic that Newcomb discerned in 1906 gained from not only Newcomb's 1892 paper but also the report of the NEA Conference on Mathematics. When Newcomb first accepted Eliot's invitation to join the ten-person Conference and help revise the mathematics curriculum of the nation's secondary schools, his old Harvard friend had written: "Your judgment will be valuable to the Conference because of your just views about teaching the elementary mathematics." Newcomb's Conference colleagues apparently concurred on the value of his judgment and elected him chairman. As chairman, he not only directed the deliberations but also had a large hand in drafting the final report.[35] His personal views colored the opinions of his nine Conference colleagues as expressed in their report. The document consisted of a general statement of conclusions and specific statements on the teaching of arithmetic, concrete geometry, algebra, and formal geometry. The discussions of the more elementary subjects of arithmetic and geometry included the same message that Newcomb had just published in the *Educational Review*, two months before the Conference convened. In teaching both arithmetic and introductory geometry, the instructor should emphasize concrete illustrations tied to sensory experiences.

In their opening statement of general conclusions, the members of the Conference unanimously recommended that the typical course in arithmetic "be at the same time abridged and enriched; abridged by omitting entirely those subjects which perplex and exhaust the pupil without affording any really valuable mental discipline, and enriched by a greater number of exercises in simple calculation and in the solution of concrete problems." Regarding the addition of exercises and concrete problems, the conferees specifically suggested that the metric system be introduced through "applications to actual measurements to be executed by the pupil himself" with "the measures and weights being actually shown to, and handled by, the pupil." In addition, "the illustrations and problems should, so far as possible, be drawn from familiar objects." The Conference members detailed these recommendations in the special section of their report on arithmetic. Again emphasizing the concrete, they wrote: "The relations of magnitudes should, so far as possible, be represented to the eye. The fundamental operations of arithmetic should not only be performed symbolically by numbers, but practically, by joining lines together, dividing them into parts, and combining the parts in such a way as to illustrate the fundamental rules for multiplication and division of fractions." This technique of manipulating actual lines is the one that Newcomb had pushed in his *Review* article and later labeled "visible arithmetic." Also in the spirit of Newcomb's article, the Conference members went on to endorse "instruction in concrete or experimental geometry."[36]

The comprehensive report of the Committee of Ten reinforced the message of Newcomb's Conference on Mathematics. The Committee not only summarized the Conference's recommendations regarding concrete instruction but also generally adopted a tone supportive of experiential forms of education. Thus, the Committee emphasized that "all the Conferences on scientific subjects dwell on laboratory work by the pupils as the best means of instruction." Supportive of the nascent, national movement toward instruction through laboratories but distrustful of teaching based exclusively on discursive, authoritarian textbooks, the Committee went so far as to suggest that pupils devote Saturday mornings to a laboratory component of their science courses. Regarding instruction in mathematics, the Committee quoted those passages of Newcomb's statement that called for enrichment of the course in arithmetic by using a greater number of "concrete problems" and by "giving the teaching a more concrete form."[37] As mentioned earlier, the recommendations of the Committee of Ten and the nine conferences had an enormous impact. In formally submitting the report to the Secretary of the Interior in late 1893, Committee member and U.S. Commissioner of Education William T. Harris characterized it as being "the most important educational document ever published in this country."[38] In mathematics, the document implicitly bore the imprint of Newcomb's linguistic, empirical methodology. The method underpinned his campaign to hasten the movement among the nation's mathematics teachers away from abstract toward concrete instruction.

CHAPTER X

Mental and Psychical Sciences
Challenging Current Beliefs

In attacking the outlooks of nonscientists like Stallo, Newcomb endorsed the basic doctrines of contemporary physical science while implicitly assuming that the doctrines reflected sound methodological practices. He made similar presumptions when instructing his fellow scientists on the precise use of technical terms such as radiance. These episodes do not mean, however, that he avoided challenging on methodological grounds specific, substantive interpretations of his scientific colleagues. Said differently, though Newcomb was confident of the conceptual foundations of modern science, he did question the reliability of some of the superstructures built on them. Thus, he attempted to discredit the common belief in "scientific materialism"—the belief that mental phenomena could be reduced to physical processes. Similarly, he sought to debunk the growing belief in what might be called "scientific spiritualism"—the belief that psychic phenomena could be attributed to new and unseen processes operating in, for example, the ether.

These probings of the superstructure of science, of scientific materialism and scientific spiritualism, provide once again merely modest examples of Newcomb operating on our first level of methodological rhetoric—the level where scientists, conversing among themselves, first articulate, argue, and appraise competing knowledge claims. They are only modest examples because, though Newcomb was employing methodological pronouncements to discredit emerging and still controversial theories of perceived scientific merit, he was neither grappling with theories of central concern to practicing natural scientists nor always directing his remarks to his professional colleagues. Indeed, both attempts can be seen as further rhetorical efforts on the public level—efforts at demarcating "true" science and guarding its image. Intent on securing broader support for science in the United States, he was seeking to distance the modern professional community from those "scientists" who held naive notions of mental and psychical phenomena.

SCIENTIFIC MATERIALISM

Newcomb wrote a five-part series on "Modern Scientific Materialism" for the *Independent*, the religious weekly that had carried his AAAS speech and the reactions of Gray and Porter. In this series he applied his methodological tools to the unsettled question of materialistic explanations of mental phenomena. Such explanations, he reported, are "very generally believed in, both by large classes of scientific investigators and by the public." Newcomb first employed his methodological criteria to divest the term "materialism" of all "metaphysical subtlety," thus arriving at the precise concept of "scientific materialism." With the issue now amenable to objective investigation, he proceeded to assail and reject even the stripped-down version of materialism.[1]

In the first installment of the series, subtitled "The Question Stated," Newcomb clarifies the terminology of materialism. "To attain precision in our conclusions," he advises, "we must begin by settling what we are to understand by the term materialism." He examines various common meanings, finding them all deficient. If a person claims, for example, that materialism is the doctrine which denies the existence of spirit, then the person faces the impossible task of giving meaning to the word spirit. Similarly, if a person claims that materialism is the doctrine that denies the existence of mind independent of matter, then the person faces the impossibility of investigating "by scientific methods" the truth of the claim of independence. Newcomb even finds fault with the more conventional statement of the doctrine by respected scientists such as Tyndall—that matter alone can account for all mental phenomena. Like a latter-day Comte or Mill, he dissects the phrase "account for phenomena" in this conventional formulation:

> But to state this particular form as a scientific materialism might give rise to incorrect ideas. We question whether there is in it anything sufficiently

> definite to seize with the instruments of the investigator. We question whether those who speak of matter accounting for everything have a clear conception of what a scientific philosopher understands by accounting for phenomena. He accounts for nothing and explains nothing in the metaphysical sense. His work is confined to the discovery of correlations between phenomena; to showing how one state of things follows another state of things. The laws which regulate these correlations and connections he calls *laws of Nature*. The larger the class of things connected by one law the more general that law is. Accounting for an event consists only in showing its connection with the events which precede it and with the laws which determine the succession. Moreover, his laws are merely generalized facts, or, rather, connected statements of facts.

As he had done in his AAAS presidential address, Newcomb illustrates this view of law using the theory of gravitation—"simply a generalized fact of observation, which does not penetrate into the nature of things at all." He concludes that, when properly expressed, the phrase scientific materialism simply denotes the doctrine "that there is a complete correlation between mental phenomena and physical processes going on in the brain."[2] Having a clear definition, he is now ready to show that the theory is untenable.

In the remaining four installments of the series, Newcomb concentrates on whether mental processes can be correlated with the actions of underlying atoms and molecules that follow the laws of classical mechanics. That is, as in his attack on Stallo, he takes it as a given that physical processes—the phenomena of physics and chemistry—are ultimately atomo-mechanical processes. "The hypothesis upon which all investigation proceeds, so far, at least, as any hypothesis is necessary, is that every attribute of a body can be explained in a scientific sense by its internal structure and by mechanical forces at play among its molecules." Can this structure and these forces account for mental phenomena? Atomo-mechanical processes, Newcomb responds, could never completely represent the phenomena because the small amount of matter in the brain does not have the physical capacity to match the overwhelming complexity of mental phenomena.[3]

But could not this objection be overcome if thought were a form of energy, which has no spatial limitation? According to the theory of conservation of energy, energy is convertible into different forms; perhaps physical energy is transformed in the brain into mental energy. Newcomb first clarifies for his readers how a seemingly abstract concept like energy—or as it was still commonly known, "force"—is admissible in the vocabulary of scientists who normally insist that concepts be defined in terms of phenomena.

> Force [energy], as such, does not enter at all into the fundamental definitions of physical science. In itself, it cannot be seen or felt, and physical science fundamentally concerns itself only with things that can be seen or felt.... But it is not at all necessary that we should enter into a discussion of this abstruse question, because science, in confining itself to phenomena, does not deny the existence of things which are not phenomena. It is safe to say that in the general theory of physical action the idea of force [energy] is merely an extremely convenient instrument for expressing and measuring relations among material things which, without it, would be beyond the power of the ordinary mind to grasp.

Although the concept of energy is an admissible instrument in science, and though the concept is useful in understanding the physiology of the brain, Newcomb goes on to say that the concept cannot be correlated with mental phenomena—"thought, passion, and other mental acts." Newcomb draws this conclusion on empirical grounds, pointing out that there simply is no evidence for the conversion of physical into mental energy. After a brief speculation on what the "hypermaterial essence of life" might be, Newcomb ends the five-part series, having challenged both atomistic and energetistic theories of materialism.[4]

A decade later, when William James published *The Principles of Psychology*, Newcomb responded to the new study by calling James's attention to his critique of materialistic interpretations of the mind. James replied that, although he had not read the series of articles, he concurred with Newcomb's general thesis:

> Thanks for your note. It is flattering to have anyone *react* on one's book, especially when the reacter is a man like you. I have never seen your articles in the Independent, nor can I get them here. But I agree with you that a lot of the discussion that goes on is logomachy from not defining terms. I think that "materialism" is very well kept with the vague meaning (said to be ascribed to it by Comte) of "the explanation of the higher by the lower."

James went on to endorse another of Newcomb's long-standing convictions—that "the word 'freedom' is deplorable from its ambiguity."[5]

PSYCHICAL RESEARCH

It was William James who previously had described as an "uncommon hit" Newcomb's election as first president of the American Society for

Psychical Research. The society had begun to take shape in early September of 1884 when a group of distinguished British scientists traveled from a special Montreal meeting of the British Association for the Advancement of Science to Philadelphia for an equally special meeting of the AAAS. Among the approximately three hundred visiting scientists who joined nine hundred Americans were some of the leading lights of the British Society for Psychical Research. A flourishing organization founded two years earlier, the British society had the aim of bringing scientific objectivity—and, thus, respectability—to the study of "that large group of debatable phenomena designated by such terms as 'mesmeric,' 'psychical,' and 'spiritualistic.' "[6]

During the Philadelphia meeting, luminaries such as physicist Oliver Lodge sparked the interest of the Americans in founding their own society for the scientific investigation of psychical effects. Later in September, in Boston, a small group that included William James met with William Barrett, another British physicist and leader of the psychical movement, and actually began to structure the American society. They convened the first formal meeting three months later in Boston. Elected members soon numbered about 80, with associate members adding about another 170 persons. Mostly from the Boston area but also representing cities extending south to Baltimore and Washington, the elected members included the cream of American science and higher education: George Barker, Henry Bowditch, Edward Cope, Asa Gray, Edwin Hall, G. Stanley Hall, William Harris, Benjamin O. Peirce, Edward Pickering, Josiah Royce, and Andrew White.[7]

William James (1842–1910), who inclined toward belief in psychical effects, was one of the prime movers of the new society. His pleasure in securing Newcomb as president reflected a strong personal desire both to have the organization flourish and to see psychical effects possibly substantiated. He disclosed these feelings in a letter to a friend who, shortly after the establishment of the society, complained that the group seemed to be staffed disproportionally by scientists having a bias against spiritualism. After assuring his friend that the group was impartial and merely aiming to determine "what the *phenomenal conditions of certain* concrete phenomenal occurrences are," James confided that having scientists as officers of the society had strategic value in garnering wider support for psychical research.

> The choice of officers was largely dictated by motives of policy. Not that scientific men are necessarily better judges of all truth than others, but that their adhesion would popularly seem better *evidence* than the adhesion of

others, in the matter. And what we want is not only truth, but evidence. We shall be lucky if our scientific names don't grow discredited the instant they subscribe to any "spiritual" manifestations. But how much easier to discredit literary men, philosophers or clergymen! I think Newcomb, for President, was an uncommon hit—if he believes, he will probably carry others. You'd better chip in, and not complicate matters by talking either of spiritualism or anti-spiritualism. "*Facts*" are what are wanted.[8]

Politically astute, James realized that Newcomb's name brought scientific credibility to a suspect society operating at the fringe of accepted scholarly practice. Indeed, the scientists who organized the American society were quite self-conscious about their professional image—a sensitivity engendered by the broader national trend among all scientists to forge a professional identity and establish disciplinary coherence.[9]

Newcomb had flirted with belief in prior years. As a boy in Nova Scotia, he had fancied that he might influence the actions of friends and relatives through his own thoughts, although he was disappointed in actual trials. His interest was also piqued during the early 1850s when, as he later described it, "the great wave of spiritualism, with its rappings, table-moving, and communications from the dead, was reaching its height." While living in Cambridge, he continued to follow reports of various mediums and spiritualists in the newspapers. By the time he moved to Washington, however, he had grown wary of psychical claims, an attitude reinforced by conversations with Joseph Henry—a staunch skeptic who once even nipped President Lincoln's budding belief in the powers of a noted spiritualist.[10]

Discussions at the AAAS meeting in 1884 rekindled Newcomb's interest in the psychical. Looking back two years later, he explained that when "the early experiments of the English psychical society were made known, it seemed to me that a strong case was made out for a new law of nature governing the transmission of thought, or some form of mental influence from person to person. . . . Under the influence of this possibility, I encouraged the formation of our own society, and accepted membership in it." Of course, his denial of materialist arguments—as in his articles in the *Independent*—also left open the possibility of telepathy and other such mental effects.[11]

In response to the September initiative to create a society but before the first official meeting in December of 1884, Newcomb published an article in *Science* warning against investigations of "psychic force" not conducted in the proper "scientific spirit." Effectively a commentary based on general methodological criteria, the article exposed recent researchers' inability to isolate actual psychical effects and draw justifiable conclusions. "The first and greatest obstacle we meet with in such investigations," he advised, "is the absence of clear ideas of what it is we are to look for, and how we are to distinguish between real relations of cause and effect and mere chance coincidences." The British, in Newcomb's opinion, were guilty of conducting inquiries into thought transference, apparitions, and other purported psychical effects that were doomed to be inconclusive in that the effects could be explained equally well "as the result of chance coincidence." Newcomb could speak with authority on chance occurrences. As an astronomer and mathematician, he had honed his skills in the theory of probability, beginning with minor publications and talks while still in Cambridge and continuing through lectures on topics like the method of least squares to astronomy students at Johns Hopkins. Drawing on his expertise in the theory of probability, he sketched a simple calculation to show the frequency with which an investigator might expect the occurrence of coincidences that might mistakenly be interpreted as psychical manifestations. Concentrating on alleged apparitions involving dying friends or relatives and taking into account the entire population of the United States, he showed that by pure chance it is likely that almost every day someone in the nation will have a dream or vision of a friend or relative at the moment of the acquaintance's death.[12]

The British, he continued, commit a further methodological blunder—specifically a logical blunder—while trying to explain effects associated with so-called haunted houses; when, after investigating many cases, they are left with a few that seem to defy standard explanations, they erroneously conclude that they have identified genuine psychical effects. The British forget that conventional natural causes may remain hidden; consequently, in their findings on haunted houses, they have actually presented nothing more than "very scientific children's ghost-stories."[13]

Newcomb shifts to a broader plane to close this article in *Science*: "The general question at issue is, whether there is any such process as what the psychists very happily denominate 'telepathy,' which may be defined as *feeling at a distance* without the intervention of any physical agent." He suggests that the only research that augurs in favor of telepathy is that involving drawings made by one person in response to images being thought about by another person. But even in this one promising branch of telepathic inquiry, stronger evidence will be needed to satisfy the incredulous.[14]

Newcomb's remarks were bound to rankle the British researchers, a group that paraded their scientific credentials and prided themselves on adhering to the empirical methods of the physical sciences.[15] Indeed, his remarks provoked no less a figure than the honorary secretary of the British Society for Psychical Research, Edmund Gurney. A little over two weeks after the appearance of Newcomb's article, Gurney responded to the challenge to the society's research by submitting a letter for publication in *Science*. In this detailed communication, Gurney acknowledged the importance of applying "the doctrine of chances" to the study of apparitions, but he went on to rebut Newcomb's probabilistic analysis, substituting his own demonstration that apparitions could not be dismissed merely as chance phenomena. When published, Gurney's rebuttal was accompanied by a short response from Newcomb, sarcastically questioning Gurney's numerical data. A week later, Newcomb presented a more formal response and elaboration of his general arguments in a short article titled "Can Ghosts Be Investigated."[16]

Curiously, some of Newcomb's thoughts in his original article again parallel those of Charles Peirce. This is not a case of telepathy but, once again, of shared interests and occasional interactions. Through early 1884, Peirce and a colleague had been conducting an experiment on humans' ability to discern faint sensations; he reported the work to the National Academy of Sciences on 17 October 1884, the same day that Newcomb's article appeared in *Science*. Peirce's finding that there was no threshold below which physical sensations abruptly ceased led him to suggest that certain supposed telepathic effects actually resulted "in large measure from sensations so faint that we are not fully aware of them." Newcomb, in warning his British colleagues that there may be hidden natural causes behind what appear to be telepathic effects, made the same point: "We must remember that the physical connection through which one mind affects another may be of the most delicate kind; may, in fact, nearly evade all investigation. The slightest look, an unappreciable motion of the muscles of the mouth or eyes, made perceptible through the light which is reflected to the eye of the second person, constitute a physical connection." Newcomb also paralleled Peirce in pioneering the introduction of probabilistic reasoning into psychical research. Peirce had broken new ground by injecting randomization into the design of his experiment on sensations; then, beginning in 1887, he entered into an exchange with Edmund Gurney, in which he presented some of the earliest commentaries on the application of probabilistic reasoning to psychical phenomena. Writing in the *Proceedings* of the American society, Peirce argued that Gurney and his colleagues had misapplied the reasoning. Specifically, he responded to Gurney's "ghost-stories" by outlining a calculation showing that apparitions associated with deaths of persons could be explained by chance. This, of course, was the same type of argument that Newcomb had earlier used in his *Science* article. Moreover, it was Gurney who had lashed back at Newcomb's charges in a letter to the editor. Newcomb's exchange with Gurney not only foreshadowed Peirce's later exchange with the British secretary but also antedated the first formal account of probabilistic reasoning to be published by either the British or American societies. This account came in the *Proceedings* of the British society for 30 December 1884 when Gurney summarized recent French work on the application of the method of probabilities to the investigation of psychical phenomena. With the publication of his article in *Science*, Newcomb thus joined Peirce in at least helping to precipitate the emergence of probabilistic thinking in British and American psychical studies.[17]

Why did this avowed skeptic accept membership in, let alone the presidency of, the American Society for Psychical Research? To be sure, the flattery of being asked and the challenge of the research problem contributed to his decision. So too did his prior interest and the enthusiasm of the British visitors.[18] But he had another motive. Like many of his colleagues, Newcomb felt that there existed a moral imperative to weigh the evidence concerning psychical effects; the objective was to sort out what, if anything, was scientifically defensible and what must be debunked as superstition or fraud. The organizing council of the society had stressed this "scientific duty" in the first circular sent out to solicit members. A lead editorial in the 17 October issue of *Science*, however, was more eloquent. Appearing a few pages before Newcomb's cautionary article on "Psychic Force," the editorial portrayed scientists as possessing the proper moral qualities to deliver the nation from the evils of raw spiritualism and, in its place, substitute systematic study of sanitized phenomena such as telepathy.

> Psychical research is distasteful to some persons; for it touches upon spiritualism, and to them seems akin to it. Now, spiritualism is an evil in the world,—in America it is a subtle and stupendous evil; a secret and unacknowledged poison in many minds, a confessed disease in others,—a

disease which is sometimes more repulsive to the untainted than leprosy. Spiritualism has two supports,—the first trickery and deceit, the second the obscurity and inexplicableness of certain psychological processes and states. The strength of spiritualism is protected by the utter mystery which screens certain mental and nervous conditions from the light of explanation. As of others, so the basis also of this superstition is, in one word, ignorance.

To those gifted with a clearer intelligence and purer moral sense, there is a moral duty in one aspect of the proposed studies. A hope that psychical research may liberate us from a baneful superstition is a stimulus to inaugurate the work of the American society.

Never one to investigate a topic halfheartedly, Newcomb accepted the moral challenge and threw himself into the thick of psychical research. Not content with merely reading psychical literature, he sought to witness the phenomena firsthand. Psychologist G. Stanley Hall, one of the founders of the society and Newcomb's colleague on the faculty of Johns Hopkins, recalled that he, university president Daniel Gilman, and Newcomb "visited every medium who advertised in Philadelphia, and later, when Dr. Gilman had withdrawn, Professor Newcomb and I made similar rounds in New York." Early in 1885, Newcomb had the opportunity to observe closely a young woman known as the "Georgia magnetic girl" who supposedly could move large objects and other persons without any excessive muscular exertion on her part. The young woman consented to test her powers before Newcomb and other "educated men" in the Washington laboratory of Alexander Graham Bell, another of Newcomb's close colleagues. Newcomb soon realized that he could account for all her feats using elementary laws of physics and specifying the ways in which she subtly achieved mechanical advantages. He subsequently reported in *Science*: "The scientific tests were productive of the usual result,—that ghosts, spirits, and occult forces absolutely refuse to perform their functions in the presence of scientific paraphernalia."[19] Though a spirited student of the psychical, by the end of his first term as president, Newcomb's skepticism had hardened. This became evident in his formal presidential address to the psychical society, read in his absence at the annual Boston meeting in January, 1886, and published with slight revisions in the society's *Proceedings*.

ADDRESS TO THE PSYCHICAL SOCIETY

Newcomb's speech to the American Society for Psychical Research was, in effect, a sustained and forceful methodological discourse that elaborated the arguments earlier touched on in his *Science* articles. It was a speech in which President Newcomb used elementary canons of scientific method—or at least his reading of the canons—to question ostensibly scientific claims that were themselves supposedly products of scientific method. His meticulous methodological analysis was a rebuke to the British researchers, especially, since their legitimacy as a scientific group was rooted in the support that method provided for their claims. Restricting his arguments to simple contentions respecting the empirical inference of natural laws and the reproducibility of results, Newcomb did not explicitly employ his usual linguistic analysis. In fact, the basic contours of his contentions date back to his Cambridge textbook statement on method and to similar, standard statements such as Henry made in 1877 to the Washington Philosophical Society. He apparently saw his task as responding to the large, extant body of purported psychic evidence, especially involving thought transference. This task required foremost a critique of how proponents used the evidence to infer possible new natural laws, not a linguistic explication of their fundamental concepts and propositions.[20]

Newcomb begins by remarking that there exist popularly accepted "laws of mental action" seemingly prohibiting thought transference, mental telepathy, mind reading, and the like. In particular, an individual mind can neither act on nor be acted upon by anything external to itself, including another mind, except through some material or physical medium, and the latter process is subject to purely physical laws. However, Newcomb quickly adds, though these laws are based on very wide experience, they are merely hypothetical laws derived through induction. Furthermore, as he similarly stressed during these same years while defending the old school of political economy, these laws are, "like all other general laws, in seeming disaccord with occasional phenomena." Newcomb is thus led to grant that, viewed a priori, it is an open question whether these "occasional phenomena" point to a new law of nature that allows for thought transference and other forms of "mental *actio in distans*." Drawing an analogy to Newtonian gravitational action at a distance, he also grants that, at least for the moment, proponents of mental action at a distance are not obliged to explain the underlying mechanism, if any, of the effect; they must only establish the empirical law that describes the effect.[21]

While acknowledging the possibility of a new law that encompasses mental action at a distance, Newcomb insists that prudent investigators must approach the subject skeptically. In particular, the probabilities are against finding such a new law and the burden of proof must fall on the proponents. Newcomb observes, however, that it so happens that the proponents—especially members of the British Society for Psychical Research—have already assembled in the preceding few years a large amount of psychical data. In light of this wealth of data seemingly weighing on the affirmative side of the psychical question, Newcomb concludes that scientific method offers the only way to proceed in further deciding the question. He writes: "On this side we have a mass of evidence so great that we cannot deal with it in detail, unless our task is facilitated by reference to those logical principles which should direct our thoughts." As in prior essays for other audiences, Newcomb offers to express these methodological principles in their most down-to-earth and rudimentary modes. "In order to avoid employing these principles in too abstract a form, I shall borrow them directly from our common-sense methods of drawing conclusions in every-day life."

At this juncture, Newcomb leads his audience through a careful review of "the frame-work which underlies these methods." In particular, he describes a methodological criterion for a sound scientific explanation: "Every explanation of natural phenomena, when complete, involves two elements,—a general law and a particular fact." The general law, itself inferred by induction and open to further generalization, is usually taken for granted and not explicitly articulated in everyday explanations. For example, to adequately "explain" hearing a certain type of sharp explosion (the "phenomenon," "effect," or "result"), it is necessary to specify both that someone fired a gun (the particular "fact," "cause," or "circumstances") and that firing a gun produces an explosive sound (the "law").

How does this twofold criterion for a well-grounded explanation relate to the large collection of phenomena seemingly involving thought transference and other manifestations of mental action at a distance? Newcomb answers that when faced with these phenomena that apparently defy conventional explanation, investigators have two options: "We may conclude either that some law of nature of which we have before remained ignorant has come into play, or that the result is due to known laws acting under particular circumstances of which we are ignorant." In other words, the present inability to explain psychic phenomena implies either flawed conventional laws of mental action or hidden facts that if exposed would substantiate the applicability of the conventional laws. How investigators might legitimately conclude that they have uncovered a new law of nature is a difficult matter, and it is over this matter that Newcomb, as in his previous articles in *Science*, parts company with his British brethren.

Members of the British Society for Psychical Research make the mistake of concluding that they have established a new law of nature when, after rigorous research involving many cases, they remain unable in a few of the cases to disclose the requisite "facts," "attendant circumstances" or "conditions" through which the phenomena could be explained in accord with conventional laws. That is, failing within the framework of traditional laws to specify the causes of certain effects, they leap to the conclusion that there must exist a new law involving mental action at a distance. Responding to this line of reasoning, Newcomb states: "I must, with all due respect to those who have applied it, express my dissent from its validity as a method of discovering such laws." He goes on to explain that it is illegitimate to infer new laws using this method and that, in fact, the British investigators have only demonstrated that they are dealing with circumstances beyond their investigative powers. It is invalid to adduce an alternative law simply because a group of "spurious" effects cannot, at least for the moment, be related to discernible causes.

Though the British method of inferring new laws was flawed, the problem of psychic phenomena remained: thought transference and kindred effects seemed to defy conventional explanations of mental action. Was this, Newcomb inquires as he returns to his twofold criterion for a sound explanation, because of concealed facts that if known would fit with conventional laws or because of entirely new laws? "The question is, Did it [thought transference] take place through some physical connection between two organisms which eludes our scrutiny, but which, had we seen it, we should have recognized as involving no new principle, or did some new law of nature come into play?" Newcomb answers with a criterion for deciding between the two alternatives. Like Whewell before him, he finds the criterion not in the axioms of cloistered philosophers but in the actual history of physical science. "The true method of investigation," he explains, "is exemplified by the whole history of physical science."[22]

History illuminates the way in which psychical researchers might determine whether they have discovered a new law of nature. Specifically, in earlier stating that a complete explanation of natural phenomena involves both a general law and a particular fact or circumstance, Newcomb had mentioned that the law itself was inferred by induction. He now elaborates, saying that the historical record provides a means of determining when investigators are justified in saying that a new law has indeed been inferred. History shows: "When the same phenomenon occurs under the same conditions time after time, we infer a law of nature. When we cannot trace its repetition to any common set of conditions [as with the British investigators], we conclude that it is due to varying circumstances, perhaps unknown to us." Thus, as he goes on to emphasize, reproducibility of results is the key: "It is a characteristic of all scientific progress, that, when we ascertain any new law connecting phenomena, we are able to produce them with continually increasing facility." Does psychic research meet this test? It does not, Newcomb responds, remarking that researchers have added nothing to their general knowledge during the prior decade. More specifically, the British investigation of thought transference "has failed to show any common feature in the ideas transferred, and has thrown no light on the question of the condition under which the phenomena can occur." It is not enough to describe "isolated facts" or establish "the mere recurrence of the phenomena"; rather, the investigators must discover the "invariable relation" between the circumstances (the cause) and the phenomenon (the effect). Only then, after determining how the phenomenon can be replicated, will psychical researchers achieve "a statement of general laws, like those which we find in books on mechanics, electricity, magnetism, or physiology, setting forth the conditions under which thought-transference can be brought about."

While doubtful that psychical researchers would ever arrive at new laws of mental action, Newcomb ends his speech on a positive note. He stresses that, though phenomena such as thought transference are probably "apparent" and not "real" phenomena, they still need to be investigated and explained. In addition, related psychological phenomena deserve serious study, including William James's current interest, hypnotism. "Whatever may be the fate of the theory of thought-transference," he concludes, "the phenomena of hypnotisms, as well as of dreams, illusions, and faults of memory, are all before us. They form a field of which the cultivation has only commenced, and which ought to prove attractive to all. I even venture to say, that, if thought-transference is real, we shall establish its reality more speedily by leaving it out of consideration, and collecting facts for study, than by directing our attention especially to it." President Newcomb had not completely closed the door through which the membership of the

American Society for Psychical Research hoped to enter new rooms of the mind. Indeed, in line with Newcomb's thoughts on the society's potential contributions to psychological research, a historian of mental science has recently judged the society to be in reality "the first organized scientific body in the country devoted to experimental psychology."[23]

When Newcomb first sent his speech to Boston to be read at the annual meeting, he had asked the society's secretary, Harvard physicist Edwin Hall, to inform him of any objections that members might have to the speech. Two and a half weeks later, Hall responded with tact, explaining that there had not been enough time to discuss the address at the general meeting but assuring Newcomb that the council still intended to consider it. "You are of course aware," Hall added in an understated appraisal of the address, "that it was well calculated to excite discussion." A few days later, Hall wrote again to tell the apostate president that the council would like to print the address in the society's *Proceedings* but that, henceforth, all issues of the *Proceedings* would carry a disclaimer denying the society's and council's "responsibility for both the facts & reasoning in addresses, reports or papers."[24]

The editors of *Science* were less circumspect in their reaction. They led off the entire issue for 29 January 1886 with a curt "Comment and Criticism" concerning the address:

> The attitude of Professor Newcomb towards the alleged discoveries in regard to thought-transference is one of extreme intellectual dissent, and will necessarily accentuate the impression of exceedingly great conservatism, which already prevails in regard to the American society for psychical research. His presidential address was essentially a frank though delicate denial, not only of the results concerning telepathy claimed by the English society, but also of the utility of pursuing any investigations upon the subject further.

While adding that "certain flaws" taint Newcomb's argument—flaws that prohibit wholesale endorsement of his position—the editors ended by seconding Newcomb's charges against the British psychical researchers. Two pages later, in a summary of the full Boston meeting, the editors repeated that Newcomb's "logic is open to criticism"; they expressed misgivings about the reasoning that led him to conclude that alleged cases of thought transference did not indicate a new law of nature but merely hidden facts that fit conventional laws.[25]

James found fault not only with Newcomb's logic but also with his conclusion. From first hearing the address, he had been particularly dismayed by Newcomb's criticism of the experiments done by the British researchers on the production of drawings through thought transference. Convinced of the integrity of the British "gentlemen," James initiated a friendly skirmish with Newcomb in an exchange of published letters in *Science* and personal notes. In a follow-up note to Newcomb half a year later, James summarized his reservation regarding Newcomb's speech:

> I feel as if the evidence for thought-transference were very good, and I must say that the a priori arguments of your presidential address were far from shaking the effect upon me of the whole body of concrete experience in favor of some thing of the kind. . . . I shall be much surprised if it does not become an orthodox scientific fact, realized like many other facts, in individuals of a certain idiosyncrasy. I am very much disposed to doubt your suspicions in this case.

Whereas Newcomb had belittled the British researchers for inferring the existence of a new law of nature without being able to replicate the relevant phenomena, James turned the tables and now faulted Newcomb on methodological grounds: for reaching a negative conclusion before weighing all the evidence. As James advised in an earlier letter, "*Conviction* either way now can only come from a much larger mass of fact observed." The British researchers themselves also remained in direct contact with Newcomb during this period. Henry Sidgwick, first president of the English society, sent him a copy of the society's compendium of psychical case studies, *Phantasms of the Living* (1886), explaining that "the book itself is the best answer that I can give to your objections."[26]

Newcomb's doubt was not enough to remove him from the presidency of the society, many members of which, according to *Science*, actually shared Newcomb's skepticism. Though he offered to step down after his first term, he was reelected. And when he himself finally declined reelection at the end of a second term, in January 1887, he continued as a member of the governing council. He remained with the society until it faded and was absorbed by the British organization in 1889. Through these final years, he remained a dissenter. Late in 1887, for example, he offered his opinion on psychical research to Richard Hodgson, an Australian scholar active in the British society who had been enlisted as secretary of the American organization. "For myself individually," Newcomb wrote, "the American and English societies have completely done their work by proving to my entire satisfaction that there is no such thing as telepathy; but as everybody is not satisfied I want to see the investigation go on to a satisfactory conclusion."[27]

Once again, the skeptic had left the door open for well-meaning colleagues such as James to continue their studies. After all, the researchers were responding to a moral imperative to expose the fraud and superstition of common spiritualism; in addition, Newcomb's historical view of how new laws are inferred precluded him from asserting unequivocally that the researchers should abandon their work. Nevertheless, Newcomb was convinced and hoped to convince others that, on methodological grounds, psychical research was a scientific dead end. In his opinion, neither the crude spiritualism of the public nor the sanitized telepathy of the British and American societies had scientific merit. When the societies' members had sought to legitimate their results by cloaking them in the garb of scientific method, Newcomb had countered with a methodological critique designed to expose the rents in the fabric. Looking back, we suspect yet again that Newcomb also had a concealed purpose, whether conscious or unconscious. He was using method to guard the popular image of science by disassociating normal practitioners from a group of enthusiasts operating at the margin of recognized scientific investigation. He was excluding what he took to be pseudoscientific intruders.[28] This demarcation seemed essential if American scientists were to enjoy fuller public support, cognitive authority, and, ultimately, professional autonomy.

CHAPTER XI

Later Years

Retirement meant for Newcomb a realignment of work, not an end or even slackening of research, writing, and public speaking. Forced by law to leave naval employ at age sixty-two, he stepped down from the superintendency of the Nautical Almanac Office on his birthday in March of 1897. A special, albeit modest, congressional appropriation and then, beginning in 1903, generous grants from the new Carnegie Institution in Washington enabled the distinguished retiree to maintain his intense schedule of research and professional interaction. He assumed the lead in a major international project to bring order to astronomical computations through the adoption of uniform constants and consistent data. At the same time, he persisted with his long-standing work on planetary tables, especially the motion of the moon. Continuing to display great drive, he also, in 1898, resumed his affiliation with the astronomy and mathematics programs at Johns Hopkins. In addition, he helped organize and, in 1899, became the first president of the Astronomical and Astrophysical Society of America (later renamed the American Astronomical Society).[1]

The creation of this society exemplifies the professional gains that the overall scientific community had made in the United States during the quarter century since Newcomb wrote his first article calling attention to the nation's scientific deficiencies. One statistical survey confirms that by about 1900, for example, academic physicists in the United States were on a par with those in France, Germany, and England at least in terms of expenditures, volume of publications, and number of practitioners. Similarly, counts of significant contributions to astronomy and of major astronomers suggest that, by about 1900, Americans had already surpassed the Germans and would in the next few decades overtake the British for the top international spot in astronomy. Yet another statistical survey, scanning the period from about 1870 to 1915 and assaying only Americans, indicates that the physics, mathematics, and chemistry communities had made dramatic advances in quantity and quality of research and researchers—advances made in alliance with accelerated growth of professional societies, special journals, and centers of teaching and research.[2] Certainly, Newcomb's steadfast campaign to improve institutional support for scientists in the United States contributed to the community-wide gains evident by the turn of the century. Said differently, he helped establish the late nineteenth-century institutional base on which scientists, especially astronomers, physicists, mathematicians, and political economists, would build in future decades. His "example and leadership" would specifically redound, Harvard's Harlow Shapley observed in 1935, to the eventual preeminence of the United States in world astronomy.[3]

FACTS AND FICTION

The international project in which Newcomb participated—to establish uniformity in astronomical work—grew out of an 1896 conference in Paris. Convened by the leaders of the world's main ephemeris offices, the conference focused on an activity basic to all of positional astronomy: establishing a standard catalog of fundamental stars to be used as celestial reference points. One of the events that triggered the conference was the appearance of Newcomb's "preliminary results" from his multidecade research program to determine as accurately as possible the orbital "elements" of the planets and the associated, basic "constants" of astronomy. Going into the conference with the support of the British, Newcomb sought to eliminate inconsistency in astronomical work around the world by urging the adoption of his highly refined tables and numerical values. He carried from the Paris conference general support for his program and a dual charge: to complete his reevaluation of the positions of fundamental stars and to update his constant for the precession of the equinoxes.[4]

Surprisingly, the only forceful opposition to Newcomb and the European conferees' call for a uniform system of fundamental constants came from a segment of the American astronomical community. Writing in the *Astronomical Journal*, Lewis Boss, director of the Dudley Observatory in Albany, New York, questioned the techniques of data analysis that Newcomb had gone on to use in determining a new value for the precessional constant. He also suggested that the entire campaign for standardization was premature.[5] Similarly, but writing in British astronomical periodicals, Williams College astronomer Truman Safford expressed doubt about the adequacy of Newcomb's treatment of the "personal equation"—that is, the troublesome differences in time that individual observers record for a given celestial transit. Safford invoked psychological experiments on reaction times to question Newcomb's solution to the problem, in effect challenging the reliability of Newcomb's standardized constants.[6]

Publicly responding only to the accusations carried by the domestic *Astronomical Journal*, Newcomb answered Boss's technical criticism by first eliminating a "misapprehension" caused by a common "defect of language"—astronomers loosely speaking of the earth's precessional motion in terms of a simple "constant" when what was really intended was a complex "numerical expression." He then addressed Boss's technical complaints point by point. Adopting a diffident air, however, Newcomb dodged Boss's further accusation that the conference's plan was premature: "I expressly disclaim any expression of opinion on the question whether the action of the Conference of 1896 in adopting in advance the new values of the precessional motion was wise. Professor Boss impugns the wisdom; I leave the conclusion to others." Unpersuaded, Boss continued to impugn the wisdom. A few months later, claiming an "infringement upon the free spirit of scientific research," he charged that Newcomb and his European cronies were using the "weight of authority" to "legislate for astronomy in general." Newcomb denied the accusation, saying that he wished "to disclaim in the strongest manner any desire to force the conclusion that my work ought to be adopted."

In countering Boss's charges regarding technical procedures and results, Newcomb made no recourse to abstract, global methodological dictums, except for the passing stricture on defining terms clearly. This is not to suggest that his counterarguments lack clarifications and justifications seemingly underpinned by concrete, particularized rules of an implicit character. As was his pattern in addressing colleagues within his own scientific circle, however, he engaged Boss in an analytic debate on the management of observational errors and calculational complexities that was unembellished by grand methodological flourishes.⁷

Though Newcomb's retirement during the thick of these disputes complicated implementation of the Paris program—and though his countrymen dragged their feet on its adoption—compilers of the world's ephemerides began, at the turn of the century, to phase in the new constants and tables. Newcomb and his allies had prevailed. Indeed, Newcomb did his job so well that many of his values would remain in official use until the coming in the 1950s of electronic computers and artificial satellites.⁸

All the while, top awards and honors were coming to him at an increasing rate from around the nation and world. In fact, his formal government career ended propitiously, coinciding with his being named in 1895 one of eight Foreign Associates of the Paris Academy of Sciences. Writing anonymously in the *Nation*, Charles Peirce explained the salience of this designation. "This is universally acknowledged to be the greatest public honor that can be conferred upon a non-French man of science. Newcomb is the first citizen of the United States to receive it (if we are right in thinking that Louis Agassiz never completed his citizenship). It has never yet been bestowed upon a native citizen of the United States, although Franklin and Rumford received it." (Peirce, of course, overlooked that Newcomb was not a native, but a naturalized citizen who had been born in Nova Scotia.) A year after retiring, Newcomb also became the first recipient of the Bruce Medal, endowed by Catherine Wolfe Bruce of New York City and awarded by the Astronomical Society of the Pacific. In describing the international process used to select the most deserving astronomer from among many brilliant candidates, society President William Alvord reported that "one name stood forward so prominently in the communications from heads of six leading observatories of the world, that the Directors of this Society could but set the seal of their approval upon the verdict of his peers, and award the first Bruce Medal to Professor Simon Newcomb."⁹

With a modicum more of leisure time, the new retiree indulged in autobiographical reflections. A series of three brief sketches published in the *Atlantic Monthly* during late 1898 grew into his four-hundred-page *Reminiscences of an Astronomer*, issued by Houghton Mifflin in 1903. In an anonymous review of the book—the same review in which he lauded Newcomb's selection as an Associate of the Paris Academy—Peirce declared the autobiography to be "pleasant reading." He elaborated: "Newcomb's powers of telling a story and of painting a situation are much beyond the mediocre, while his light, pleasant style is quite remarkable."[10] In fact, Newcomb so fancied himself a storyteller that he also tried his hand at fiction, publishing two short stories and a novel.

The novel, which appeared in 1900 with an edition of more than fifteen hundred copies, traced the successful quest of a Harvard professor of molecular physics to disarm the armies and navies of the world. Newcomb set the story, titled *His Wisdom the Defender*, in the distant year of 1941 but on the familiar turf of Cambridge, Washington, rural Maryland, and various foreign capitals. At the heart of the plot is the professor's discovery of a new substance called "etherine." When activated by burning coal, etherine interacts with the luminiferous ether to enable various vehicles to overcome gravity and fly through space. The professor uses a large fleet of these vehicles, built of oak and staffed by college athletes, to subdue the militaries of all nations and thus ensure international peace. Besides constructing this fantasy out of nineteenth-century props—the physics of atoms and ether, the technology of oak and coal, and the bodies of wholesome college gymnasts and oarsmen—Newcomb interjected other, more personally familiar themes into his novel. He created a protagonist who was, for example, well versed in the subtleties of political economy; throughout the novel, the professor ponders the impact of his weightless vehicles on the world's labor force and financial markets, especially the stocks and bonds of railway and transportation companies. Similarly, he portrays the professor as a political liberal, a man who repeatedly voices the conviction that human progress is best served through the promulgation of individual liberty. Newcomb even makes his hero a crusader for clear language. On ensconcing himself as "Defender of the Peace of the World," the professor enacts a fivefold set of governing statutes. "The first," Newcomb tells his readers, "consisted of definitions showing the exact sense in which various expressions occurring in the statutes should be construed."[11]

Newcomb took seriously his speculation on etherine, the new type of matter that facilitated flight by canceling the pull of gravity. About the time that he was concocting the tale of the Harvard professor's exploits with etherine, Newcomb was actively speaking out against the possibility of conventional heavier-than-air flying machines of the type being tested by Smithsonian Secretary Samuel Langley. Newcomb was adamant that human flight was impossible if the would-be aviators relied on available technology and established scientific principles. Heavier-than-air flight would come to pass only if, in the future, someone discovered either an extremely lightweight but strong alloy or a new material that reacted with the ether to overcome gravity.[12] Thus, the fictional professor provided a mouthpiece for Newcomb to express his feelings on the future of flight as well as his thoughts on political economy, liberalism, and clear language.

REFLECTIONS ON METHOD

Though busy in retirement with his research, autobiography, and fiction, Newcomb still found time for the "scientific philosophy" that he had forged while still a young man and then fully developed during midcareer. While in his later years he never matched the continuous public comment and controversy of midcareer, he did carry on his exploration of general methodological themes. In fact, he experienced a resurgence of interest. To be sure, many of his later writings on the nature of science and its relationships remained in manuscript form, unpublished and often incomplete. Nevertheless, these manuscripts attest to Newcomb's continued commitment to the scientific use of language or, more specifically, a linguistic, empirical method and the application of that method primarily on the public level. Often ambitious in their proposed coverage of topics and issues, these manuscripts also attest to Newcomb's continued seriousness of intent.

In 1896, on the eve of retirement, for example, Newcomb produced a polished essay titled "The Creed of Naturalism," the first sixteen pages of which still exist. Intended for publication or perhaps as the text of a speech, the essay generalized the arguments that Newcomb had developed during and immediately after his 1878 presidential address to the AAAS—provocative arguments on the demarcation of natural science and Christian religion. Now more broadly seeking to contrast the "scientific philosophy" of "naturalists" with the "supernaturalism" of all metaphysical thinkers, he began by identifying the problems the two groups have in communicating with each other. He turned particularly to "the difficulty of assigning a specific meaning to general statements of any subject whatever." He explained: "In all our general statements about Nature we are obliged to use a nomenclature borrowed, to a large extent, from the language of common life. Without a careful definition of the sense in which the terms thus used are to be understood when applied to subjects outside the field in which they originated we are liable to be led into discussions which have no higher character than that of logomachy." Still intending to avoid these aimless discussions characterized by the reckless use of terms, he proceeded to define carefully the differences between a metaphysical thinker and an advocate of the scientific philosophy. Regarding the latter, he wrote: "The naturalist sees in Nature only a complex collection of phenomena and therefore considers her only in those aspects which, from their nature, may be apprehended in our experience or concerning which we may imagine ourselves to have experience." Newcomb emphasized that the naturalist, in limiting himself to only the domain of phenomena, completely suspends judgment on other possible domains: "If he should claim that the sphere thus defined comprises the whole of existence he would be passing quite out of his proper field. It is unnecessary that he should even claim this to be the only aspect of Nature about which we can know anything." Nondogmatic—and politic—as ever, Newcomb was still allowing for the complementarity of naturalistic and metaphysical modes of knowing.

As in his 1878 address and follow-up articles when he formally introduced his "fundamental postulate of the scientific philosophy"—a postulate dating to his Cambridge years and reflecting the influence of Mill and Wright—Newcomb further drew "a sharp distinction" and marked "the dividing line" between the two groups of thinkers by stating what he now labeled the "fundamental principle of Naturalism." That is, he restated the postulate that only antecedent causes operate in the world of phenomena—causes that are independent of consequences but in accord with natural laws that are uniform over time and place. Newcomb illustrated this postulate and the dissimilarities between the naturalist and metaphysical thinker by, as in earlier years, scrutinizing accepted notions of free will and stripping them of their metaphysical veneers. "The naturalist," he pointed out, "concerns himself only with cognisable actualities and is in no way concerned either with necessity on the one hand or liberty on the other, unless these words are so used as to imply objective certainty or uncertainty." If metaphysical thinkers of all ilk would only appreciate this emphasis on phenomena—in the present case, on concrete human actions—then "illusory questions" such as those concerning freedom of the will would vanish.

Newcomb seemed optimistic about this general prospect. Writing eighteen years after his AAAS address, he sensed an inevitable "growth of naturalistic doctrine" and its "extension into new fields." This increase in the cognitive authority of science, he reasoned, "is not so much the direct work of any set of men as a part of the movement of the age, carrying with it men of all schools of thought however unwilling they may be to admit the change." He cautioned, however, that the metaphysical school continues to challenge the naturalistic. As in the heyday of natural theology, the challenge comes from blurring the creeds of the two schools. "Attempts to lessen the significance of the movement towards Naturalism by obliterating the distinction between its views and those of Supernaturalism are so popular," he concluded, "it becomes imperative to enquire whether there is necessarily a fundamental difference of conception." Newcomb's personal response to this query was to seek clear definition of the terms, and hence tenets, of the two schools by writing this essay on the "creed" of naturalism.[13]

Another example of his sustained allegiance to his linguistic, empirical method appears in the "rough notes" for a speech he presented in 1898 in New Haven, presumably to a Yale audience. Again, the notes hark back to the 1870s when he first publicly underscored the "practical" character of scientific inquiry and the necessity of providing experiential definitions of propositions and terms. After briefly tracing the emergence in history of "the scientific idea," the title of the talk, he summarized the defining characteristics of this "idea":

> All men at the present day have some notion of a line of distinction between those conceptions which may be called indefinite, vague or visionary, and those which are properly termed practical. In its relations to progress, the first work of science is to eliminate from our mental activities everything that has not the attribute which may be called practical. It does not look at propositions and assertions from a distance to admire them, or criticize them according to what they seem to be; on the contrary, the first thing it wants to do is to get hold of the proposition, pull it to pieces, see what it means, see whether it is true. If, as is probably the case, it is sometimes true and sometimes false, it wants to know when it is true and when it is false. It rejects all vague speculations, all meaningless forms of language; it uses no words which do not admit of exact definition in terms of actual experience.

Newcomb completed this section of the speech by endorsing, as in prior decades, William Clifford's description of science as "organized common sense." Though the common sense of the layperson has the benefit of being based on concrete experience, the common sense of the scientist has the further advantage of being based on the shared experiences of other researchers in different places and times. Again, as in his 1896 manuscript "The Creed of Naturalism," Newcomb appeared optimistic about the spread of scientific thinking and its organized common sense. "As its scope widens," he advised, "we find that there is no field of knowledge to which science is not ready to extend its activities. Everywhere it desires to replace the vague speculations and doubtful conclusions of the untrained man by the precise methods and exact conclusions of the investigator."[14] At age sixty-three, Newcomb was still campaigning to extend the domain of scientific method and to use the method to demarcate scientific from other forms of thought.

Examples such as these from Newcomb's later, unpublished writings are numerous. Unfortunately, certain of these writings, while attributable to his later years,[15] are neither arranged systematically in the Newcomb archives nor identified by date. The titles alone of these disordered and often fragmentary works, nonetheless, further indicate Newcomb's continued ambitions in even the realm of pure philosophy of science. For instance, one work in progress carries the title "Logic," with preliminary sections containing discussions of the complex relationship between actual sensations, mental perceptions, and the formation of conceptions. Another manuscript has the label "The Epistemology of Science: A Study of the Sources and Limitations of Our Knowledge of Nature"; chapters include "The World of Sense" and "The Universe of Perception." Early in this manuscript, in a passage once again reminiscent of Charles Peirce's pragmatic maxim and anticipatory of Percy Bridgman's operational outlook, Newcomb suggests that for both the astronomer and the layperson the meaning of words resides in practical actions. For an illustration, he turns to his own everyday life: "In all my motions while going to the train, crossing crowded streets, avoiding passing vehicles and watching my progress toward the station, I am at every moment acting in response to sense impressions, which, although I interpret them at every moment as exhibitions of material objects, they only result in actions on my part susceptible of being defined by language without any reference to material things." That is, the words of a language find meaning through concrete actions and associated sensory effects, not in a so-called external or material reality.[16]

Finally, there is a lengthy but untitled typescript on the philosophy of science that opens with a chapter titled "The Analysis of the Processes of Nature." Near the beginning of the chapter, Newcomb succinctly reiterates his goal of developing "right thinking" through a scientific perspective on terminology and language. He also reiterates his views on the appropriateness of hypothetical propositions in science and the corresponding need to suspend judgment on issues such as the reality of a fourth spatial dimension or the ultimate constitution of atoms and ether. Similarly, he again emphasizes that for scientists "laws of nature" have an abstract and hypothetical character, that "anthropomorphic" concepts such as "force, action and agent" are merely heuristic aids to thought "whether or not there is any reality corresponding to them," and that "cause" is a metaphysically neutral term to be used in accord with its everyday meaning.[17] This manuscript, which goes on to discuss "Inductive Inference" and "Probability in Natural Phenomena," is possibly a fragment of—or, at least, a derivative of—a major treatise on the application of mathematical theories of probability to not only astronomy but also meteorology and human vital statistics. In the last months of his life, Newcomb commented that of all his unfinished works, the most important was this book on extending "the theory of Probable Inference" to "human affairs." In an extended fragment definitely from the treatise and written after 1906, he introduced the topic of probability by marking its main characteristic: "No particular conclusion reached by probable inference is categorically expressible in terms of certainty, and can only be described as having some degree of probability." The limitations and imperfections of scientific knowledge remained all too apparent to the aging astronomer and mathematician.[18]

Whereas a portion of Newcomb's later writings on the general workings and ties of science remained unpublished, many did get into print. He reiterated his basic methodological precept, for example, in 1904 at age sixty-nine while serving as president of the Saint Louis Congress of Arts and Science. Held with the world's fair that commemorated the Louisiana Purchase and bringing together foremost scholars from around the world, the congress was, in the words of a contemporary, "doubtless the most memorable and impressive scientific gathering ever held in America."[19] In a *Popular Science Monthly* article announcing the upcoming congress, Newcomb expressed the hope that this international gathering would help to unify the numerous branches of present-day science. He was optimistic about this integration occurring, particularly if the practitioners in the diverse fields adhered to sound methodological doctrines. In words recalling his midcareer writings, he spoke of the natural tendency of every science, when pursued by the best methods, to become more precise in the expression of its laws, and thus to bring mathematical conceptions to the aid of its investigators. When we have not only assigned a name to an object of study, but have made measurement of its size, or of the intensity of any ascertained properties it exhibits, we have taken a great step toward giving precision to our results, and making them comprehensible to a wider body of investigators.

He also touched on method in his presidential address on the evolution of science through history, a topic that Charles Peirce was "gratified" to find taken up by "a man so prominent in the world of science." Again confronting the problem of the proliferation of disciplines, Newcomb reiterated his view on method's unifying capacity: method provides a common element in otherwise diverging branches of science.[20]

Newcomb's time-consuming involvements as president of the Saint Louis Congress disrupted his methodological musings. In a letter to Peirce in 1904, he wrote: "I am so much occupied with the Congress and other work that I have wholly dropped out of the field of logic." Futhermore, he complained that he saw little chance of getting various existing manuscripts published: "So far as getting things printed is concerned, I am in much the same boat with yourself, having a great number of unfinished works, but it would not pay for a publisher to bring them out, and no institution, so far as I yet know, is ready to undertake the publication."[21]

Though frustrated in his efforts to conclude his more sweeping methodological studies, Newcomb did continue to write, speak, and occasionally publish on more particular topics. As mentioned earlier, he renewed his efforts in mathematics instruction, speaking in 1906 to the National Educational Association on the advantages of his system of "visible arithmetic" that built on concrete student experiences. The following year, he pursued the subject further through a detailed correspondence with a younger but kindred spirit, Edward L. Thorndike. Professor of educational psychology at Teachers College within Columbia University and a prior student of William James, Thorndike was attempting to show through rigorous experiments and quantitative measurements that learning occurred through the formation of "connections" between stimuli influencing a pupil's life and responses in thought, attitude, and action. "The one thing which I would like best to do," Thorndike told Newcomb, "would be to have a share in turning discussions of education and teaching from a set of quarrels and opinions, no matter how eminent, into something really like a science."22

Newcomb also continued to write, speak, and occasionally publish on particular topics that were more controversial. The most familiar of these were the state of American science, political economics, psychical research, and religion. It was in this nexus of fields, of course, that Newcomb found during midcareer both the means and ends for his pronouncements on method. To achieve the end of increased public appreciation of science and, in turn, increased institutional support, Newcomb used the means of not only lauding the usefulness of scientific method but also demonstrating its merits, especially in the realm of political economics. Similarly, to protect the professional standing of American scientists and extend their public province, he used the means of disassociating truly scientific practitioners from nonscientific dabblers in the areas of political economy, psychical research, and natural theology. Of course, these ends that served the aggregrate of American scientists also inextricably reflected Newcomb's personal beliefs as an individual citizen-scientist seeking to serve the commonweal.

OLD ISSUES REJOINED

On the occasion of the nation's centennial in 1876, Newcomb had published a comprehensive critique of American science in the January number of the *North American Review*. Three decades later, to the month, he issued in the same journal an overview of how scientific research could be better organized in the United States. Though no longer conveying a sense of urgency regarding the nation's research failings in this era of increased support and productivity, he suggested that scientific work in the United States could still profit from institutional restructuring. He had in mind a specific organizational form: "If we aim at the single object of promoting the advance of knowledge in the most effective way, and making our own country the leading one in research, our efforts should be directed towards bringing together as many scientific workers as possible at a single centre, where they can profit in the highest degree by mutual help, support, and sympathy." Merely wishing to show the advantages of such a centralized facility to his "fellow-citizens interested in the promotion of American science," Newcomb did not dwell on the national deficiencies that necessitated the institution.23 He had written more candidly four years earlier, however, in an article titled "Conditions Which Discourage Scientific Work in America."

In this prior essay, published as a lead article in the *North American Review* in 1902, Newcomb began by highlighting Americans' increased stature in the international scientific community. Within the past decade or two, he explained, his countrymen in both the physical and biological sciences had advanced greatly. Nevertheless, refusing to be appeased, he yet found fault. In the world scientific hierarchy, he contended, American science remained in a "backward condition" and still held an "inferior place." His explanation for this alleged lag was a variation of that offered in the 1870s. On the one hand, the American people—including private citizens and government officials—were interested in science and were willing to finance facilities for learning and research. On the other hand, these same citizens and officials did not comprehend the nuances of how best to encourage actual scientific inquiry. Newcomb observed that "our public, with all its appreciation of learning in the abstract, has only the vaguest and most imperfect idea of the true spirit of science and of the peculiar conditions under which the most advanced research is possible." While supportive of "science" as a professional grouping of faceless practitioners, neither private citizens nor government officials provided ample encouragement or recognition to the individual scientist, particularly the individual "leader in science, the divinely inspired explorer of nature." Compared specifically to Europeans, Americans did not adequately acknowledge outstanding scientists. Neither did the American government maintain close enough ties with scientific leaders; indeed, during the prior three decades, Congress had essentially ignored and let languish the National Academy of Sciences.[24]

Though these complaints were similar to the ones he had voiced in the 1870s, Newcomb's personal experiences as a world-class researcher in government employ had contributed to a fixation on the plight of the outstanding individual and the dilemma of government inattention. His concern extended beyond himself, however, to all highly talented researchers. At the dedication of the Yerkes Observatory a few years earlier, in a widely reprinted speech, he rebuked patrons and the public for presuming to foster astronomy merely by providing observatory buildings, large telescopes, and related apparatus. These material resources are necessary but not sufficient for the successful prosecution of research. "A great telescope," he counseled, "is of no use without a man at the end of it, and what the telescope may do depends more upon this appendage than upon the instrument itself." He went on to recommend the nurturing of those individual astronomers who, because of innate gifts, are "destined to advance the science by works of real genius."[25]

To begin to remedy the "want of touch" between politicians and scientists—and thus also to help close the gap between the wider public and scientists—Newcomb in the 1902 article again urged an institutional solution: the establishment in Washington of a research facility that would bring together the nation's best scientists. As he would in 1906 when detailing this centralized facility, he made the suggestion in the spirit of "one who desires to see our country pre-eminent in science and learning." Other American researchers, continuing through the opening decades of the twentieth century, would repeat this call for "sheltered enclaves," responding to what they felt were inadequate recognition and support. In the meantime, Newcomb was also privately promoting a more specific remedy for a particular manifestation of government inattention. Though retired, he was still actively pressing for professional, civilian astronomers to take control of the Naval Observatory (which had expanded its domain by absorbing the Almanac Office in 1894).[26]

Through most of his career, as part of his effort to redress the institutional failings of American science, Newcomb had worked to improve communication between policymakers and scientists. One of his initiatives had been to bring scientific method to bear on current political and economic issues. Beginning in the 1890s, in the aftermath of the ideological fray between the old and new schools, his involvement in political economy had dropped sharply. His last thrust in the field came in 1905 when he addressed the section on Social and Economic Science at the national meeting of the AAAS. The title of his talk, "The Basis of Economics as an Exact Science," revealed that Newcomb still sought to demarcate the disciplinary boundaries of a mathematical science of political economy. Similarly, the first sentence of the talk, as later abstracted in *Science*, showed that he still felt that the initial step in creating the science entailed the adoption of scientific methodology and, specifically, a clarification of language and meaning. "One of the first things to strike us in the effort to apply scientific methods to economics," he began, "is the absence of nomenclature." As in earlier years, he went on to point out specific instances of "defective definition." He then moved to his more general concern, the "method of inquiry" best suited to economics. Drawing analogies to physics and recalling past applications of mathematical techniques to economics, he maintained that economics could be made an exact science even though its practitioners had to contend with the "vagaries of human nature." Speaking in the opening decade of the twentieth century, he was advocating a goal increasingly embraced by a growing cadre of Progressives—reformers seeking social and economic improvements, often through the use of supposedly disinterested scientific methods. Of course, he had not deviated from the goal for political economy that he had first enunciated at the end of the Civil War.[27]

Neither had he deviated in his opinion of psychical research—a field not only in which he had challenged basic claims but also from which he had endeavored to disassociate professional scientists. Unlike his final thrust into economic territory, however, his final foray into psychical precincts precipitated a protest. An old controversy reignited when in January 1909, half a year before his death, Newcomb published a rehash of his views on the failings of psychical studies. As when presiding in the 1880s over the American Society for Psychical Research, Newcomb argued, on mainly methodological grounds, that no scientific justification existed for alleging the presence of psychical effects such as mental telepathy. He brought his argument up to date by contrasting the continuing failure of researchers to determine the conditions for replicating psychical effects with the rapid success of scientists in establishing the principles of the seemingly equally elusive phenomena of radioactivity. The reaction from the psychical community to Newcomb's adverse appraisal was immediate. It was also strong, coming from the pens of Sir Oliver Lodge, the distinguished English physicist, and James Hyslop, an ex-professor of philosophy at Columbia who had revived the American psychical society a few years earlier. (Centered in New York City and catering to the avocational interests of lay people, the new society no longer enjoyed the support of the wider and now more self-consciously "professional" scientific community.[28]) Besides directly countering Newcomb's charges, both men chided the astronomer and mathematician for intruding into and passing judgment on a branch of inquiry where his knowledge was limited or, at least, out of date.[29] By now, Newcomb was accustomed to this complaint, having heard it from adversaries in political economy, religion, and the philosophy of physics.

The exchange between Newcomb, Lodge, and Hyslop, while neither adding much to the psychical debate as first joined in the 1880s nor to our understanding of Newcomb's rhetorical uses of method, provides a window into the shift that had occurred in attitudes toward science. Specifically, the exchange illuminates the extent to which Newcomb's type of "scientific philosophy" had taken hold in American culture by 1909. His pronouncements on method and its utility had contributed to this broader cultural shift, even though his motives were narrower: on the one hand, to cultivate public support of basic science and to strengthen science's institutional framework; on the other hand, to press his personal political and religious beliefs for the benefit of society at large. Hyslop, in responding to Newcomb's charges, complained that scientific skepticism was now so in vogue that Newcomb's iconclastic arguments enjoyed a credibility beyond what they deserved. "What interests me primarily, at present," he began, "is the authority which Professor Newcomb will have with a certain class of Philistines, who do not think for themselves, but like to stand behind the skirts of any man with a scientific reputation, as an excuse to deny." He continued, in more detail: "The conservative and sceptical spirit of the article will have the antecedent probabilities in its favor with the scientific minds of this age, as doubt and denial are regarded as the marks of intelligence in this much deluded period of human reflection. It is only one kind of truth of which men are in search, and they think themselves qualified to substitute a scientific for a religious dogmatism, and no one can be treated as sceptic who dares to question the authority of doubt. All this is in the favor of Professor Newcomb's article." To counter Newcomb's advantage in "this much deluded period of human reflection," Hyslop adopted a strategy that, ironically, further illuminates the extent to which Newcomb's particular type of empirical, linguistic methodology had taken hold (again, at least in part, through Newcomb's own efforts). He accused Newcomb of committing the very error that Newcomb had spent his career warning against—the unscientific use of language, in this case, the imprecise use of the term "telepathy." "Professor Newcomb does not mention any clear definition or delimitation of the term," Hyslop explained, "and so cannot be indorsed or disputed. If the term were recognizably defined, and if it were not so equivocally employed by various disputants in these problems, we might find no occasion for analysis. But such is not the case, and it is time to call attention to the responsibilities of writers on this subject when employing the term." As Newcomb had done many times before in other contexts, Hyslop proceeded to evaluate the alternative usages, arriving finally at "the only conception of the term that will stand scientific consideration."[30]

Americans in general had witnessed by century's end a rise of positivist science. And as the twentieth century opened, they were further witnessing the beginning of the Progressive movement with its attachment to scientific method.[31] They were also increasingly encountering, as we will see, pragmatic perspectives. Thus, Hyslop was not alone among Americans in noticing that general society had shifted into a scientific age of "doubt and denial." Newcomb himself, in his various writings from the period, commented on this expansion of science's cognitive authority, the "growth of naturalistic doctrine." Writing in 1909, he particularly marveled over the transformation in attitude of Christians toward Darwinian doctrines: "There is perhaps no one of my experiences more interesting than the change which I have witnessed in the attitude of the world generally, especially of the world of orthodox christianity toward these doctrines during the past thirty years." William James detected a similar transition in attitudes regarding the relationship between science and religion. In 1902, with his usual aptness, the Harvard psychologist and philosopher remarked: "Though the scientist may individually nourish a religion, and be a theist in his irresponsible hours, the days are over when it could be said that for Science herself the heavens declare the glory of God and the firmament showeth his handiwork."[32]

From Newcomb's personal perspective, the days had never existed when it could be said that scientists and theologians were partners in a quest to elucidate God's design in the universe. In his debate during the late 1870s with Gray, Porter, and McCosh, he had invoked the criterion of scientific method to sever natural science from natural theology, and he had gone on to express, albeit anonymously, his personal misgivings about Christian teachings. Now, in the early summer of 1909, he remained convinced that scientific inquiry must be kept distinct from traditional natural theology. We know this because, in the weeks preceding his death on July 11, he was drafting a chapter on his religious views to add to an intended new edition of his autobiography. The resulting manuscript was part of a flurry of works, including the final portions of his study of the moon's motion, that he dictated to stenographers as he faced imminent death from cancer of the bladder.[33]

Toward the end of the fifteen-page manuscript, this religious skeptic, ailing and facing certain death, acknowledged that he had always allowed for the possibility of a "Great First Cause." He pointed out that he had touched on this possibility, for example, while discussing the origin of the universe in his *Popular Astronomy*, a book that had gone through many editions and translations since publication three decades earlier. (He might have had in mind a quotation that he had endorsed from John Milton's *Paradise Lost* in which the poet has "the almighty Maker" create the world from chaos.)[34] "Such being the case," he reasoned, "I can scarcely be classed as an agnostic, though I am such as regards current views of religious subjects. The agnostic simply knows nothing; I hold at least the beginning of a positive creed in [conceding that evidence exists for] the action of an infinite cause." True to form, Newcomb clarified his position on a first cause by returning to the methodological critique of language on which he had depended throughout his career. "But if I am asked whether I regard this cause as an intelligent one I am unable to answer until the word 'intelligent' is defined. This term implies a certain mental quality belonging to the human race and it seems to me a belittling of a great universal cause to apply any such term to it." "All we can say," he added in his final words on the subject, "is that the cause exists and must be capable of preceding the result, which is the universe as we find it to be. But I have never indulged in vain speculation on subjects which I found it impossible to form a clear and definite conception, and so shall not pursue the subject further."[35] He died a few weeks later.

Newcomb's closest colleagues, friends, and relatives would have understood his refusal, even on his deathbed, to indulge in religious speculations. They would have further appreciated that this refusal followed from not only a general disposition traceable back to childhood but also his particular vision of scientific method. In a memorial address prepared for the Philosophical Society of Washington, Herbert Putnam, the United States Librarian of Congress and an old friend, captured the extent to which that method provided Newcomb with personal and professional guidance. In recalling his friend's personality, Putnam wrote:

> He was never, I say, intolerant against contrary opinion which he supposed sincere; yet, a master of an exact science, he required of others that exactitude in expression which he imposed upon himself. While therefore patient before assertions of fact or of principle opposed to his own impressions or convictions, I have often remarked his insistence upon a clear expression of them, and have as generally observed that the inability of the interlocutor to frame such an expression concluded the argument. In an address as former President of this Society, after asserting "the greatest want of the day to be the more general introduction of the scientific method and the scientific spirit into the discussion of our political and social problems," he explained that under the scientific method must be included dis-

cipline in the scientific use of language. To such a discipline he had doubtless subjected himself. Indeed, there was evidence that he was doing so daily. The result was a style, both written and spoken, of extraordinary lucidity. He has been called "a master of clear thought and of good English," but he was more than a master of good English: he was a master of clear expression, and he could have been a master of clear expression only *because* he was a master of clear thought. His capacity for simplicity of expression amounted to genius. . . . To Dr. Newcomb the universe was either intelligible or unknown. If intelligible, it must be capable of clear and exact expression; if unknown, it must be clearly distinguished as such.[36]

In other words, Newcomb put his basic methodological precept into everyday practice. His skill in using language sprang in part from his insistence on the scientific use of language. In turn, his call for the scientific use of language benefited from his cogency as a speaker and writer. By joining precept to practice regarding his linguistic, empirical method, Newcomb thus enhanced his effectiveness in his various public, disciplinary, and technical campaigns. He spoke with strength when he spoke in advocacy of science.

PART III
Commentary

CHAPTER XII

Newcomb and American Pragmatism

Simon Newcomb embraced and then sought to propagate an enticing vision of scientific method. After tapping contemporary beliefs in a seemingly singular method capable of producing results both in and out of natural science, he went on to proclaim his version of the beliefs, doggedly advocating a distinct outlook on method. In particular, he found in his linguistic, empirical method a sharp polemical weapon. He wielded the weapon with confidence in a range of professional and public arenas as he spoke for either the American scientific collective or himself as an individual citizen-scientist. An appreciation of the centrality of method in Newcomb's thought provides insight into the mind of one of late nineteenth-century America's most prominent scientists and crusaders for science. Furthermore, his comments and exchanges help illuminate contemporary issues involving not only the institutionalization of natural science in America but also the state of political economy, religion, philosophy, the foundations of physics, education, and psychical research. His fascination with method also tells us much about what traditionally was considered one of the nation's most penetrating cultural and intellectual movements—pragmatism.

The topic of pragmatism has fallen into disfavor among some American historians. In part, this is because of earlier excessive claims about its role in the nation's development. Seeking a fresh perspective on the subject, David Hollinger has recently reappraised the place of pragmatism in American thought and culture. Through his study of the individuals and the writings that have come to be associated with pragmatism, Hollinger identifies three ways to construe the movement. That is, pragmatism had three distinctive but interrelated presences in American history. On the most basic plane, it consisted of a set of characteristic theories of meaning and truth that served to identify and inform the movement. This refers to the much studied philosophical core of pragmatism as variously articulated by Charles Peirce, William James, and John Dewey during the late nineteenth and early twentieth centuries: for example, Peirce's seminal essay of 1878 suggests that the meaning of a concept is to be determined through its concrete, practical effects. In contrast to this strictly philosophical presence, pragmatism also consisted of a collection of general images or stereotypes of American thought. Here Hollinger has in mind the sweeping claims of commentators who identified pragmatism with a set of supposedly distinctive American traits. To these persons, pragmatism seemed "a style of thought characterized by voluntarism, practicality, moralism, relativism, an eye toward the future, a preference for action over contemplation, and other traits of the same degree of generality."

Hollinger feels that studies restricted to the philosophical core are too narrow to shed much light on the place of pragmatism in American history when considered in its full cultural and intellectual scope. Scholars interested just in Western philosophy, however, still legitimately can find much to analyze regarding the variegated technical core. As for the attempts to associate pragmatism with a package of characteristic American traits, Hollinger agrees with most other recent students of American history that such attempts are overambitious, exceeding the bounds of valid historical generalization. "If our understanding of American history can be enriched by the study of pragmatism," he reasons, "that enrichment can scarcely be expected to come about so long as pragmatism is either stretched to cover all of America or confined to those of its formulations sufficiently fruitful philosophically to have found places in the history of Western philosophy." Instead, he suggests that we can add to our understanding of American history by recognizing a third, somewhat intermediary way in which pragmatism displayed itself during the late nineteenth and early twentieth centuries. In this view, pragmatism consisted of "a cluster of assertions and hopes about the basis for culture in an age of science."[1] What does Hollinger mean?

PRAGMATISTS IN HISTORICAL CONTEXT

Expressed most simply, Hollinger's position is that the pragmatists are significant because they reflected and contributed to Americans' preconceptions about and aspirations for a modern, scientific culture. Said differently, Hollinger is seeking to do for pragmatism what other scholars are seeking to do for methodology in general: to place pragmatism within its full historical context. He is seeking to move beyond another "internalist" analysis of its technical core or another "Whiggish" analysis of its characteristics that can be associated with a preconceived set of cultural or intellectual attributes. To be sure, he remains alert to the importance of the philosophical core in flagging the persons and texts associated with the pragmatic movement—that is, in identifying them for further study. He also remains alert to differences in the outlooks of Peirce, James, Dewey, and their followers as well as to the dangers of categorizing them all under one label. Nevertheless, having identified the principal pragmatists, he finds that they "emerge as reflectors of, and powerful agents for, a distinctive cluster of assertions and hopes about how modern culture could be integrated and energized." More specifically, he distinguishes three elements of this cluster, all of which embody the pragmatists' preoccupation with science's place in modern life.[2]

The first element in this cluster involved the pragmatists' fixation on the process of inquiry itself rather than the actual acquisition of knowledge. Envisioning the world as a place of change where knowledge often seemed uncertain and transient, the pragmatists turned specifically to the scientific method of inquiry for something certain and enduring. They believed that the scientific method was sufficient by itself to sustain modern culture. This tendency appears in Dewey's recurrent calls to rescue civilization from detrimental trends by adopting "the scientific habit of mind" and in James's insistence on bringing the process of free scientific inquiry to even the sanctioned beliefs of orthodox science (including some scientists' dogmatic rejection of psychical effects). Hollinger also sees the element in Peirce's writings. "Peirce was unusual—even in an age of extravagant 'scientism'—in the extremity and singularity with which he identified goodness and progress with science, and he was among the first admirers of science to focus this adulation explicitly on the community of investigators and on the common methodological commitments that enabled members of this community to correct both each other and the stock of propositions they took to be true." This tendency to accentuate the scientific method of inquiry was already evident in Peirce's important paper of 1878 with its pragmatic principle. In particular, the principle "connoted in all its formulations a willingness to treat knowledge as temporal and to treat method as both primary and enduring." Of course, as Hollinger adds, Peirce was "the Melville of American philosophy" in that his writings were largely neglected for many years.

Whereas method was certain, knowledge was problematic for the pragmatists. Although most comfortable with modern empirical propositions, which they viewed as having a high degree of reliability, they generally perceived human knowledge to be uncertain and transient. And though sometimes eager to take advantage of available knowledge and apply it to social and religious issues, they generally approached the body of existing knowledge with caution. Thus, James viewed scientific laws as heuristic "approximations" rather than absolute transcripts of reality. Dewey similarly stressed the "instrumental" nature of truth in all realms. Peirce, while more optimistic about the trustworthiness of knowledge because he envisioned the eventual coalescence of ideas, joined Dewey and James in acknowledging its temporal limitations. Although they debated the specific nature of truth, the pragmatists agreed "that truth was a condition that happened to an idea through the course of events as experienced and analyzed by human beings." With human knowledge perceived to be fallible and temporal, the upshot was that Peirce, James, Dewey, and their followers believed that the broader culture should assign priority to method.[3]

The second element in the triad of assertions and hopes was the conviction that in all its physical and social manifestations the world was responsive to human purpose—specifically, that the process of inquiry itself could alter the world. Hollinger elaborates: "In contrast to the moralists who hailed or lamented the scientific enterprise as the exploration of a one-way street, down which orders for belief and conduct came from 'nature,' the pragmatic tradition carried a faith in inquiry's reconstructive capabilities in the most rigorous of the sciences and in everyday life." The pragmatists believed that such transformations could occur not merely on the simple level of technological improvements, but also on social, moral, and even cognitive levels. Though Dewey was the main popularizer of this "reconstructive vision" and though he began popularizing the vision around 1900, it was not until the late 1930s that he clarified what he meant by the elusive claim that the process of inquiry itself allowed investigators to reconstruct or transform aspects being studied. During the early decades of the twentieth century, he merely reassured the public that science and technology illustrate the extent to which the world could be purposively altered and that similar alterations should be attempted in hitherto untouched social and moral realms. Besides this belief in "social engineering" (which meshed with certain Progressive ideals), a corresponding conviction also surfaced in James's voluntarist statements regarding the power of human will to affect the world both internal and external to the self. The will, which James distinguished from outside forces, could affect an individual's life and the world in which he or she lived. "James was particularly reassuring," Hollinger adds, "about the role of purpose in the inquiries carried out by scientists; the phenomena of nature would ultimately decide an issue, but all the more decisively if scientists brought to their investigation—critically, to be sure—their own most intense hope for a given outcome." As for Peirce, he shied away from the notion that inquiry could transform the world in any fundamental sense.[4]

The final element of pragmatism was the belief in the accessibility of inquiry to the general citizenry of an educated, democratic society. Dewey, and James to a lesser extent, stressed that common-sense modes of inquiry were similar to scientific modes and that public education should be used to draw common-sense modes even more into line with the scientific. Citizens having the appropriate education could bring scientific inquiry to bear on social, ethical, and philosophical issues. "Not only were social scientists encouraged," Hollinger explains, "to extend to the social realm the search for facts pioneered by practitioners of the physical sciences; persons of any station confronting issues in politics and morals were encouraged to face them 'scientifically.' " One upshot of the call for global application, as seen in the writings of James and especially Dewey, was to question the relevance of traditional philosophical issues. Hollinger adds that such a readily employable pragmatic outlook actually was democratic in two senses: "Not only was its method announced as widely accessible and as an engine of improvement; the very practice of that method was supposed to be open, undogmatic, tolerant, self-corrective, and thus an easily recognized extension of the standard liberal ideology articulated by Mill and cherished by so many late-nineteenth-century Americans."[5]

To what degree did Newcomb participate in this American pragmatic tradition? Was he a reflector of, and an agent for, a distinctive triad of assertions and hopes about how the method of scientific inquiry could integrate and energize modern culture? With certain qualifications, the answer is yes. This does not mean, however, that he was a pragmatist, but merely that he shared with the pragmatists certain preconceptions and aspirations. Indeed, we need to remember that Hollinger does not intend his three elements to constitute a checklist for identifying pragmatists. In undertaking his study, he actually depended on distinctive components of the philosophical core to flag the pragmatists; only after having so identified the primary members of the group did he examine their broader beliefs. In this process, he found that Peirce, James, Dewey, and their followers shared a particular combination of ideas about the role of science in culture. Hollinger does not deny that individual elements in the cluster appeared often in the writings of science enthusiasts unaffiliated with pragmatism. He merely contends that the full complement of elements was advanced with a high degree of persistence and national visibility by only the pragmatists.[6] Thus, Newcomb could be allied with the pragmatists and even play an important role in their history, but still not be a pragmatist. In fact, as we will see in the next chapter, Newcomb's strong affinities with the pragmatists can be explained by noticing that both he and they were part of a larger cultural and intellectual tradition—one that subsumed them both. For now, let us look more closely at his relation to the three pragmatic elements.

If there is a central message running through Newcomb's general essays and speeches, it is that the scientific method of inquiry rather than the acquisition of knowledge should have priority in contemporary culture. The primacy of method is, of course, the first element in the pragmatic triad. From as early as his Cambridge days when drafting a physics textbook, Newcomb took method to be the key to progress in natural science. Moreover, from the mid 1870s on, he publicly argued that training in the method of science, rather than mastery of the substantive content, should be the foundation of a national program of liberal education. Like Dewey, he emphasized that human progress could be enhanced and disaster avoided through the application of scientific method to areas such as political economy and religion; indeed, he could exclaim that there is "no field of knowledge" to which science is not ready to extend its method. Like James, he was even willing to redirect the method back at prevailing scientific opinion, criticizing doctrines such as scientific materialism. And though he and James ended on opposite sides of the psychical question, he had joined James in seeking an open, scientific hearing for the controversial question. Newcomb also paralleled Peirce regarding the primacy of method. In particular, Hollinger could have been describing Newcomb when he marked Peirce for "the extremity and singularity with which he identified goodness and progress with science." Similarly, he could have been talking about Newcomb when he placed Peirce "among the first admirers of science to focus this adulation explicitly on the community of investigators and on the common methodological commitments that enabled members of this community to correct both each other and the stock of propositions they took to be true." For Newcomb, scientific method, especially the proper use of language, was essential for creating a "common basis" or "community of understanding" among practitioners.[7]

The first pragmatic element, the belief in the primacy of scientific method, has a corollary: human knowledge is fallible and truth is temporal. As he repeatedly attests in his essays and speeches, Newcomb mistrusted the received knowledge in fields such as political economy, religion, philosophy, and psychical research. Moreover, he joined the pragmatists in affirming that, though the propositions, theories, and laws of natural science constitute the most reliable form of knowledge, even they are not true, complete, or explanative in any ultimate, metaphysical sense. Thus, while defending his scientific approach to political economy, he could assert that "all scientific propositions are in their very nature hypothetical"; similarly, he could grant "the necessary imperfections of all scientific statement." Theories have come and gone, with some withstanding the tests of evidence better than others. When a scientist labels a theory as true, it does not mean "that it bears any recognized seal of truth," but simply that the available evidence weighs in its favor—a point that Newcomb, beginning as early as his Cambridge days, expressed also in terms of the probabilistic character of knowledge. Truth, therefore, is not established through revelation or authority, but "decided by the human judgment." More generally, all scientific conclusions require the "exercise of judgment on the part of the individual." Like James and Dewey, Newcomb acknowledged the provisional and hypothetical character of theories and granted that a concept such as energy functioned as "merely an extremely convenient instrument for expressing and measuring relations."

Newcomb does not reflect the second pragmatic element to the degree that he reflects the first and, as we shall see, third elements. The second element is the assertion that the purposive action of human inquiry can change the world on personal, social, and physical levels. Newcomb displays this belief only in its simplest sense. Specifically, he agrees wholeheartedly with Dewey that persons can draw on scientific method to reshape areas of human discourse such as political economics, religion, and philosophy. Indeed, the conviction that progressive change can occur through scientific inquiry is at the heart of Newcomb's entire program. "I make bold to say," he begins his 1880 address to the Philosophical Society of Washington, "that the greatest want of the day, from a purely practical point of view, is the more general introduction of the scientific method and the scientific spirit into the discussion of those political and social problems which we encounter on our road to a higher plane of public well being." And to the extent that this reshaping involves value judgments, he realizes along with Dewey that the application of scientific method ultimately has ethical and moral implications. But Dewey and James, each in his own distinctive manner, believed the world could also be reconstructed in more elemental senses, with James even asserting that human will could actually alter the world both internal and external to the self. Newcomb, in contrast, was adamant that volition is "powerless" in altering the basic course of nature, whether physical, social, or personal. In accord with Mill's compatibilism, he was fond of arguing that, while human actions can be "free" in the sense of not coerced, they are still "determined" in the sense of subject to external causal influences. More generally, his well-worn "fundamental postulate" of the "scientific philosophy" precluded James's voluntarist conception of the will. That is, in further accord with Mill, Newcomb avowed that all natural phenomena are "conditioned solely by antecedent causes," not by imagined "consequences."

Newcomb was not alone in reflecting only a delimited version of the second pragmatic element. Neither Peirce nor, for that matter, Wright—a believer in the universality of physical as opposed to mental or metaphysical causation—deemed that nature in its essence was responsive to human purpose.[8] Both merely held that persons could use scientific method to restructure areas of human discourse such as politics. The second element in its full form thus seems to be an attribute of pragmatism as formulated variously and more exclusively by James and Dewey, especially around and after the turn of the century. Indeed, in drafting his influential essay "The Will to Believe" in 1895, James was reacting against his late nineteenth-century predecessors—

responding explicitly to agnostic scientists such as William Clifford and Thomas Huxley and responding implicitly to the agnosticism of his old mentor, Wright. Furthermore, Dewey's "reconstructive vision" remained at the level of obscure and oblique pronouncements from the late 1890s until its fuller explication in the late 1930s.[9] Newcomb, a product of an earlier period, joined Wright and Peirce in believing that scientific method could be used to transform areas of human discourse but did not concur with either James or Dewey in believing further that inquiry could fundamentally change the world. To this extent, he manifested the second element.

The third element of the pragmatists' assertions and hopes consists of the belief that scientific inquiry is accessible to all citizens of an educated, democratic society. As with James and especially Dewey, Newcomb repeatedly called attention to the accessibility of scientific method by highlighting its similarities to the "common sense" of "practical" persons such as businessmen. He was convinced that the method, as opposed to the content, of science could and should be taught in the nation's schools, thus creating a true program of liberal education. Dewey, who dominated educational thought in the United States during much of the first half of the twentieth century, later gave wide currency to this notion, continually calling for the teaching of "scientific method in its largest sense."[10] Newcomb particularly urged the public and the nation's leaders to adopt the method in order to bring it to bear on muddled thinking in political economy. His vision of method was democratic and in harmony with prevailing liberal sentiment, not only in the sense of the method being accessible to the nation's citizens but also in the sense of its practice being unconstrained and undogmatic. Accessibility also extended to persons interested in philosophy and religion, with Newcomb identifying uses for method in both areas. Again as with James and Dewey, one result was to question the basic legitimacy of traditional philosophical perspectives such as those involving freedom of the will and traditional Christian doctrines such as those associated with natural theology.

Clearly, Newcomb's beliefs about scientific method are akin to the pragmatists' triad of assertions and hopes. Before considering the possible historical significance of the affiliation, it will be useful to consider Newcomb's further connection to the philosophical core of pragmatism.

THE PHILOSOPHICAL CORE OF PRAGMATISM

It is difficult to isolate the technical base of the method publicly advocated by Newcomb and compare it to pragmatic theories of meaning and truth without sounding like a strict internalist or a predisposed Whig historian. There exists, nevertheless, a sound historical reason for focusing on the technical foundation. It happens that Newcomb variously joined Wright and Peirce in endorsing certain philosophical precepts that were as much reflections of the times as were the three types of cultural assertions and hopes. Specifically, in his view on meaning, Newcomb joined Peirce and the pragmatists in going beyond Wright's related but more restrictive view. In his outlook on truth, however, he merely prefigured but never articulated a pragmatic conception of truth. Thus, whereas his philosophical beliefs do not mark him as a pragmatist, they do reveal him again to be a close ally. His view of method entailed key aspects of the pragmatic theories of meaning and truth.

Recall that Newcomb's combined linguistic and empirical emphasis translated into one central methodological rule: scientists, in the process of inquiry, should use only those concepts that can be derived from or explicated in terms of particular sensory perceptions. Viewed from a slightly different perspective, this rule establishes a means of determining which concepts a scientist accepts as legitimate. In other words, Newcomb's basic methodological rule can be reexpressed as a rule for evaluating meaning—what philosophers designate a meaning criterion. According to this reformulation, Newcomb is claiming that the meaning of a concept lies in its relation to particular sensory perceptions. What type of relation and what type of terms, propositions, and concepts did Newcomb intend?

Newcomb applied his meaning criterion in a variety of fields to concepts ranging from the highly abstract to the empirically simple. He used the criterion in his 1878 AAAS presidential address to distinguish between the language of science and religion. Only in the former do all terms "have exact literal meanings, and refer only to things which admit of being perceived by the senses, or, at least, of being conceived as thus perceptible." In his 1880 speech "The Relation of Scientific Method to Social Progress" he went on to argue that the terms and propositions of political economics, especially, should have "connections with special objects of sense." These included "abstract terms" and "words expressive of mental states," although such terms and words had only indirect sensory ties. For examples of proper specifications of meaning, Newcomb cited the fields of business and physics, including modern physicists' definition of "force"; for examples of inadequate specifications of meaning, he turned to philosophy, particularly traditional philosophers' mishandling of "freedom."

During 1880, in his series on scientific materialism, Newcomb elaborated on natural scientists' use of abstract terms such as "energy." Although science limits itself to phenomena, it "does not deny the existence of things which are not phenomena"; that is, it does not deny the existence if the concept (such as "energy") provides a "convenient instrument for expressing and measuring relations among material things which, without it, would be beyond the power of the ordinary mind to grasp." In his 1882 rebuttal to Stallo, he insisted that even the simple concepts of "hardness" and "elasticity" must refer precisely to "qualities of sensible masses" to be meaningful; in addition, words such as "active" and "passive" lacked meaning if they did not refer to "effects which follow when matter is placed under certain conditions." His review of Stallo's book also contained an operational variant of his criterion: the meaning of a scientific proposition or term, in this case "mass" or "length," lies in its laboratory method of measurement. In his technical economic writings from the decades bracketing 1880, he also showed how concepts such as "value" could be defined using quantitative measurements and how concepts such as "capital" were often left improperly defined.

During his later years, Newcomb continued to sound these familiar themes. In his New Haven speech, for instance, he again points out that science uses only words that "admit of exact definition in terms of actual experience." And in his "Epistemology of Science," he explains more generally that words acquire meaning through concrete actions and associated sensory effects, not through an independent material reality.

To what extent is Newcomb's meaning criterion cut from the same philosophic cloth as Wright's or Peirce's? To help answer this question, we can draw on Edward Madden's comparative analysis of Wright's writings and Peirce's 1878 essay, Madden being the leading authority on Wright's philosophy. He argues that Wright's and Peirce's criteria of meaning, while having much in common, have a "crucial" difference. The difference suggests that Wright had helped set the stage for a pragmatic theory of meaning but had not expressly prefigured it. In particular, though Wright and Peirce both endorse criteria of meaning based on verification through sensuous consequences, they differ on what actually is to be verified. Whereas Wright seeks only to verify "transcendental or ideal notions or theories," Peirce and the pragmatists expand the scope of what is to be verified to encompass even "simple empirical sentences." Thus, both men would want to have the meaning of a concept such as "force" explicated through a specification of sensuous consequences. Only Peirce, however, would further insist that the meaning of a concept such as "hard" needs to be explicated through "its conceived effects"; Wright would be content to treat this concept as if it had manifest meaning. Madden adds that the common feature of Wright's and Peirce's criteria of meaning—verification by sensuous consequences—is not distinctive to pragmatism. This feature characterizes much of traditional empirical philosophy, going back at least to David Hume.[11]

Madden offers an alternative statement of this similarity and difference between Wright and Peirce. Here the starting point is to notice that traditional empiricists and some pragmatists agree that abstract words or propositions are linked through a chain of definitions to simpler terms—terms such as "hard." For the traditional empiricist, each of these simpler terms is self-contained in the sense that it denotes particular sensuous elements and requires no further definition. Some pragmatists, however, deny "the self-containedness of a sensuous 'given' " and insist that even apparently simpler terms require explicit definition through their effects. Again, whereas both Wright and Peirce required that abstract terms such as "force" be tied to sensuous consequences, only Peirce went on to apply this criterion to simpler terms such as "hard."[12]

From as early as his Cambridge days when he invoked Mill's emphasis on experienced sensations to fault even Comte's positivism, Newcomb shared Wright's traditional empirical conviction that abstract terms be linkable to sensuous consequences. In his 1880 address on scientific method and social progress, for example, he explained that abstract words require an intermediate chain of supportive definitions ultimately grounded on terms corresponding directly to "sensible objects." However, as with Peirce, Newcomb later asserted that apparently simple terms such as "hard" did not qualify as final anchors in the definitional chains. Specifically, he objected to Stallo's use of "hard" and "elastic" as simple, self-contained, empirical terms. Newcomb advised: "All our conceptions of hardness and elasticity are derived from the qualities of sensible masses.... We call a body hard when we cannot compress it, and elastic when it rebounds on being struck." In this manner, Newcomb's meaning criterion displays the very feature that distinguishes Peirce's "pragmatic" criterion from Wright's traditional, empirical criterion. Whereas all three men require that abstract terms be tied to sensuous consequences, Newcomb concurs with Peirce in further applying the criterion to simpler terms. Peirce states that by "calling a thing *hard*" we mean that it cannot be scratched by other substances; Newcomb, in a parallel formulation but with a different test for hardness (and, therefore, presumably an alternative meaning of the term), states that we "call a body hard" when it cannot be compressed. We should also notice that, unlike Newcomb who applied his criterion of meaning in a variety of intellectual contexts ranging from physics to religion and from political economy to philosophy, Wright applied his criterion only to terms in natural science.[13]

Madden identifies one other important difference in the meaning criteria of Wright and Peirce. Peirce and later pragmatists, most notably Dewey, in maintaining that the meaning of a proposition lies in its sensuous consequences, associated these consequences with physical manipulations. For instance, Peirce variously associated the sensuous consequences with "laboratory," "experimental," or "practical" actions. Dewey likewise stressed manipulative procedures in science. Wright, however, did not display this emphasis on active manipulation in his analysis of meaning, but limited himself to a more passive specification of sensuous consequences.[14] Once again, Newcomb's approach corresponds to Peirce's. To be sure, in his earlier methodological statements, he called generally for defining concepts through sensory experiences, an inclusive criterion that allowed for passive observation and active experimentation. In his methodological advice to Stallo, however, he explicitly called for definitions using actual laboratory manipulations, specifically, concrete measurements. To give an expression meaning, he warned in the passage that anticipated Bridgman's operationalism, we must first define the method in which the constituent terms (such as mass or length) are to be measured; then, "in so many ways as we choose to make this measurement, just so many meanings may the expression have." Newcomb displayed a similar emphasis on measurement in his various technical articles on defining physics terms such as "radiance" and economic terms such as "value."

Having even additional relevance for pragmatism is Newcomb's admonition to Stallo—the admonition just mentioned—that a given expression can have different meanings depending on what measurements are used to define the expression's constitutive terms. This notion dovetails with another version of a pragmatic meaning criterion that Madden further contrasts with Wright's version. In this version, "the pragmatic view is presented as a criterion of synonymity of meaning rather than a criterion of propositional meaning." Thus, James states that "the tangible fact at the root of all our thought-distinctions, however subtle, is that there is no one of them so fine as to consist in anything but a possible difference of practice."[15] James, who was actually building on an idea from Peirce's 1878 paper when he formulated this "quieter of controversy," meant that the meanings of what appear different concepts are actually synonymous if the practical consequences are identical; conversely, what appear identical concepts actually have different meanings if the practical consequences are not the same. Newcomb similarly meant that apparently identical expressions actually have different meanings if the constituent terms are defined using different methods of measurement. This is what he intended when he stated that a particular expression can have many meanings, with the number of meanings being a function of the different methods of measurement that can be used to define the expression's terms. This notion led Newcomb to conclude: "The very first necessity of any exact scientific proposition is a definition, without ambiguity, of a precise method in which every quantitive [sic] measure brought in shall be understood. The conclusions are then valid, assuming that particular method of measurement, but they are not valid on any other method."

Madden, after differentiating the views on meaning, highlights a basic similarity between Wright and later pragmatists, namely Peirce and James. Indeed, the similarity is simply that they all chose to emphasize the meaning of terms and the warrant for specific propositions rather than the origins of scientific knowledge, whether mental or experiential. Part of the general trend during the second half of the nineteenth century away from Baconian inductive reasoning toward hypothetico-deductive reasoning, this emphasis on verification specifically represented a departure from the traditional empiricism of Hume and Mill. It also signaled a demise of the debate over origins of knowledge between idealists of Whewell's camp and sensationists of Mill's. Madden explains: "Wright, more than any other nineteenth-century empiricist, reoriented empiricism from concern over the origin of concepts and hypotheses to concern for their test or verification. Empiricists tried hard to show that every concept and hypothesis was learned through experience.... Wright insisted that the origin of a concept or hypothesis was irrelevant to the empiricist's position." According to Madden, it did not matter to Wright whether concepts originate "through imagination, dreams, hallucinations or intuitions"; it mattered only that "the hypotheses in which they occur must yield consequences which can be checked or tested in experience." Dewey later identified this viewpoint with James, finding in his writings the insight that "validity is not a matter of origin nor of antecedents, but of consequents."[16]

As early as his Cambridge years, when he intentionally sidestepped the debate between Whewell and Mill by concentrating on the "necessity" rather than the "source" of scientific propositions, Newcomb subscribed to this view. Specifically, in emphasizing the role of the senses in science, he did not deny that the mind might also contribute to the origin of scientific knowledge; he merely chose to suspend judgment on this difficult issue of origins and concentrate solely on the more tractable question of the meaningfulness of language. "Whatever theory we may adopt of the relative part played by the knowing subject, and the external object in the acquirement of knowledge," he wrote in 1880, "it remains none the less true that no knowledge of the meaning of a word can be acquired except through the senses, and that the meaning is, therefore, limited by the senses." Adherence to this general view, of course, did not preclude taking an epistemic stand on the origins of particular types of knowledge. In advancing his program for teaching mathematics, Newcomb sided with Mill in saying that mathematical knowledge is acquired through the senses.

Moving from the realm of meaning to truth, we find that only a few of Newcomb's published writings pertain directly to this second philosophical cornerstone of pragmatism. As a practicing physical scientist, Newcomb's primary interest was in what he perceived to be the actual method used by successful scientists. In making the linguistic, empirical method the focal point of his perspective on science, he accentuated the issue of meaning while somewhat disregarding the issue of truth.

Again, Madden's distinctions regarding Wright and later pragmatists are helpful. As with the other components of pragmatism, the pragmatic notion of truth varies with the individual pragmatist. In general, Madden reports: "Some pragmatic theories of truth amount to definitions of 'true'—i.e., 'true' *means* 'the useful,' or 'the workable,' etc.—while other pragmatic theories of truth simply offer the useful, the workable, etc. as criteria or tests of what is true." While Wright hinted at the latter pragmatic formulation in one brief mention, he held a more traditional view of truth. He endorsed what in effect is a correspondence theory of truth, saying that the truth of a thought consists in its agreement with the particulars of experience.[17] Neither did Peirce explicitly envision truth in the forms James and Dewey would later frame; in his 1878 paper, he postulated that truth is what results from the inevitable convergence over time of seemingly disparate scientific investigations within the community of inquiry. "The opinion which is fated to be ultimately agreed to by all who investigate, is what we mean by the truth, and the object represented in this opinion is the real."[18]

Newcomb's notion of truth reflects his general conviction, traceable to Comte, Mill, and others, that the role of the scientist is not to provide ultimate, metaphysical explanations but to establish correlations and connections between phenomena. As such, his notion of truth had aspects in common with those held by Wright and Peirce. Like Wright, he viewed truth in terms of a correspondence between a proposition and the particulars of experience. For Newcomb, to say that a hypothesis is true is simply to say that it agrees with experiential evidence; as he expressed it around 1860 in his physics textbook, when a scientist obtains agreement between hypothesis and evidence, "there will be a high probability, or a reasonable certainty, of the truth of the hypothesis." But like Peirce, he further viewed truth as emerging over time through consensus within the community of investigators. "A countless host of theories have thus been demolished and forgotten with the advance of knowledge," he wrote in his AAAS address, "but those which remain, having stood the fire of generations, can show us a guarantee of their truthfulness which would not be possible under any other plan of dealing with them." In his commentary on psychical research, he similarly argued that the history of science shows that true laws are those that enable scientists to replicate results with "continually increasing facility."

Unlike Peirce, however, Newcomb did not believe that this movement toward truthfulness was an inevitable process acting in some sense independently of individual humans; neither did he hold that the process led to an ultimate reality. Newcomb thus added in his AAAS address that the truth of a scientific proposition is not arrived at through "revelation" but through "human judgment" based on a careful weighing of all pertinent evidence.[19] This emphasis on the temporality of truth is the closest that Newcomb comes to the pragmatic attitude toward truth. That is, like Wright and Peirce, Newcomb does not identify truth explicitly with the useful or the workable; but he does stress its contingency, its dependence on human judgment. This particular emphasis, according to Hollinger, is integral to the pragmatic approach to truth; that is, "whatever else the pragmatist theory of truth entailed, it carried with it the sense that truth was a condition that happened to an idea through the course of events as experienced and analyzed by human beings."[20]

NEWCOMB AS A CATALYST

Our examination of the historical triad of assertions and hopes along with our analysis of the two components of the philosophical core suggest that Newcomb had much in common with the founders of American pragmatism. To be sure, he was not literally a pragmatist in the mode of Peirce, James, or Dewey—persons who after 1898 self-consciously but variously embraced something called "pragmatism" or "pragmaticism." Neither was he a pragmatist in the strict sense of an individual committed to a distinctive package of philosophical precepts. Nevertheless, he had many of the same cultural aspirations and philosophical predilections as Peirce, James, Dewey, and their precursor, Wright. Recall that throughout his writings and speeches, he fully displayed two key elements of Hollinger's distinctively pragmatic cluster of assertions and hopes. He believed adamantly in the primacy of scientific method in a world where human knowledge was fallible and temporal. Equally heartfelt was his conviction that the scientific method of inquiry was accessible to rank-and-file members of the broader, democratic society. To a degree, he also manifested the remaining element of the threefold cluster. That is, he shared with the pragmatists the belief that scientific method empowered persons to reconstruct areas of human discourse such as politics, economics, religion, and philosophy.

Newcomb also advocated central components of the technical, philosophic core of pragmatic belief as summarized by Madden. In his view on meaning, Newcomb was in step with Peirce and the pragmatists who extended Wright's related but narrower view. Newcomb's meaning criterion, among other characteristics, had domain both in and out of the framework of natural science and over both abstract and simple terms. In his view on truth, he concurred with pragmatists regarding its temporality; but like Wright and Peirce, he never expressly defined truth as later pragmatists would in terms of what is useful or workable.

In light of the cultural aspirations that Hollinger distinguishes and the philosophical predilections that Madden pinpoints, what is Newcomb's significance for American pragmatism? In short, Newcomb contributed to the formulation around the 1870s of these pragmatic aspirations and predilections as well as to their transmission up to and following the time of their broader dissemination and fuller development beginning around 1898. The latter occurred with William James's rediscovery of Peirce's original 1878 essay and the subsequent popularization and elaboration of various pragmatic attitudes and positions by James, Peirce himself, John Dewey, and others. Stated differently, we must reevaluate James's contention that Peirce's 1878 pragmatic principle "lay entirely unnoticed by any one for twenty years" until about 1898 when "the times seemed ripe for its reception."[21] Regarding the first half of this contention, there exists indirect evidence that Newcomb had "noticed" Peirce's principle. More important—in fact, much more important—there exists direct evidence that Newcomb was operating in the same cultural and intellectual milieu as Peirce. This was the milieu fostered by Wright and influenced by Europeans such as Comte, Darwin, and Mill and by working American scientists such as Henry. From the teachings and practices of these men, and perhaps Peirce, Newcomb was able to synthesize certain aspirations regarding scientific method and certain predilections regarding questions of meaning and truth—"pragmatic" aspirations and predilections germane to diverse domains of human experience. Regarding the second half of James's contention, part of the reason for the time seeming "ripe" around 1898 for pragmatism was that Newcomb had been helping—unknowingly, of course—to prepare the way during the prior "twenty years." In the dual process of seeking to strengthen American science and to enhance the public well-being, Newcomb crusaded throughout his career for beliefs and positions that equate with the cultural and philosophical attributes of pragmatism. That is, through his multifaceted commentaries on science in American culture, he helped catalyze the diffuse movement later labeled pragmatism.[22]

James's reconstruction of the early history of pragmatism, with its twenty-year gap, is understandable given his concentration on Peirce and, at times, Wright. He focused on them for obvious reasons. Not only was Wright his old mentor and Peirce a friend, but both men were brilliant philosophical thinkers, much more subtle and informed in their philosophical thought than other late nineteenth-century American commentators on science, including Newcomb.[23] However, as their colleagues realized and historians have repeatedly remarked, Peirce and Wright were both relatively obscure scholars during their lifetimes. Also, their writings that foreshadowed pragmatism were either few, as with Peirce, or abstruse, as with Wright. When we include Newcomb in the history of pragmatism, James's twenty-year gap diminishes. In contrast with Peirce and Wright, Newcomb had become by the late 1870s one of the most prominent and influential natural scientists in America, if not the world; he also had developed into a respected political economist and national spokesman on topics ranging from religion to psychical research. Moreover, as opposed to Peirce and Wright, Newcomb repeatedly and lucidly emphasized the "scientific philosophy" in his speeches before important groups like the AAAS and in his articles in widely read publications such as the *North American Review*. Periodically, he actually riveted the educated public's attention through spirited exchanges with the nation's intellectual elite: Ely and Edmund James in economics; Gray, Porter, and McCosh in religion; Stallo regarding the foundations of physics; and William James in psychical research. Put most simply, Simon Newcomb played a key role during the late nineteenth century in formulating and disseminating among Americans a pragmatic perspective toward nature, human beings, and society.

CHAPTER XIII

Pragmatism and Methodological Rhetoric

The intellectual and cultural world of the American pragmatists provides a telling historical context in which to place Newcomb's pronouncements on method; each sheds light on the other. Beyond the customary reaches of the history of philosophy and general American history, however, there exists an alternative historical context in which we can locate both Newcomb and the early pragmatists. This is the context of the history of science, specifically the politics and rhetoric of scientific method. Just as Newcomb both reflected and fostered key elements of the cluster of pragmatic assertions and hopes, so too did the pragmatists and allied thinkers such as Newcomb reflect and foster key elements of the rhetorical and political tradition.

Recall that John Schuster and Richard Yeo, in delineating this tradition, distinguish three levels of interaction on which representatives of natural science use claims about scientific method: the internal level where method serves in forming, rationalizing, interpreting, promoting, or discrediting the technical content of science; the disciplinary level where method figures into questions of institutional organization and control; and the public level where methodological claims are directed outward, usually to legitimate a particular image of science, to demarcate intellectual territories, or to proselytize nonscientists. Recall also that Newcomb throughout his career operated principally on the public level. His frequent forays into political economics, religion, philosophy, and the foundations of physics served variously to enhance science's image and to demarcate its borders, thus pressing American scientists' claims to expanded public support and cognitive authority. Even when he occasionally moved to the disciplinary and internal levels, his pronouncements still promoted public ends. In particular, his efforts to discredit "scientists" having suspect political-economic and religious credentials as well as his attempts to pan current mental and psychical research all fostered the goal of protecting the popular image of science and expanding the public role of true scientists. Of course, in his multifarious public initiatives, he often mixed collective and personal agendas. As we saw earlier, he blended two personas: the group spokesman garnering fuller national support for the overall scientific enterprise and the individual citizen-scientist seeking to improve society at large by pushing forward a personal ideology. Though overlap exists, we need to follow Schuster and Yeo's lead and, for the moment, restrict ourselves to applications of method serving the scientific collective in its manifold aspects.

Taking this tack, we find Newcomb tracing a pattern in his uses of scientific method that parallels the pattern displayed by Wright, Peirce, James, Dewey, and their supporters. In their pronouncements on method, these pragmatists often spoke for the overall scientific enterprise; and when they did, they operated especially on the public level and only to a limited extent on the disciplinary and internal levels. Though personal agendas once again spilled into agendas for the scientific collective, the pragmatists stand with Newcomb as primarily a subspecies of that segment of Schuster and Yeo's apostles and apologists having public aims. That is, Newcomb as well as Wright, Peirce, James, and Dewey directed most of their methodological statements not toward disciplinary and technical issues but toward the broader social and cultural relations of science, variously trying to legitimate a distinctive image of science, demarcate its boundaries, and proselytize the lay public. "The recognition that the intellectuals who rallied to pragmatism were preoccupied with the place of science in modern life," Hollinger comments more generally, "is indeed the point at which to begin an assessment of pragmatism's role in the lives of Americans who cared about it."[1]

Not only did the pragmatists match these public apologists in directing their pronouncements on method mainly outward, but both groups shared similar sets of basic assumptions about scientific method. As we will see, the assumptions endorsed by the apologists correlate with the triad of assertions and hopes endorsed by the pragmatists. They correspond, that is, if we continue as in the last chapter to qualify or restrict the second element of the pragmatic triad, the one involving method's transformative or reconstructive capabilities. Thus, both groups believed in and urged others to accept the primacy of scientific method in a world where knowledge is fallible, the power of inquiry to reconstruct areas of human discourse such as politics (not to alter the fundamental structure of the world), and the accessibility of the scientific method of inquiry to members of the broader society.

As mentioned, Hollinger readily acknowledges that all three elements of his cluster individually have precedents or nonpragmatic manifestations. He even lists science enthusiasts who displayed isolated components of the cluster. He maintains, however, that the full combination of the three elements, and the combination's emergence within the American context, give pragmatism its essential quality—its distinctiveness and significance in the national culture. He writes: "The particular elements in this cluster were often articulated singly and in relation to other ideas by other moralists of the period, including some critical of pragmatism, but the combination of elements found in the writings of the pragmatists and their popularizers was nowhere else advanced more persistently and with more notice from educated Americans."[2]

Granting the accuracy of this judgment regarding the collective force of the three elements, it remains that two complete elements of the triad—and the remaining element when delimited—do not merely have separate precedents or unconnected histories. Rather, as we will see, the three are conjugate components of a long-standing tradition, especially as manifested in Britain, of using scientific method as a rhetorical resource in the social arena. Newcomb, Wright, Peirce, James, and Dewey were part of that broader tradition, even though their personal and national circumstances conditioned their responses to it. To the extent that they were part of it, American pragmatism can be viewed as a variant of the rhetorical tradition. This leaves as the distinguishing feature of pragmatism—the distinguishing feature, that is, from the perspective of cultural rather than exclusively philosophical developments—the triad's second element in its full form: the belief not only that scientific method empowers persons to reconstruct areas of human discourse but also that the purposive action of inquiry can fundamentally change the world on personal, social, and physical planes. But recall that this element in its unrestricted form manifested itself only after the period around 1900 and only in the often disparate writings of James, Dewey, and their followers, not in the earlier writings of Wright, Peirce, and, of course, Newcomb. We must cede to other scholars the problem of deciding if the notion of purposive change is an essential cultural attribute of pragmatism. Our present goal is to show that we can further our understanding of American pragmatism, including Newcomb's contribution, by viewing it within the historical context of the politics and rhetoric of scientific method.

PRAGMATISTS AS APOSTLES OF SCIENTIFIC METHOD

The relationship between the beliefs of the pragmatists and the apologists becomes clearer if we examine Yeo's elaboration of the third, or public, level of method claims. In the same book that he and Schuster edited and introduced, but in a later, separate chapter, Yeo details a case involving the social relations of science that is especially germane to our analysis of late nineteenth-century America. Specifically, he concentrates on Great Britain from 1830 to 1917, recounting how "certain assumptions and statements" about the method as distinguished from the content of science were used strategically in debates about the character of the scientific enterprise. In general, he finds that

> there were three assumptions about method which were conditioned by, and served as rhetorical resources in, the social relations of science. At various levels of debate, scientific method was represented as *accessible, single* and *transferable*. These three characterizations respectively claimed that the method of science could be understood and practised by a large number of people; that there was a single method common to all branches of science; and that this method could be extrapolated from natural science to other subjects. Statements deriving from these assumptions were designed to promote the cause of science by presenting it as founded upon a well-defined and successful method, accepted by all practising researchers, accessible to laymen, and capable of being extended beyond the study of nature to the study of society.[3]

Yeo's three "assumptions and statements" do not map in a one-to-one manner onto Hollinger's triad of "assertions and hopes." But, as will be explained in a moment, various combinations of the three assumptions collectively entail each element of the triad—the triad involving method's primacy, ability to reconstruct human discourse, and public accessibility.

That the assumptions held by the apologists incorporate those held by the pragmatists should come as no surprise if we realize that some of the main British apologists for scientific method number among the main influences on Wright and Newcomb as well as Peirce, James, and Dewey. As we saw with Newcomb, Mill is perhaps the chief transatlantic link. James, for example, dedicated a 1907 book that highlights pragmatism's connections, *Pragmatism: A New Name for Some Old Ways of Thinking*, to the memory of Mill, "from whom I first learned the pragmatic openness of mind and whom my fancy likes to picture as our leader were he alive to-day."[4] Of course, the pragmatists also embraced certain views of an earlier British commentator who formerly had dominated methodological landscapes in the United States, Bacon, whose canon had been conveyed to a prior generation of Americans by the Scottish Realists.

The first element of the pragmatic triad, the belief in the primacy of scientific method rather than the acquisition of knowledge, shares aspects with the apologists' notions of the "unity" and "transferability" of method. Spokesmen for science in nineteenth-century Britain, distinguishing between the content and method of science, believed that the progress of all scientific inquiry stemmed from the application of a common method and that this single method could be successfully extended beyond natural science. In other words, the notions of singularity and transferability in method translate into the pragmatists' emphasis on a universal method taking precedence over particular bodies of knowledge. Whereas the ideas of singularity and transferability have long and complex histories, Bacon had the largest influence, at least among the British, in establishing them. During the Enlightenment, these Baconian beliefs flourished, setting the stage for early nineteenth-century commentators on method such as John Herschel, Whewell, and Mill. Herschel maintained, for example, that method was the essential, defining feature of science and that scientists, while often in disagreement over particular theories, were in consensus on method. Method's success in science led Herschel to advocate its extension to social and political realms, a suggestion seconded by Mill and even Whewell (a critic of many Baconian ideas). For his part, Mill felt that progress in the "moral and social sciences" would be achieved not through "the truths" that physical science reveals but through "the process by which it attains them." "These were early expressions of the claim to intellectual and cultural authority, beyond the realm of natural knowledge," Yeo explains, "which later become a major element in the rhetoric of public science."[5] Indeed, these calls for extending the method or "process" of science to wider realms of reasoning were expressions of method's primacy. Though more than years and an ocean separated Americans such as Peirce from Herschel and Mill, they at least would have agreed that method is primary and that shared methodological commitments allow members of the scientific community to correct one another's ideas.

Having a history nearly as long as the primacy of method is the concomitant conviction that human knowledge is fallible and temporal, although knowledge derived through science is the most reliable. As the nineteenth century progressed, most scientists relinquished the aim of certain knowledge and resigned themselves to "a more modest program of producing theories that were plausible, probable, or well tested."[6] Indeed, the roots of the displacement of certain knowledge by probable knowledge extend to experimental natural philosophers of the seventeenth century. While granting that their knowledge of the empirical world was only probable, these natural philosophers still felt, Yeo reminds us, that "this form of knowledge was far more secure than alternative sources of information about nature because it derived from the application of rigorous, publicly repeatable, procedures which gave a *moral* certainty to its discoveries." In fact, the very definition of science came to reflect this belief that knowledge derived through set procedures was morally preferable to other types of knowledge of nature. In particular, science's definition became joined to notions about the efficacy of a single scientific method. Additionally, assertions about method entered the justifications and defenses used to promote the value of science in political, social, and cultural arenas.[7]

This linkage to political, social, and cultural agendas leads us to the second element of the pragmatic cluster: belief in method's power to effect change in various realms of human discourse. In explicitly endorsing Bacon's call for the extension of scientific method to the study of society and politics, British writers such as Herschel were implicitly endorsing the belief that human inquiry could transform these areas. Even antagonists Whewell and Mill concurred on method's relevance to political and moral concerns. Though divided on epistemic issues, they agreed that human progress depended on the reconstruction of areas such as political economy and moral philosophy through the application of scientific method. Later in the nineteenth century, Huxley, Spencer, and other British representatives of the natural sciences elaborated the same theme.[8]

Finally, the pragmatists' notion of method's accessibility to the public not only parallels the apologists' idea of "accessibility" to lay people but also includes aspects of their idea of method's "transferability" from natural science to other subjects. That is, the pragmatists believed that the scientific mode of inquiry was open in general to educated citizens and usable in particular by persons confronting issues in the social sciences, politics, and morals. Once again, Bacon and his followers fostered the image that the scientific method of investigating nature—specifically, the inductive method—was accessible to a wide segment of the public. The image of accessibility persisted into early nineteenth-century Britain. Although he reserved a niche for scientific specialists, Herschel generally held a democratic or egalitarian vision of natural science, maintaining that the lay public could participate in scientific inquiry. Other early nineteenth-century writers reinforced this view by emphasizing the features that method shared with "common sense." Still others such as Mill argued that study of the method rather than the content of science be made part of students' liberal education.[9]

To recapitulate, just as Newcomb reflected and contributed to the pragmatic tradition, so too did the pragmatists and allied thinkers such as Newcomb reflect and contribute to the rhetorical tradition. In other words, Newcomb and the pragmatists were part of a larger cultural and intellectual movement that manifested itself through a well-established and ongoing exchange, both written and spoken, on method. Specifically, the pragmatists' belief in the primacy of method, its use in reconstructing areas of human discourse, and its public accessibility echoed the nineteenth-century British rhetoricians' belief in the accessibility, unity, and transferability of an efficacious method.

PRAGMATISM'S EMERGENCE IN THE LATE NINETEENTH CENTURY

While these conclusions enable us to locate Newcomb and the pragmatists in a broader historical context, we are left with a basic question: Why did this methodological discourse in general and pragmatism in particular take hold among American scientists and their representatives with such intensity in the late nineteenth century? Why did this chiefly public methodological rhetoric emerge so compellingly only then, especially after having existed in Britain for decades?

Cultural lag, the delay between the inception and possible assimilation of a trend in different communities, does not provide a full answer. As we saw earlier, Americans lagged but little in their endorsement of Baconian method as mediated by Thomas Reid and Dugald Stewart. Well versed in Baconian inductivism from at least early in the nineteenth century, they quickly appropriated Baconian rhetoric and exploited Baconian assumptions. In fact, a review of Baconian thought among Presbyterians from 1820 to 1860 indicates that many influential Americans shared their British cousins' belief in the accessibility, unity, and transferability of method. Also, a glance back at American scientists from 1815 to 1845 discloses that many of them were using at least the language of the method as they purported to adapt the Baconian model to areas of active research.[10] Even young Newcomb, around 1860 while still in Cambridge, fashioned a textbook statement on method having a decided Baconian hue. Also, as early as the decade following the Civil War, he revealed a commitment to the accessibility, singularity, and transferability of method when he sought to apply method to pressing economic issues.

Though American scientists displayed Baconian trappings during the century's first half, they did not enter public arenas and advance a methodological campaign for the scientific collective with the persistence and polemical fanfare of their British contemporaries or their American descendants. After all, unlike their British colleagues—or Americans in later years—these early nineteenth-century Americans possessed only the most rudimentary institutional and professional props. Similarly, they possessed only the most tentative perceptions of themselves as a cohesive national community. Bruce reminds us it was not until the third quarter of the century that science "came to see itself, and society came to see it, as an established profession." Merely beginning to experience the advantages of fuller support, to acquire consequential platforms for swaying public opinion, and to develop a sense of professional cohesion, these scientists of the first half century employed a methodological rhetoric lacking the punch and focus of their British contemporaries and their own American descendants. Indeed, the most avid and influential apologist for Baconian method during the period was Samuel Tyler, not a scientist but a Maryland lawyer and philosopher who sought primarily to reconcile Baconianism with Protestantism.[11] In this era of nascent professionalism, the rhetoric of method mainly provided individual citizen-scientists with a way to rationalize their specialized practices and to legitimate personal political and religious ideologies. To be sure, a flourishing methodological tradition among individual American scientists during these early years helps explain the pervasiveness of personal commitments among later nineteenth-century citizen-scientists to the power of *the* scientific method and to the ease with which it could be grasped and applied. The individualistic temper of the tradition also helps explain the dearth of early initiatives in the name of American science as a whole. We still lack, however, a full answer to our original question: Why did American scientists in the late nineteenth century, and those who spoke on their behalf, carry the rhetoric into the public arena with such regularity and fanfare?

A more complete answer to the question of timing arises if—while continuing to restrict ourselves to pronouncements having significance for the collective agenda of the scientific community rather than merely for the personal agendas of separate citizen-scientists—we return to Newcomb. Specifically, Newcomb's perceptions of his institutional and conceptual milieu help us understand why the rhetoric of scientific method emerged with such consistency and vigor in late nineteenth-century America, even though the roots of the rhetoric extend to the Baconianism of the first half century. In addition, his perceptions help us understand why the rhetoric was deployed primarily on the public level and only secondarily on the disciplinary and internal levels.[12]

As we have seen, Newcomb affirmed a definite image of his institutional and conceptual environment. On the one hand, he felt that he and his American scientific colleagues, especially in the physical sciences, were burdened by an institutional framework only beginning to take shape. Repeatedly, he complained about inadequate educational programs, technical journals, and professional societies—deficiencies that he attributed ultimately to a general lack of public support for basic research. On the other hand, he felt that he and his colleagues, again particularly in the physical sciences, were favored by secure conceptual foundations—secure in the sense that members of the scientific community generally agreed that all physical phenomena could be represented in terms of the tested formalisms of classical mechanics. Specifically, Newcomb assumed that Newton's theory of motion, as elaborated in the eighteenth and nineteenth centuries, could be applied with equal advantage to celestial orbits and to the hypothetical world of atoms and ethers—a world that lay behind phenomena such as heat, light, electricity and magnetism, and the behavior of gases. In astronomy, he even suggested that researchers had left the era of fundamental conceptual breakthroughs and entered a calmer period of elaborating or systematizing basic Newtonian research programs.[13] And though not professionally involved in the life sciences, he was similarly sanguine about the Darwinian theory of evolution.

Newcomb's sense that the conceptual foundation was secure while the institutional framework was insecure was well grounded in actual circumstances. His view, which he articulated so fully in the 1870s and 1880s, derived from his own experiences in especially physical science beginning in the 1850s and extending through the Civil War. Although classical, mechanical outlooks were neither monolithic nor unchallenged in the United States during the last half of the nineteenth century, they did prevail. The Americans' loyalty to mechanical research programs and outlooks mimicked a similar commitment by many of the mid-century European leaders of physical science.[14] In addition, especially in the physical sciences, American institutional structures were still relatively weak and ineffective during the last half of the nineteenth century; they were, however, in the process of strengthening and expanding, having already advanced well beyond the first half century's glimmerings of professionalism and specialization. Nathan Reingold, in fact, interprets the recurrent laments voiced by Newcomb and other American scientists near the centennial year of 1876—a period of national self-appraisal—as signs of a vigorous and healthy scientific community that was experiencing "growing pains" as "support for the expanding sciences probably tended to lag behind needs." Reviewing the same group of centennial laments—and responses such as Newcomb received from Benjamin Silliman, Jr., John LeConte, and James Dana—Bruce draws a similar conclusion. American science fit the image of "a newborn foal, callow, ungainly, yet fully formed and swiftly gaining strength. It still had a long way to go. But it had some definite ideas of how to get there."[15] To say that Newcomb's disparaging view of the institutional framework was well grounded in current conditions thus does not mean that he simply was reporting evident deficiencies in schools, journals, societies, and public support, but that he also sensed the potential for further improvement.

Why then did the rhetoric of scientific method emerge among American scientists and their representatives with such consistency and conviction in the late nineteenth century? And why did the Americans deploy the rhetoric primarily on the public level and only secondarily on the disciplinary and internal levels? We approach answers to both questions by observing that Newcomb's pronouncements on method came at the moment in the history of American science when, on the one hand, institutional supports for the pursuit of natural science seemed deficient but promising while, on the other hand, the conceptual props of natural science seemed reliable and compellingly ready to carry additional loads. Eager to strengthen the institutional underpinnings of American scientists so that they could take fuller advantage of the opportunities offered by the established and robust research programs of the day, Newcomb sought to construct and diffuse in the democratically inclined nation a favorable image of science, thus increasing the level of public support for scientists. He did this by attempting to convince politicians, economists, educators, and other presumably indifferent but influential Americans of the public value of an essential attribute of modern science, its method. Also in an attempt to consolidate institutional gains, encourage the further professionalization of the scientific community, and promote the distinctiveness of science's social role, he drew on method to demarcate the boundaries of science. When it was advantageous, he variously distinguished science from and associated it with philosophical, theological, commercial, and other nonscientific realms of culture. In other words, Newcomb's perceptions of his conceptual and institutional environment led him to operate primarily on the public level of claims about scientific method. Even his occasional excursions onto the disciplinary and internal levels usually involved only less established branches of science and furthered his basic goal: enhancing science's popular image, fixing its professional boundaries, and, ultimately, legitimating the claims to wider cognitive authority and, thus, fuller public support of those scientists having proper credentials. Indeed, Newcomb saw little to criticize in the recognized research programs of natural scientists, but much to fault in intellectually unsettled fields such as political economy and psychical research—fields themselves in the first throes of possible professionalization.

Yeo makes similar points in focusing on nineteenth-century British history and the methodological initiatives taken for science as a collective enterprise. He judges that in Great Britain the deployment of the rhetoric of scientific method on the public level correlated with the institutional status of science. Examining the mid-nineteenth-century shift in allegiance from the inductive method (with its rules of empirical discovery as enunciated especially by Bacon) to the hypothetico-deductive method (with its criteria for the verification of hypotheses as pioneered by Galileo and others), Yeo relates the shift to "changes in the social conditions under which science was pursued." Specifically, commitment to the Baconian method correlated with the period of relative institutional weakness early in the nineteenth century while commitment to the hypothetico-deductive method correlated with the period of relative institutional strength later in the century. That is, as the scientific community consolidated in mid-century Britain, the need diminished for such traditional rhetorical resources of Baconianism as invoking the possibility of lay participation and stressing the cultural relevance of science. The earlier deference to Baconianism had sprung, in part, from "the need to advertise the sober, empirical character of scientific inquiry and to distance it from speculation and hypotheses which had controversial political or theological overtones." In contrast, the mid-century rise of the hypothetico-deductive method reflected a new generation's "comparative independence from religious control, and the greater confidence of a more professionalized community." With the coming of institutional security, practicing scientists and their representatives no longer needed to placate lay people by including them in their discourse through claims of an accessible inductive method; instead, they could mandate a methodology that was solely the domain of specialists skilled in verifying often abstruse hypotheses. Method, here the hypothetico-deductive method, took on a new rhetorical role in the social relations of British science—justifying the existence of "the trained scientific expert pursuing an exciting intellectual quest which lay beyond the reach of the public." Though apologists now envisioned method being restricted to an elite, they continued to portray it as transferable to social and political spheres and, thus, as still serving a public function.[16]

Newcomb, of course, was formally operating within the hypothetico-deductive framework when he enunciated his linguistic, empirical method; that is, like his British colleagues in the second half of the nineteenth century, he had moved beyond Baconian inductivism and was emphasizing the verification of hypotheses whatever their source. But significantly, he was using his version of the hypothetico-deductive method under social conditions and for ulterior purposes similar to those associated with the Baconian method in early nineteenth-century Britain. In particular, Newcomb's pronouncements on method came at a time in American scientific development of institutional deficiency but promise. The statements served important purposes: to build bridges between the scientists and the lay public and between science and other areas of culture, while at the same time maintaining the scientists' distance from theological, moralistic, psychical, and other potentially compromising forms of inquiry. That is, the pronouncements aided the campaign for enlarged scientific authority in American intellectual and cultural life and expanded public support for scientific research and education.

Newcomb's coupling of a variant of the hypothetico-deductive method to these public uses seems to contradict Yeo's analysis of British patterns. After all, in Great Britain, the rhetoric of public accessibility flourished in conjunction with Baconian, not hypothetico-deductive, methodology; furthermore, the demise of both this rhetoric and the Baconian methodology occurred in conjunction with the rise of the hypothetico-deductive method as British science became more securely institutionalized. To understand Newcomb's deviation from the British pattern, we need to draw back from Yeo's specific case study and return to his and Schuster's joint, more general commentary on the character of methodological rhetoric. From this broader perspective we see that, while Newcomb's particular posture stands in contrast to the posture of later nineteenth-century British scientists, the disparity actually reinforces a more basic point: that scientific method, being extremely pliable, is capable of promoting a broad compass of rhetorical and political ends. Specifically, the disparity illustrates the extent to which similar sets of methodological rules can be put to different strategic uses under different circumstances or—conversely—the extent to which different sets of methodological precepts can be put to similar strategic uses under similar circumstances. In Britain, scientists and their spokesmen used Baconian method as an argumentative resource to strengthen the public image of science during the early nineteenth-century period of relative institutional weakness; the Baconianism gave way to the hypothetico-deductive method in the subsequent period of increasing institutional security. In contrast, in the United States during the late nineteenth-century period of institutional weakness but potentiality, Newcomb's type of hypothetico-deductive method provided the resource to strengthen the public image of science. Said differently, the institutional status of science is not inevitably linked to the rhetorical use of a *particular* method (or, for that matter, any rhetorical use of method at all). In the late nineteenth-century American case, we merely need to recognize a tie between institutional status and a *particular* rhetorical use of a method currently preferred.

Ultimately then, Newcomb's perceptions of the institutional and conceptual environments within the United States from the middle through the close of the nineteenth century enable us to understand why he discoursed on scientific method with such conviction during this period. His perceptions also help us grasp why he saved the rhetoric primarily for the public level. To the degree that Newcomb reflected and contributed to a pragmatic perspective in the United States, we can generalize from his case, thus gaining insight regarding the timing of the perspective's appearance. The rise of pragmatism, so distinctive of the late nineteenth century, correlated with the common notions among scientists and their representatives that the institutional framework of natural science within the United States was deficient, but promising, whereas the conceptual foundation was secure. Wright and Peirce shared these notions with Newcomb, their old Cambridge colleague; Wright expressed the views throughout his mature years, while Peirce did so into at least his middle years. James and Dewey—though active not in the natural sciences but in the even more institutionally deficient social, psychological, and pedagogic sciences—were weaned at Harvard and Johns Hopkins on similar ideas.

The extent to which the insights regarding Newcomb apply to Wright, Peirce, James, and Dewey remains a question for future historians to settle. The question is complex in that the four men paralleled Newcomb in mixing statements serving the scientific collective with statements mirroring their own personal agendas—that is, statements in which they invoked method to retail individual opinions on social, cultural, and intellectual issues. Looking at their pronouncements involving science as a collective enterprise, however, we can conclude that pragmatism's emergence reflected scientists' impressions of their institutional and conceptual settings and their associated public deployment of the rhetoric of method. Indeed, viewed from the perspective of the history of science rather than the customary outlook of either the history of philosophy or general American history, pragmatism *was* the deployment of the rhetoric of scientific method on the public level. Although in their campaigns for science Wright, Peirce, James, and Dewey recited their own variations of the litany, they shared Newcomb's zeal as they heralded method's accessibility, unity, and transferability. And though it was James and Dewey who helped formalize the pragmatic gospel, it was Newcomb, the vocal and visible preacher of scientific method, who helped frame and promulgate pragmatism's underlying litany.

Notes

PREFACE

1. Frederick E. Brasch, "Einstein's Appreciation of Simon Newcomb," *Science* 69 (1929): 248–249.
2. Robert S. Woodward, "Address of Mr. R. S. Woodward," in "Simon Newcomb: Memorial Addresses," *Bulletin of the Philosophical Society of Washington* 15 (1910): 148.

CHAPTER 1: METHOD, RHETORIC, AND NEWCOMB

1. My survey of scientific methodology in this and the next three paragraphs derives especially from Larry Laudan, *Science and Values* (Berkeley, Los Angeles, London: University of California Press, 1984), 1–140. For a succinct summary of the same topic, see "Methodology," in *Dictionary of the History of Science*, ed. W. F. Bynum, E. J. Browne, and Roy Porter (Princeton, N.J.: Princeton University Press, 1981), 267.
2. Helge Kragh, *An Introduction to the Historiography of Science* (Cambridge: Cambridge University Press, 1987), 39.
3. Larry Laudan, "Some Meta-Methodological Preliminaries," paper presented at the Workshop on Scientific Methodology, April 1984, Blacksburg, Virginia, pp. 3–4.
4. John A. Schuster and Richard R. Yeo, eds., *The Politics and Rhetoric of Scientific Method: Historical Studies*, Australasian Studies in History and Philosophy of Science, vol. 4 (Dordrecht, Holland: D. Reidel Publishing Co., 1986), x–xi, xiv. For a brief, general review of recent analyses of "scientific discourse as a rhetorical construction," see Jan Golinski, "The Theory of Practice and the Practice of Theory: Sociological Approaches in the History of Science," *Isis* 81 (1990): 494–500.

5. Schuster and Yeo, *Politics and Rhetoric*, xi, xiii. Richard Yeo, "Scientific Method and the Rhetoric of Science in Britain, 1830–1917," in *Politics and Rhetoric*, 261–263.

6. Paul Forman, "Independence, Not Transcendence, for the Historian of Science," *Isis* 82 (1991): 73–77.

7. Frank M. Turner, "Public Science in Britain, 1880–1919," *Isis* 71 (1980), 589–590. Thomas F. Gieryn, George M. Bevins, and Stephen C. Zehr, "Professionalization of American Scientists: Public Science in the Creation/Evolution Trials," *American Sociological Review* 50 (1985): 392–409.

8. Schuster and Yeo, *Politics and Rhetoric*, xi–xii. For an overview of recent scholarship in rhetoric, see Herbert W. Simons, "Introduction: The Rhetoric of Inquiry as an Intellectual Movement," in *The Rhetorical Turn: Invention and Persuasion in the Conduct of Inquiry*, ed. Simons (Chicago: University of Chicago Press, 1990), 1–31. In this same book, for a useful discussion of the "distinction between strong and weak versions of the idea that rhetorical practices are an integral part of science," see Robert E. Sanders, "Discursive Constraints on the Acceptance and Rejection of Knowledge Claims: The Conversation about Conversation," 145–148; for a useful distinction between "discursive practices, both internal and external" in science, see Dilip P. Gaonkar, "Rhetoric and Its Double: Reflections on the Rhetorical Turn in the Human Sciences," 552–553.

9. Ironically, in championing empirically literal and presumably nonrhetorical uses of language over more customary figurative uses, Newcomb was unknowingly using the *rhetoric* of scientific method to discredit rhetoric as traditionally conceived (i.e., "mere rhetoric"). As John Christie points out in discussing the shaping of language in early modern science and philosophy, it was through rhetorical persuasion in the seventeenth and eighteenth centuries that the sharp distinction between "literal" and "figural" discourse was itself "built into modern philosophical and scientific sensibility." Christie, "Introduction: Rhetoric and Writing in Early Modern Philosophy and Science," in *The Figural and the Literal: Problems of Language in the History of Science and Philosophy, 1630–1800*, ed. Andrew E. Benjamin, Geoffrey N. Cantor, and John R. R. Christie (Manchester: Manchester University Press, 1987), 1–9. See also Steven Shapin, review of *The Figural and the Literal*, *Isis* 79 (1988): 127–128.

10. For an analysis of the shift "from methods of enumerative and eliminative induction to the method of hypothesis," see Laudan, *Science and Values*, 81–82. See also Laudan, "Why Was the Logic of Discovery Abandoned?" in his *Science and Hypothesis: Historical Essays on Scientific Methodology* (Dordrecht, Holland: D. Reidel Publishing Co., 1981), 181–191.

11. Albert E. Moyer, *American Physics in Transition: A History of Conceptual Change in the Late Nineteenth Century* (Los Angeles: Tomash, 1983), 3–118.

12. Daniel Kevles, *The Physicists: The History of a Scientific Community in Modern America* (New York: Alfred A. Knopf, 1978), 25–74. Kevles, "The Physics, Mathematics, and Chemistry Communities: A Comparative Analysis," in *The Organization of Knowledge in Modern America, 1860–*

1920, ed. Alexandra Oleson and John Voss (Baltimore: Johns Hopkins University Press, 1979), 147. Robert Bruce, *The Launching of Modern American Science, 1846–1876* (New York: Alfred A. Knopf, 1987), 3, 93, 101–105, 279–305, 348–352. Stephen Brush, "Looking Up: The Rise of Astronomy in America," *American Studies* 20 (1979): 50. For the positive contributions of American institutional structures in the late nineteenth century, see Moyer, "American Physics in 1887," in *The Michelson Era in American Science: 1870–1930*, ed. Stanley Goldberg and Roger Stuewer, AIP Conference Proceedings, no. 179 (New York: American Institute of Physics, 1988), 102–110. See also Kevles, Jeffrey L. Sturchio, P. Thomas Carroll, "The Sciences in America, Circa 1880," *Science* 209 (1980): 27–32.

13. Thomas Haskell, *The Emergence of Professional Social Science: The American Social Science Association and the Nineteenth-Century Crisis of Authority* (Urbana: University of Illinois Press, 1977). Mary Furner, *Advocacy and Objectivity: A Crisis in the Professionalization of American Social Science, 1865–1905* (Lexington: University Press of Kentucky, 1975).

14. He engaged in, that is, what Donald McCloskey has labeled "the rhetoric of economics." McCloskey, *The Rhetoric of Economics* (Madison: University of Wisconsin Press, 1985), 20–35.

15. Karl Popper quoted in Kragh, *Historiography of Science*, 53.

16. Newcomb, "Evolution and Theology: A Rejoinder," *North American Review* 128 (June 1879): 661. Similarly, Percy Bridgman wrote in 1938 that his operational "method" or "technique of analysis" definitely "does not attempt to set up a theory of meaning or to be an epistemology." Bridgman, "Operational Analysis" (1938), reprinted in *Reflections of a Physicist*, by Bridgman, 2d ed. (New York: Philosophical Library, 1955), 1–2, 25–26. Compare Bridgman's comment to Ernst Mach's warning that "above all, there is *no* Machian philosophy, but at most a scientific methodology and a psychology of knowledge." Quoted in Larry Laudan, "Ernst Mach's Opposition to Atomism," in his *Science and Hypothesis*, 207.

17. Chauncey Wright, "Evolution of Self-Consciousness," *North American Review* (1873), reprinted in *Philosophical Discussions by Chauncey Wright*, ed. Charles E. Norton (New York: Henry Holt, 1877), 244.

CHAPTER 2: FORMATIVE YEARS

1. Newcomb, *Reminiscences of an Astronomer* (Boston: Houghton Mifflin Co., 1903), 20; Newcomb, "Formative Influences," *Forum* 11 (April 1891): 186.

2. In discussing Newcomb's childhood and adolescence, I follow throughout Newcomb's own detailed account in his *Reminiscences*, 1–61, and "Formative Influences," 183–187. To a lesser degree, I draw on Sara Newcomb Merrick (Newcomb's sister), "John and Simon Newcomb: The Story of a Father and Son," *McClure's Magazine* 35 (Oct. 1910): 677–687; W. W. Campbell, "Simon Newcomb," *Memoirs of the National Academy of Sciences* 17, 1st mem. (1924): 1–18; and Raymond C. Archibald, "Simon Newcomb," *Science*, n.s., 44 (1916): 871–878. I also draw on two items in the

Simon Newcomb Papers, Manuscript Division, Library of Congress (hereafter cited as SNP). These items are Simon Newcomb, MS, "George Eggley: A Story of a Life" (a thinly veiled account, using satirical names, of Newcomb's early years, written ca. 1863), SNP, Box 101; and information assembled by Anita Newcomb McGee (Newcomb's daughter) in the folder labeled "Biographical Material," SNP, Box 61.

3. John Newcomb to Simon Newcomb, 8 June 1858, reprinted in Newcomb, *Reminiscences*, 10–14, and "Formative Influences," 184; the original letter is in SNP, Box 111.

4. Three unpublished manuscripts exist in which Newcomb focused exclusively on his religious development; they are located in Box 104, SNP. See Newcomb, untitled MS in folder on "Religious Autobiography," ca. Aug.–Sept. 1879, p. 1; "A Religious Autobiography," ca. 1879–1880, pp. 1–8; and "Development of My Religious Views," ca. 1909, pp. 1–8. The date of the first draft is determined through a note attached to a series of letters written by Newcomb to his wife, Mary, in early September 1879, SNP, Box 12; the type of paper and handwriting for both the draft and the note confirm the date. (The note, unsigned and directed either to himself or an intimate associate, has the heading "Here is something for you to burn up when read.") "A Religious Autobiography" seems to be a second draft intended for publication. The date of "Development of My Religious Views" is determined by a note inserted into the manuscript by Newcomb's daughter, Anita Newcomb McGee: "This 'chapter' on SN's religious views was written within a few weeks of his death, as a contribution to a revised edition of his autobiography." Newcomb also commented on "Religious Influences" in a section of the unpublished MS "Autobiography of My Youth," ca. 1880, p. 12, SNP, Box 100.

5. Roger Cooter, *The Cultural Meaning of Popular Science: Phrenology and the Organization of Consent in Nineteenth-Century Britain* (Cambridge: Cambridge University Press, 1984), 1–11, 101–133, 156, 262. There are parallels between Combe's and Newcomb's Calvinist upbringings.

6. John Newcomb to Simon Newcomb, 8 June 1858; Newcomb, "Development of My Religious Views," 3.

7. Newcomb, *Reminiscences*, 7, and "Formative Influences," 184. Newcomb, MS, "Stray recollections of a teacher South" (a thinly veiled account, using satirical names, of Newcomb's first year of teaching, written ca. 1863), SNP, Box 101. Throughout the remaining discussion of Newcomb's years in Maryland, I follow his own detailed account in his *Reminiscences*, 49–61, and "Formative Influences," 187–188. William Cobbett, *A Grammar of the English Language*, the 1818 New York 1st ed. with passages added in 1819, 1820, and 1823, Costerus New Series, vol. 39, ed. Charles C. Nickerson and John W. Osborne (1818; reprint, Amsterdam: Rodopi, 1983): 31, 33–34, 76.

8. Jean-Baptiste Say, *A Treatise on Political Economy*, new American ed., trans. from the 4th French ed. by C. R. Prinsep (Philadelphia: Grigg & Elliot, 1844), xvii, xxxix, xlvi.

9. Newcomb, MS, "Essay on Human Happiness," SNP, Box 97; for the section on teaching, see untitled MS in folder titled "On Education," SNP, Box

114. Later notes that Newcomb attached to these two portions of the MS indicate that it was drafted in 1854.
10. Henry to Newcomb, 16 Oct. 1855, SNP, Box 26.
11. Newcomb, "Letter," *National Intelligencer*, 26 May 1855, p. 1, col. 2.
12. Henry to John Rodgers, 24 April 1878, Joseph Henry Papers, Smithsonian Institution Archives, quoted in Arthur Norberg, "Simon Newcomb's Early Astronomical Career," *Isis* 69 (1978): 212.
13. Nathan Reingold, "Joseph Henry," *Dictionary of Scientific Biography*.
14. Henry, *Scientific Writings of Joseph Henry* (Washington, D.C.: Smithsonian Institution, 1886), 1: 297–305; 2: 6–10, 34–38, 89, 310–311; these 2 vols. were issued as vol. 30 of *Smithsonian Miscellaneous Collections* (Washington, D.C.: Smithsonian Institution, 1887).
15. Norberg, "Newcomb's Early Astronomical Career," 212–217. Bruce, *Launching of Modern American Science*, 32–33, 39–41, 87, 102–103, 175, 231–232. A. Hunter Dupree, *Science in the Federal Government* (Cambridge, Mass.: Harvard University Press, 1957), 105–109. Bessie Z. Jones and Lyle G. Boyd, *Harvard College Observatory: The First Four Directorships, 1839–1919* (Cambridge, Mass.: Harvard University Press, 1971), 66.
16. Henry to Newcomb, 6 May 1857, SNP, Box 26. For Henry serving as a role model, see Norberg, "Newcomb's Early Astronomical Career," 209.
17. Winlock to Capt. D. N. Ingraham, Naval Observatory Records, National Archives and Records Service, quoted in Norberg, "Newcomb's Early Astronomical Career," 214.
18. Newcomb, Diary for 1858, 2 Feb. 1858, 27 Sept. 1858, SNP, Box 1.
19. See, e.g., Diary for 1859, 14 April 1859, SNP, Box 1.
20. *Catalogue of the Officers and Students of Harvard University for the Academical Year 1856–57, Second Term* (Cambridge, Mass.: John Bartlett, 1857), 67, 69–73. Newcomb, "Formative Influences," 189–190; Newcomb, *Reminiscences*, 74–75.
21. Newcomb, "Formative Influences," 189–190. Norberg, "Newcomb's Early Astronomical Career," 212–213. Diary for 1858, 16 July 1858, SNP, Box 1. MS, "Alphabetical Catalogue of Books of Simon Newcomb," SNP, Box 115. Carolyn Eisele, "Benjamin Peirce," *Dictionary of Scientific Biography*. Helena M. Pycior, "Benjamin Peirce's *Linear Associative Algebra*," *Isis* 70 (1979): 249–251.
22. Newcomb, *Reminiscences*, 195–197. Campbell, "Simon Newcomb," 4.
23. Newcomb, "Development of My Religious Views," 8–10. Newcomb, Diary for 1858, 25 April 1858; Diary for 1859, 2 Oct. 1859; Diary for 1860, 8 Jan. 1860, 22 Jan. 1860, and 3 June 1860; Diary for 1861, 7 June 1861, SNP, Box 1. Newcomb, "A Religious Autobiography," 10–12. Newcomb, "Autobiography of My Youth," 59, 74. One of Newcomb's sources on Hume's argument might have been John Stuart Mill, *A System of Logic* (New York: Harper & Brothers, 1858), 379–382.
24. Newcomb, Diary for 1860, 19 April 1860, SNP, Box 1. Newcomb, "Alphabetical Catalogue of Books." Newcomb, "Development of My Religious Views," 10.

25. George Santayana, *The Last Puritan: A Memoir in the Form of a Novel* (New York: Charles Scribner's Sons, 1935), 7, 438. Maila Walter finds similar traits in operationalist Percy Bridgman; see Walter, *Science and Cultural Crisis: An Intellectual Biography of Percy Williams Bridgman (1882–1961)* (Stanford, Calif.: Stanford University Press, 1990), 13–14, 259–264. See also Albert E. Moyer, "A Puritan of Science," review of *Science and Cultural Crisis*, by Walter, *Science* 251 (1991): 815. For general background, see Alan Heimert and Andrew Delbanco, eds., *The Puritans in America: A Narrative Anthology* (Cambridge, Mass.: Harvard University Press, 1985), 12–15; Perry Miller and Thomas H. Johnson, eds., *The Puritans*, rev. ed. (New York: Harper & Row, 1963), 1: 2–5.

26. Newcomb, "Alphabetical Catalogue of Books."

27. James Walker, "Editor's Notice," in *Essays on the Intellectual Powers of Man*, by Thomas Reid, 9th ed. (Boston: Phillips, Samson, & Co., 1859), iii–iv.

28. Richard Olson, *Scottish Philosophy and British Physics, 1750–1880: A Study in the Foundations of the Victorian Scientific Style* (Princeton, N.J.: Princeton University Press, 1975), 42–48, 94–95.

29. Theodore D. Bozeman, *Protestants in an Age of Science: The Baconian Ideal and Antebellum American Religious Thought* (Chapel Hill: University of North Carolina Press, 1977), 21, 29–30, 160–162. See also Herbert Hovenkamp, *Science and Religion in America: 1800–1860* (Philadelphia: University of Pennsylvania Press, 1978), 3–36.

30. George H. Daniels, *American Science in the Age of Jackson* (New York: Columbia University Press, 1968), 63–66, 71, 118–119, 198–200. John C. Greene questions aspects of Daniels's analysis in "An Earlier Day in American Science," review of *American Science in the Age of Jackson*, by Daniels, and *Science in Nineteenth-Century America*, by Nathan Reingold, *Science* 160 (10 May 1968): 638–640; see also A. Hunter Dupree, review of *American Science in the Age of Jackson*, by Daniels, *American Historical Review* 74 (Oct. 1968): 281.

31. Bruce Kuklick, *The Rise of American Philosophy: Cambridge, Massachusetts, 1860–1930* (New Haven, Conn.: Yale University Press, 1977), 5–27. See also Edward H. Madden and Peter H. Hare, review of *Rise of American Philosophy*, by Kuklick, *Transactions of the Charles S. Peirce Society* 14 (1978): 53–72.

CHAPTER 3: INFLUENCES OF COMTE, DARWIN, AND MILL

1. Henry James, *Charles W. Eliot* (Boston: Houghton Mifflin Co., 1930), 1: 250; for a fuller discussion of this episode from Fiske's junior year of 1861–62, see Samuel E. Morison, *Three Centuries of Harvard, 1636–1936* (Cambridge, Mass.: Harvard University Press, 1936), 308–309. For Comte's general impact among Unitarians in Cambridge and Boston, see Charles D. Cashdollar, *The Transformation of Theology, 1830–1890: Positivism and*

Protestant Thought in Britain and America (Princeton: Princeton University Press, 1989), 93–111, 281–283.

2. Edward H. Madden, personal letters, 14 and 24 April 1981; Madden, *Chauncey Wright and the Foundations of Pragmatism* (Seattle: University of Washington Press, 1963), 8–9, 91; Madden, *Chauncey Wright* (New York: Twayne, 1964), 12–18, 123–124.

3. Newcomb, *Reminiscences*, 95; Norberg, "Newcomb's Early Astronomical Career," 214.

4. Diary for 1858, 11 Oct. 1858 and 12 Oct. 1858, SNP, Box 1. *Catalogue of the Officers and Students of Harvard University for the Academical Year 1858–59, First Term* (Cambridge, Mass.: John Bartlett, 1858), 40–41. This and subsequent Harvard *Catalogues* list Newcomb as a "resident graduate" from the first term of the academic year 1858–59 through the first term of 1861–62.

5. See esp. the first two chaps. of the first vol. of Auguste Comte's *Cours de philosophie positive* (Paris, 1830), in *Introduction to Positive Philosophy*, trans. Frederick Ferré (Indianapolis: Bobbs-Merrill Co., 1970). John Stuart Mill, *Autobiography* (1873; reprint, London: Oxford University Press, 1971), 164. Neal C. Gillespie, *Charles Darwin and the Problem of Creation* (Chicago: University of Chicago Press, 1979), 53–54.

6. *Catalogue of the Officers and Students of Harvard University for the Academical Year 1859–60, First Term* (Cambridge, Mass.: Sever & Francis, 1859), 43. In a personal letter to the author of 17 Oct. 1980, Clark A. Elliott reports that the Harvard University Archives has no record of a submission by Newcomb.

7. In folder titled "Foreknowledge of the Deity—Fatalism and Free Will," SNP, Box 94. The handwriting, type of paper, and content of the three undated and untitled drafts indicate early dates of composition; the handwriting is most easily compared with that in Newcomb's dated diaries.

8. Mill, *System of Logic* (1858), 38–41, 289–290. Ernest Nagel, ed., "Introduction," in *John Stuart Mill's Philosophy of Scientific Method* (New York: Hafner, 1950), xxxii–xxxiii. R. F. McRae, ed., "Introduction," in *A System of Logic*, by Mill, vol. 7 of *Collected Works of John Stuart Mill* (Toronto: University of Toronto Press, 1973), xxxix–xliv.

9. Gillespie, *Charles Darwin*, 41–66, 146–156; Cashdollar, *Transformation of Theology*, 182–198. See also Ernest Mayr, "Introduction," in *On the Origin of Species*, by Charles Darwin, 1st ed. (1859; reprint, Cambridge, Mass.: Harvard University Press, 1964), xi–xxiii. Morse Peckham, "Darwinism and Darwinisticism," and John H. Randall, "The Changing Impact of Darwin on Philosophy," both in *Darwin*, ed. Philip Appleman, 2d ed. (New York: W. W. Norton & Co., 1979), 302–303, 324–325. Cynthia E. Russett, *Darwin in America: The Intellectual Response, 1865–1912* (San Francisco: W. H. Freeman & Co., 1976).

10. Diary for 1859, 21 Feb. 1859 and 22 Feb. 1859, SNP, Box 1.

11. A. Hunter Dupree, *Asa Gray: 1810–1888* (Cambridge, Mass.: Harvard University Press, 1959), 255–256; Newcomb, *Reminiscences*, 70.

12. Dupree, *Asa Gray*, 285–288; Newcomb, MS, "Autobiography of My Youth" (see chap. 2, n. 4), 64. See also Francis Bowen, review of *On the Origin of Species*, by Darwin, *North American Review* 90 (1860): 474–506.

13. Diary for 1860, 27 March 1860 and 1 May 1860, SNP, Box 1. See also Norberg, "Newcomb's Early Astronomical Career," 216.

14. Newcomb, "A Religious Autobiography" (see chap. 2, n. 4).

15. Newcomb, untitled MS in folder titled "Religious Autobiography," pp. 5–6, SNP, Box 104.

16. Mill, *Autobiography*, 134. See also Nagel, "Introduction," xv–xlviii, and McRae, "Introduction," xxii–xlviii.

17. Mill, *System of Logic* (1858), 174, 183–186, 194–197, 337–339.

18. Diary for 1860, 11 Jan. 1860 and 18 April 1860, SNP, Box 1. For Bowen's opposition to Mill, see Kuklick, *Rise of American Philosophy*, 28–45. Newcomb, "Alphabetical Catalogue of Books" (see chap. 2, n. 21). Mill acknowledged his debt to Whewell in his "Preface," *System of Logic*, iv. Notice that all citations involving Mill's *System of Logic* refer to the edition published by Harper & Brothers in New York in 1858, possibly the edition that Newcomb owned.

19. Undated and untitled fragment of MS, in folder titled "Misc. Writings," SNP, Box 110. The small, neat handwriting suggests an early date of composition (see n. 7).

20. For Mill's explication of his disagreement with Whewell, see his chapter "Of Demonstrations and Necessary Truths," *System of Logic*, 148–161. See also Harold T. Walsh, "Whewell and Mill on Induction," *Philosophy of Science* 29 (1962): 279–284.

21. Madden, *Chauncey Wright*, 62–63, 124–125; see also Madden, *Chauncey Wright and the Foundations of Pragmatism*, 16.

22. Mill, *System of Logic*, 36, 153. Nagel, *Mill's Philosophy of Scientific Method*, xxxiii–xxxiv, xxxix.

23. Laudan, *Science and Values*, 81–82; see also Richard R. Yeo, "Scientific Method and the Rhetoric of Science," in *Politics and Rhetoric*, ed. Schuster and Yeo, 269–272.

24. Newcomb, Diary for 1860, 8 May 1860, SNP, Box 1; "The objections raised by Mr. Mill . . . ," *Proceedings of the American Academy of Arts and Sciences* 4 (1860): 433–440; MS, "Ideas for the introductory essay on probabilities," SNP, Box 57. Cf. Mill's chapter "Of the Calculation of Chances," *System of Logic*, 319–327. Also cf. Newcomb, "The Method and Province of Political Economy," review of *The Character and Logical Method of Political Economy*, by J. E. Cairnes, *North American Review* 121 (1875): 260. For Mill's criticism of Laplace, see Theodore M. Porter, *The Rise of Statistical Thinking, 1820–1900* (Princeton, N.J.: Princeton University Press, 1986), 82–83.

25. Newcomb, "Alphabetical Catalogue of Books." See also Newcomb, MS, "Theory of Probable Inference" (ca. 1905), SNP, Box 108.

26. Newcomb, "The objections raised by Mr. Mill . . . ," 434, 439. Cf. Mill's chapter "Of Chance, and Its Elimination," *System of Logic*, 312–319.

Notes to Pages 44–49

27. Newcomb, *A Critical Examination of Our Financial Policy during the Southern Rebellion* (1865; reprint, New York: Greenwood, 1969), 12–14, 19. Newcomb, "Method and Province of Political Economy," 264. Newcomb, "Abstract Science in America, 1776–1876," *North American Review* 122 (Jan. 1876), 122–123. Cf. Mill, *System of Logic*, 565–567.

28. Diary for 1870, 29 Nov. 1870, SNP, Box 1. Newcomb, *Reminiscences*, 272.

29. Mill, *Autobiography*, 163. Newcomb, Diary for 1870, 22 Nov. 1870 and 23 Nov. 1870, SNP, Box 1. Newcomb, *Reminiscences*, 293–294.

30. Newcomb, Diary for 1868, 29 Feb. 1868, SNP, Box 1; he never published this particular paper, although a decade later he dealt with a similar topic in his "The Course of Nature," *Proceedings of the AAAS* 27 (1878): 1–28. Newcomb, "Autobiography of My Youth," 63–64, and *Reminiscences*, 70.

31. Newcomb, *Reminiscences*, 273. In an essay written under the pseudonym of "An Evolutionist," Newcomb had earlier responded to W. S. Jevons's critique of Mill; see "An Advertisement for a New Religion," *North American Review* 127 (1878): 48–49.

32. Newcomb, *Critical Examination of Our Financial Policy*, 12–19, 45–46. Newcomb, *Principles of Political Economy* (1886; reprint, New York: A. M. Kelley, 1966), 14–31.

33. Newcomb, *Reminiscences*, 95–96. See also Norberg, "Newcomb's Early Astronomical Career," 217. Newcomb, MS in folder titled "Reminiscences of an Astronomer [Unpublished Portion]," p. 76, SNP, Box 104. Among Newcomb's miscellaneous papers, there are many undated notes and pages pertaining to physics; see, e.g., SNP, Boxes 95, 96, 97, 109, 143.

34. Gould to Newcomb, 19 Aug. 1861, SNP, Box 24. Newcomb, *Reminiscences*, 97–114. Bruce, *Launching of Modern American Science*, 176–186, 217–224, 285, 296–297. Steven J. Dick, "How the U.S. Naval Observatory Began, 1830–65," in *Sky with Ocean Joined: Proceedings of the Sesquicentennial Symposia of the U.S. Naval Observatory, December 5 and 8, 1980*, ed. Steven J. Dick and LeRoy E. Doggett (Washington, D.C.: U.S. Naval Observatory, 1983), 167–181. Dupree, *Science in the Federal Government*, 62, 105–109, 118–119, 135–137. Naturalization document, 16 June 1864, SNP, Box 143.

35. Newcomb, Diary for 1861, 18 March 1861, SNP, Box 1. Newcomb, MS, "A Textbook of Physics," pp. 1–4, SNP, Box 109.

36. Bozeman, *Protestants in an Age of Science*, 160. Daniels, *American Science in the Age of Jackson*, 63.

37. Newcomb, MS, "Chapter third statics," pp. 15–16, SNP, Box 109.

38. Reingold, "Joseph Henry," 279. Henry, *Scientific Writings* 1: 298–302; 2: 10–11, 34–37. Mill, *System of Logic*, 264–270, 290–299. For other parallels discussed in this section between Newcomb's textbook introduction to method and Mill's views, see *System of Logic*, 194–210, 216–222, 256–259, 337–343. For other parallels with Henry's views, see *Scientific Writings* 1: 259; 2: 37–38, 86–89, 310–311. Among the Scottish philosophers, Stewart had also adopted a softer stance against hypothesis than

his predecessor, Reid; see Olson, *Scottish Philosophy and British Physics*, 94–98.

CHAPTER 4: INTERACTIONS WITH WRIGHT AND PEIRCE

1. Charles Peirce, "Concerning the Author," MS (ca. 1897), reprinted in *Philosophical Writings of Peirce*, ed. Justus Buchler (New York: Dover Publications, 1955), 2.
2. See, e.g., Diary for 1859, 11 March 1859 and 16 March 1859; Diary for 1860, 10 Jan. 1860, SNP, Box 1.
3. Diary for 1859, 14 April 1859; Diary for 1861, 15 May 1861 and 10 Sept. 1861, SNP, Box 1.
4. Diary for 1861, 9 Jan. 1861 and 13 Jan. 1861, SNP, Box 1.
5. Kuklick, *Rise of American Philosophy*, 28–45. For Wright's disregard for orthodox philosophers like Bowen, see Madden, *Chauncey Wright*, 116–120. See also Wright, review of *A Treatise on Logic*, by Francis Bowen, *North American Review* 99 (1864): 592–605.
6. Bowen, review of *On the Origin of Species*, by Darwin, *North American Review* 90 (1860): 474–506. For Wright's attack on Bowen's anti-Darwinian views, see Dupree, *Asa Gray*, 289–293.
7. Newcomb, *Reminiscences*, 70; see also, Newcomb to James B. Thayer, *Letters of Chauncey Wright*, ed. Thayer (1878; reprint, New York: Burt Franklin, 1971), 71; and Newcomb, "Autobiography of My Youth" (see chap. 2, n. 4), 64. Newcomb, "Abstract Science in America" (see chap. 3, n. 27), 109–110.
8. Newcomb to Thayer, *Letters of Chauncey Wright*, ed. Thayer, p. 70. Newcomb, "Autobiography of My Youth," 63–64. Cf. to Newcomb, *Reminiscences*, 70.
9. For Newcomb's visits to Wright, see e.g., Newcomb, Diary for 1867, 27 Aug. 1867, SNP, Box 1. For evidence of Newcomb sending his scientific papers to Wright, see Diary for 1866, "Memoranda" page, SNP, Box 1.
10. Newcomb to Wright, 24 Feb. 1865, SNP, Box 4. The book, published at Newcomb's own expense, was *Critical Examination of Our Financial Policy*.
11. Wright to Newcomb, 28 Feb. 1865, SNP, Box 44. Newcomb to Wright, 11 March [1865] and 1 April 1865, SNP, Box 4; see also Newcomb, Diary for 1865, 15 March 1865 and 1 April 1865, SNP, Box 1. Wright to Newcomb, 5 April 1865, SNP, Box 44; see also Diary for 1865, 7 April 1865, SNP, Box 1. Newcomb to Wright, 7 April 1865 and 21 July 1865, SNP, Box 4; see also Diary for 1865, 14 July 1865, SNP, Box 1.
12. See Mill's chapter "Of Liberty and Necessity," *System of Logic*, 521–526; see also 207–208n. For a concise explication of Mill's position, see Paul Edwards and Arthur Pap, eds., *A Modern Introduction to Philosophy: Readings from Classical and Contemporary Sources* (New York: Free Press, 1965), 6–9, 44–50.

13. Newcomb to Thayer, *Letters of Chauncey Wright*, ed. Thayer, pp. 70–71. Joseph Henry, in contrast to Newcomb, leaned toward a deistic determinism; see Henry, *Scientific Writings* 2: 6.

14. Newcomb, "Formative Influences" (see chap. 2, n. 1), 188–189. Newcomb, "Autobiography of My Youth," 64. Arthur Norberg doubts that Newcomb led Wright to Mill's view; see Norberg, "Newcomb's Early Astronomical Career," p. 216 n. 28.

15. *Letters of Chauncey Wright*, ed. Thayer, p. 71. Cf. Madden, *Chauncey Wright and the Foundations of Pragmatism*, 18, 90–91.

16. Newcomb to Wright, included with letter of 29 April 1865, SNP, Box 4. Cf. Mill, *System of Logic*, 523–525. For analyses of free will similar to that in his 1865 letter to Wright, see two apparently early MS fragments by Newcomb in the folder "Foreknowledge of the Deity—Fatalism and Free Will," SNP, Box 94. One fragment contains the following relevant paragraph: "Every term implying freedom, power, possibility, &c in contradistinction to determination, necessity, certainty, &c has its origin in circumstances of common life which have no relation to questions of pure philosophy and therefore entirely loses its signification when applied to the latter. In any branch of physical science the greatest care is taken accurately to define all terms which are applied to new things; how much greater necessity for such precaution in metaphysics where, from the extreme sublety [sic] of the analysis, the mind is much more likely to substitute a common place phrase for a refined idea."

17. Wright to Newcomb, 18 May 1865, *Letters of Chauncey Wright*, ed. Thayer, pp. 71–75. The original letter, with some portions torn away, is in SNP, Box 44. Quotations are from the original version when possible. Cf. Chauncey Wright, "Limits of Natural Selection," *North American Review* (1870), reprinted in *Philosophical Discussions by Chauncey Wright*, ed. Norton, 119–125, and "Evolution of Self-Consciousness," *North American Review* (1873), also reprinted in *Philosophical Discussions*, 233, 244. Also cf. Charles S. Peirce, "How to Make Our Ideas Clear," *Popular Science Monthly* (Jan. 1878), reprinted in Christian J. W. Kloesel, ed., *Writings of Charles S. Peirce: A Chronological Edition*, vol. 3 (Bloomington: Indiana University Press, 1986), 267.

18. Simon Newcomb to Mary Newcomb, 27 May 1865, SNP, Box 8. Newcomb to Wright, 21 July 1865, SNP, Box 4. Newcomb, "The Relation of Scientific Method to Social Progress" (Washington, D.C.: Judd & Detweiler, 1880), 9–10, 13.

19. Newcomb to Norton, 18 Feb. 1868, Charles E. Norton Papers, Houghton Library, Harvard University, Item bMS Am 1088 (4946). Newcomb, Diary for 1868, 29 Feb. 1868, SNP, Box 1. William James, "Chauncey Wright," *Nation* (1875), reprinted in Madden, *Chauncey Wright and the Foundations of Pragmatism*, 143–144.

20. Norton to Newcomb, 19 Nov. 1875, SNP, Box 34.

21. Newcomb, "Autobiography of My Youth," 57; Newcomb, *Reminiscences*, 71–72; Newcomb, "Formative Influences," 188.

22. Philip P. Wiener, *Evolution and the Founders of Pragmatism* (Cambridge, Mass.: Harvard University Press, 1949), 18–96.

23. For the overlap in the careers of Peirce and Newcomb, see Carolyn Eisele, "The Charles S. Peirce–Simon Newcomb Correspondence," *Proceedings of the American Philosophical Society* 101 (1957): 409–433; reprinted as chap. 5 of Eisele, *Studies in the Scientific and Mathematical Philosophy of Charles S. Peirce* (The Hague: Mouton, 1979), 52–93.

24. Eisele, "Peirce–Newcomb Correspondence," 410.

25. Carolyn Eisele, "Peirce the Scientist," in *Historical Perspectives on Peirce's Logic of Science: A History of Science*, ed. Eisele (Berlin: Mouton Publishers, 1985), 1: 31.

26. Diary for 1868, 6 Jan. 1868, SNP, Box 1; Newcomb used Charles Peirce's full, formal name perhaps to distinguish him from his father. The five papers are reprinted in Edward C. Moore, ed., *Writings of Charles S. Peirce: A Chronological Edition*, vol. 2 (Bloomington: Indiana University Press, 1984), 12–97; see also "Editorial Notes," 501–503. The papers are items P 00030, 31, 32, 33, and 34 in *A Comprehensive Bibliography and Index of the Published Works of Charles Sanders Peirce . . .*, ed. Kenneth L. Ketner, et al. (Greenwich, Conn.: Johnson Assoc., 1977). Also see Murray G. Murphey, "Peirce, Charles Sanders," *Encyclopedia of Philosophy* (New York: Macmillan, 1967), 6: 71–72.

27. "Tuck Memorandum" book (inside frontispiece reads "Simon Newcomb, U.S. Navy"), SNP, Box 1. In his notes specifying the observation sites of expedition members, Newcomb impersonally wrote of "Benj. Peirce & Son" being located in Sicily; he listed by name the many other members throughout the region.

28. See "Index to Names of Contributors," *Bulletin of the Philosophical Society of Washington* 1 (March 1871–June 1874): 47–48. See also J. Kirkpatrick Flack, *Desideratum in Washington: The Intellectual Community in the Capital City, 1870–1900* (Cambridge, Mass.: Schenkman, 1975), 60–65. Newcomb to Peirce, 25 Oct. 1872, Charles S. Peirce Papers, Houghton Library, Harvard University, Section L 314; this Peirce collection contains the originals of many of the letters cited in Eisele, "Peirce–Newcomb Correspondence."

29. Charles Peirce to Zina Peirce, 17 Dec. 1871, reprinted in Eisele, *Studies*, 367–368; also see p. 253. Peirce to Newcomb, 17 Dec. 1871, SNP, Box 10; reprinted in Eisele, "Peirce–Newcomb Correspondence," 414. Max H. Fisch, "The Decisive Year and Its Early Consequences," in Moore, ed., *Writings of Charles S. Peirce*, xxxv–xxxvi.

30. Eisele, "Peirce–Newcomb Correspondence," 411 n. 17. Daniel C. Gilman to Newcomb, 7 Feb. 1876, SNP, Box 24. See also Max H. Fisch and Jackson I. Cope, "Peirce at the Johns Hopkins University," in *Studies in the Philosophy of Charles Sanders Peirce*, ed. Philip P. Wiener and Frederick H. Young (Cambridge, Mass.: Harvard University Press, 1952), 277–311; Daniel C. Gilman, *The Launching of a University* (New York: Dodd, Mead, 1906), 55, 62.

31. Newcomb to Peirce, 8 July 1882, SNP, Box 5. Newcomb had in mind the ether-drift experiment of his younger colleague, Albert A. Michelson. Nathan Reingold has reprinted the Newcomb-Michelson correspondence on this topic and others in *Science in Nineteenth-Century America: A Documentary History* (New York: Hill & Wang, 1964), 275–306.

32. Newcomb to Henry Rowland, 8 July 1882, SNP, Box 5.

33. In a personal letter to the author of 24 Sept. 1980, Arthur Norberg documented the disclosure, referring to letters from Simon to Mary Newcomb dating from 14 Oct., 21 Nov., and 25 Dec. 1883. For a related note from Newcomb to either the trustee or Gilman, see Max H. Fisch's 1975 transcription of a letter dated 22 Dec. 1883, Daniel Coit Gilman Papers Ms. 1, Special Collections, Milton S. Eisenhower Library, Johns Hopkins University. See also Paul Conkin, *Puritans and Pragmatists* (New York: Dodd, Mead & Co., 1968), 199.

34. Eisele, "Peirce–Newcomb Correspondence," 414–428. See also "Correspondence," *Nation* 48 (1889): 488, 504–505, 524.

35. Newcomb to Peirce, 3 Jan. 1894, reprinted in Eisele, "Peirce–Newcomb Correspondence," 426. Two weeks later in another letter on Peirce's proposed philosophical volumes, Newcomb explained: "I quite coincide with your expression of the spirit in which you treat the subject, although I fear my philosophy would diverge a good deal from yours." Newcomb to Peirce, 16 Jan. 1894, reprinted in Eisele, "Peirce–Newcomb Correspondence," 426.

36. [Peirce], review of *Reminiscences*, by Newcomb, *Nation* 78 (1904): 237. See also Eisele, "Peirce–Newcomb Correspondence," 429–430. For a more negative judgment written in the heat of argument—that Newcomb's outlooks were "very narrow both on the philosophical and on the mathematical side"—see Peirce to Baldwin, 26 Dec. 1900, quoted in Eisele, "Peirce–Newcomb Correspondence," 415–416.

37. Wright, "The Philosophy of Herbert Spencer," *North American Review* (Jan. 1865), reprinted in *Philosophical Discussions by Chauncey Wright*, ed. Norton, pp. 43–96. Wright discusses concepts such as "force" on, for example, p. 78; he discusses "sensuous verification" on pp. 44–47. For Edward Madden's doubts that Wright fully anticipated pragmatism, see Madden, *Chauncey Wright and the Foundations of Pragmatism*, 73–81. For a more positive interpretation of Wright's anticipation of pragmatism, see Wiener, *Evolution and the Founders of Pragmatism*, 31–69. For support of Madden's views on Wright, see Robert Giuffrida, Jr., "The Philosophical Thought of Chauncey Wright: Edward Madden's Contribution to Wright Scholarship," *Transactions of the Charles S. Peirce Society* 24 (Winter 1988): 33–64.

38. Wright, "Speculative Dynamics," *Nation* (June 1875), reprinted in *Philosophical Discussions by Chauncey Wright*, ed. Norton, 385, 388–389.

39. For the first paper, see Peirce, "The Fixation of Belief," *Popular Science Monthly* (Nov. 1877), reprinted in Kloesel, ed., *Writings of Charles S. Peirce*, 254. See also Francis E. Reilly, *Charles Peirce's Theory of Scientific Method* (New York: Fordham University Press, 1970).

40. Peirce, "How to Make Our Ideas Clear," *Popular Science Monthly* (Jan. 1878), reprinted in Kloesel, ed., *Writings of Charles S. Peirce*, 266. Philip Wiener is typical of many past scholars when he acclaims this 1878 essay, saying that "perhaps no earlier and clearer presentation of the 'operationalist' theory of meaning can be found in American philosophy." See *Charles S. Peirce: Selected Writings*, ed. Philip Wiener (New York: Dover Publications, 1960), 113. For more cautious appraisals, see: Richard Smyth, "The Pragmatic Maxim in 1878," *Transactions of the Charles S. Peirce Society* 13 (1977): 93–111; Bruce Altshuler, "The Nature of Peirce's Pragmatism," *Transactions of the Charles S. Peirce Society*, 14 (1978): 147–175.

41. Peirce, "How to Make Our Ideas Clear," 265–266, 270–271.

42. For this distinction, see Madden, *Chauncey Wright and the Foundations of Pragmatism*, 80; see also n. 37. For an overview of Wright and Charles Peirce as scientist-philosophers, see Daniel J. Wilson, *Science, Community, and the Transformation of American Philosophy, 1860–1930* (Chicago: University of Chicago Press, 1990), 12–36.

CHAPTER 5: MIDCAREER

1. Frederic C. Howe, *The Confessions of a Reformer* (New York: Charles Scribner's Sons, 1925), 30. William Thomson to P. G. Tait, 11 July 1876, quoted in Harold I. Sharlin, *Lord Kelvin: The Dynamic Victorian* (University Park: Pennsylvania State University Press, 1979), 187. Newcomb, MS, "Physico-biographical summary in Case of S. Newcomb," SNP, Box 124. Newcomb, "Alphabetical Catalogue of Books" (see chap. 2, n. 21). Merrick, "John and Simon Newcomb," 687. Herbert Putnam, "Address of Mr. Herbert Putnam," in "Simon Newcomb: Memorial Addresses," *Bulletin of the Philosophical Society of Washington* 15 (1910): 155–158. Campbell, "Simon Newcomb," 18. William Alvord, "Address of the Retiring President of the Society, in Awarding the Bruce Medal to Professor Simon Newcomb," *Publications of the Astronomical Society of the Pacific* 10 (1898): 54.

2. Merrick, "John and Simon Newcomb," 687. Putnam, "Address," 155. Irving Fisher, "Obituary: Simon Newcomb," *The Economic Journal* 19 (Dec. 1909): 641–644. Charles G. Abbott, "Simon Newcomb," *Dictionary of American Biography*.

3. Newcomb, MS, "Physico-biographical summary." Newcomb's superintendent at the Naval Observatory, Capt. James Gilliss, died in 1865, perhaps from malaria contracted at the observatory site; see Dick, "How the U.S. Naval Observatory Began," 180.

4. Alexander Agassiz to Newcomb, 22 April 1878, SNP, Box 14; Newcomb to Agassiz, 13 May 1878, SNP, Box 4. Henry eventually recommended that Smithsonian colleague Spencer Baird succeed him, which Baird did; see Bruce, *Launching of Modern American Science*, 325.

5. Mary Newcomb to Joseph Henry, 4 March 1878, SNP, Box 26. Diary for 1867, 2 Feb. 1867, SNP, Box 1. Newcomb, "An Investigation of the Orbit of Uranus with General Tables of Its Motion," *Smithsonian Contributions to*

Knowledge, vol. 19, art. 4 (Washington, D.C.: Smithsonian Institution, 1874), iv. Henry to M. Le Verrier, 20 Jan. 1874, SNP, Box 26. See also Newcomb, *Reminiscences*, 225, 235.

6. Eliot to Newcomb, 7 Oct. 1875 and 11 Oct. 1875, Charles W. Eliot Papers, Harvard University Archives, Box 90; see also Newcomb to Eliot, 7 Oct. 1876, and Charles S. Peirce to Eliot, 18 Oct. 1876, Box 69. Benjamin Peirce to Newcomb, 4 Feb. 1876 and 16 Feb. 1876, SNP, Box 34. For another colleague's regrets on Newcomb not taking the Harvard position, see Charles E. Norton to Newcomb, 19 Nov. 1875, SNP, Box 34, and Newcomb to Norton, 30 Nov. 1875, Norton Papers, Houghton Library, Harvard University, Item bMS Am 1088 (4947). Newcomb, *Reminiscences*, 74, 211–214. For a possible position at the Harvard Observatory in the mid 1860s, see Norberg, "Newcomb's Early Astronomical Career," 219–220. For the unsuccessful attempts of Charles Peirce's friends to get him the position, see Jones and Boyd, *Harvard College Observatory*, 176–177.

7. Eisele, "Peirce–Newcomb Correspondence," 410; reprinted in Eisele, *Studies*, 54. Dupree, *Science in the Federal Government*, 185. Newcomb, *Reminiscences*, 211–214.

8. Norberg, "Newcomb's Early Astronomical Career," 219–224. Newcomb, *Reminiscences*, 213–214, cf. p. 273. Alexander Agassiz to Newcomb, 9 May 1881, and Newcomb to Agassiz, 15 May 1881, SNP, Box 14. Newcomb, "Abstract Science in America, 1776–1876" (see chap. 3, n. 27), 112. Newcomb, "Conditions Which Discourage Scientific Work in America," *North American Review* 174 (1902): 156.

9. Cayley, "Address Delivered by the President on Presenting the Gold Medal of the Society to Professor Simon Newcomb," *Monthly Notices of the Royal Astronomical Society* 34 (1874), reprinted in *The Collected Mathematical Papers of Arthur Cayley* (Cambridge: Cambridge University Press, 1896), 9: 181. Newcomb, "An Investigation of the Orbit of Venus," iii–iv, 1–5. In discussing Newcomb's astronomical researches in this section, I follow throughout Norberg, "Newcomb's Early Astronomical Career," 210, 217–225; Brian G. Marsden, "Simon Newcomb," *Dictionary of Scientific Biography*, ed. Charles C. Gillespie (New York: Charles Scribner's Sons, 1970–76), 10: 33–36; Campbell, "Simon Newcomb," 4–16; G[eorge] W. Hill, "Professor Simon Newcomb as an Astronomer," *Science* 30 (17 Sept. 1909): 353–357; P. H. C[owell], "Simon Newcomb, 1835–1909," *Proceedings of the Royal Society of London*, ser. A, 84 (8 July 1910): xxxii–xxxviii; Newcomb, *Reminiscences*, 97–144, 151–181, 195–233, 285–288, 321–328; and Raymond C. Archibald, "Simon Newcomb, 1835–1909, Bibliography of His Life and Work," *Memoirs of the National Academy of Sciences*, 17, 1st mem. (1924): 32–53.

10. Newcomb, "Autobiography of My Youth" (see chap 2, n. 4), 94–95. Bruce, *Launching of Modern American Science*, 285.

11. Cayley, "Address," 184. H. P. Hollis, "The Decade 1870–1880," in *History of the Royal Astronomical Society, 1820–1920*, ed. J. L. E. Dreyer and H. H. Turner (1923; reprint, Oxford: Blackwell Scientific Publications, 1987), 185.

12. Marcus Benjamin, "Simon Newcomb: Astronomer (1835–1909)," in *Leading American Men of Science*, ed. David Starr Jordan (New York: Henry Holt & Co., 1910), 384–386. Archibald, "Simon Newcomb," *Science*, 871–878.

13. Newcomb, "The Story of a Telescope," *Scribner's Monthly* 7 (Nov. 1873): 44–55. Donald E. Osterbrock, John R. Gustafson, and W. J. Shiloh Unruh, *Eye on the Sky: Lick Observatory's First Century* (Berkeley, Los Angeles, London: University of California Press, 1988), 19. Helen Wright, *James Lick's Monument* (Cambridge: Cambridge University Press, 1987), 35–36. John Lankford, "Amateurs versus Professionals: The Controversy over Telescope Size in Late Victorian Science," *Isis* 72 (1981): 15–16.

14. Paul M. Janiczek, "Remarks on the Transit of Venus Expedition of 1874," in *Sky with Ocean Joined*, ed. Dick and Doggett, 52–73.

15. Howard Plotkin, "Astronomers versus the Navy: The Revolt of American Astronomers over the Management of the U. S. Naval Observatory, 1877–1902," *Proceedings of the American Philosophical Society* 122 (1978): 385–399. See also Bruce, *Launching of Modern American Science*, 318. For Newcomb's early discontent with "incompetent" administrators, see Newcomb to Charles Norton, 30 July 1867, Norton Papers (4944); for his own interest in directing the observatory, see Newcomb to Alexander Agassiz, 15 May 1881, SNP, Box 14.

16. Campbell, "Simon Newcomb," 10. Brasch, "Einstein's Appreciation," 248–249. Abraham Pais, *'Subtle Is the Lord . . .': The Science and the Life of Albert Einstein* (Oxford: Oxford University Press, 1982), 253–254. For Jean Chazy's critique in the 1920s of Newcomb's work on the advance of the perihelion of Mercury, see Pierre Costable, "Jean Cheazy," *Dictionary of Scientific Biography*.

17. Mary Ann James, *Elites in Conflict: The Antebellum Clash over the Dudley Observatory* (New Brunswick, N.J.: Rutgers University Press, 1987), 10–12. Simon Schaffer, "Astronomers Mark Time: Discipline and the Personal Equation," *Science in Context* 2 (1988): 119, 126–131. Newcomb, *Reminiscences*, 114, 116. For Newcomb's general comments on this "atmosphere of sentiment," see Newcomb, "The Lick Observatory of California," *Harper's Magazine* 70 (Feb. 1885): 399–406. See also Bruce, *Launching of Modern American Science*, 116–117.

18. Archibald, "Simon Newcomb, Bibliography," 32–53. Newcomb, "Some Talks of an Astronomer," *Harper's New Monthly Magazine* 49 (Oct. 1874): 693–707, (Nov. 1874): 825–841. Hill, "Newcomb as an Astronomer," 356. Milton Updegraff, "Address of Mr. Milton Updegraff," in "Simon Newcomb: Memorial Addresses," *Bulletin of the Philosophical Society of Washington*, 141. John W. Abrams, "William Wallace Campbell," *Dictionary of Scientific Biography*.

19. Flack, *Desideratum in Washington*, 60–65.

20. William James to Thomas Davidson, 1 Feb. 1885, *Letters of William James*, ed. Henry James (Boston: Atlantic Monthly Press, 1920), 249–250. Newcomb, *Reminiscences*, 410–416.

21. Gilman to Newcomb, 7 Feb. 1876, SNP, Box 24. Gilman, *Launching of a University*, 55, 62. Hugh Hawkins, *Pioneer: A History of the Johns Hopkins University, 1874–1889* (Ithaca, N.Y.: Cornell University Press, 1960), 72–73, 333.

22. Alexander Agassiz to Newcomb, 9 May 1881, and Newcomb to Agassiz, 15 May 1881, SNP, Box 14. Newcomb to Gilman, 29 March 1886, Daniel Coit Gilman Papers Ms. 1, Special Collections, Milton S. Eisenhower Library, Johns Hopkins University. Archibald, "Simon Newcomb," *Science*, 872–874; Campbell, "Simon Newcomb," 13; Newcomb, *Reminiscences*, 407. Hawkins, *Pioneer*, 74, 86, 136–137.

23. Newcomb, *Reminiscences*, 144–147, 182–194. Newcomb, "Story of a Telescope," 44–55. Osterbrock, Gustafson, and Unruh, *Eye on the Sky*, 15–27, 86, 119. Helen Wright, *James Lick's Monument*, 15–16, 35–38, 79–82, 187–188. Campbell, "Simon Newcomb," 7, 16.

24. Merrick, "John and Simon Newcomb," 687.

25. Archibald, "Simon Newcomb, Bibliography," 16–19.

26. In line with his wish, his family sold the collection intact after his death. The purchaser was the City College in New York. See "Presentation of the Newcomb Library," *The City College Quarterly* 6 (1910): 233–243. See also "Professor Newcomb's Library to Be Sold," *Publishers' Weekly*, no. 1956 (24 July 1909): 217.

27. "Simon Newcomb Papers" (Finding Guide, Manuscript Division, Library of Congress, 1963), 1–9. Besides keeping letters that he received, Newcomb usually preserved his own, outgoing letters as letterpress copies in bound volumes. (All of these copies are deteriorating and losing legibility.)

28. "Invitations and Cards," SNP, Boxes 127–134. Newcomb to His Excellency the President [Rutherford B. Hayes], 13 Nov. 1880, SNP, Box 5. See also the file labeled "President of the U.S.," SNP, Box 36.

29. Archibald, "Simon Newcomb," *Science*, 872–874; Campbell, "Simon Newcomb," 15–16; Hill, "Newcomb as an Astronomer," 356–357; Newcomb, *Reminiscences*, 275; C[owell], "Simon Newcomb," xxxvi.

30. Archibald, "Simon Newcomb," *Science*, 871; Alvord, "Address of the Retiring President," 56; William Harper Davis, "The International Congress of Arts and Science," *Popular Science Monthly* 66 (1904–1905): 8. Obituary notices, SNP, Boxes 124, 125. See also, "Simon Newcomb: Memorial Addresses," *Bulletin of the Philosophical Society of Washington*, 133–167. For a recent appraisal of Newcomb's standing among astronomers, see Brush, "Looking Up," 46–47.

CHAPTER 6: AMERICAN SCIENCE, SCIENTIFIC METHOD, AND SOCIAL PROGRESS

1. Adams to Lodge, June 1874, *Letters of Henry Adams (1851–1891)*, ed. Worthington C. Ford (Boston: Houghton Mifflin Co., 1930), 259–260.

2. "Editor's Table: Professor Newcomb on American Science," *Popular Science Monthly* 6 (Dec. 1874): 238. Silliman to Newcomb, 7 Oct. 1874, SNP, Box 39. See also Bruce, *Launching of Modern American Science*, 342.

3. Newcomb, "Exact Science in North America," *North American Review* 119 (Oct. 1874): 286–299. Reprinted with brief commentary in *Science in America: Historical Selections*, ed. John C. Burnham (New York: Holt, Rinehart & Winston, 1971), 204–221. See also [Newcomb], "Note on the American Association for the Advancement of Science," *Nation* 19 (Aug. 1874): 123. Cf. Newcomb, "Conditions Which Discourage Scientific Work in America" (see chap. 5, n. 8), 156.

4. Newcomb, "Exact Science," 301–308.

5. Ibid., 293.

6. John L. LeConte to Newcomb, 26 Oct. 1874, SNP, Box 29.

7. Adams to Newcomb, 15 Aug. 1875 and 22 Aug. 1875, SNP, Box 141. Letters also reprinted in *Henry Adams and His Friends: A Collection of His Unpublished Letters*, ed. Harold D. Cater (1947; reprint, New York: Octagon Books, 1970), 68–70.

8. Adams to Newcomb, 25 Oct. 1875, 23 Dec. 1875, SNP, Box 14. Reprinted in *Henry Adams and His Friends*, ed. Cater, 71–72.

9. Silliman to Newcomb, 31 Jan. 1876, SNP, Box 39; Dana to Newcomb, 29 Feb. 1876, SNP, Box 20.

10. Newcomb, *Reminiscences*, 403. Gilman, *Launching of a University*, 115.

11. Newcomb, "Abstract Science in America" (see chap. 3, n. 27), 109–117. For deficiencies involving "academies, institutions, and journals," see also Newcomb, "Biographical Memoir [of Joseph Henry]," *Smithsonian Miscellaneous Collection* 21 (1881): 467.

12. Newcomb, "Abstract Science in America," 112–123. Henry, *Scientific Writings* 2: 8–10, 86–87. See also Bruce, *Launching of Modern American Science*, 115–134, 192–195.

13. Newcomb, "Abstract Science in America," 91–92, 115–116.

14. Ibid., 122–123. See also Newcomb, "What Is a Liberal Education?" *Science* 3 (1884): 435–436. Cf. Mill's view in *Inaugural Address at the University of St. Andrews* (1867), in *John Stuart Mill on Education*, ed. Francis W. Garforth (New York: Teachers College Press, Columbia University, 1971), 184–198.

15. Plotkin, "Astronomers versus the Navy," 387. Schaffer, "Astronomers Mark Time," 127.

16. Diary for 1870, 6 April 1870, SNP, Box 1. *Bulletin of the Philosophical Society of Washington*, 1 (March 1871–June 1874): viii, xiv, 19–20. The founding of the society is discussed in Flack, *Desideratum*, 59–69. Bruce, *Launching of Modern American Science*, 297, 343.

17. "Special Meeting: May 14, 1878," *Bulletin of the Philosophical Society of Washington* 2 (Oct. 1874–Nov. 1878): 196. "149th Meeting: Nov. 9, 1878" and "168th Meeting: Nov. 1879," *Bulletin of the Philosophical Society of Washington* 2 (1874–78): 17, 51; "187th Meeting: Nov. 6, 1880" and "189th Meeting: Dec. 4, 1880," *Bulletin of the Philosophical Society of Washington* 4 (Oct. 1880–June 1881): 29–30, 39–40, 52.

18. The pamphlet was "The Relation of Scientific Method to Social Progress" (Washington, D.C.: Judd and Detweiler, 1880). Reprinted in *Bulletin*

of the Philosophical Society of Washington 4 (Oct. 1880–June 1881): 40–52; *Smithsonian Miscellaneous Collections*, 25, Article 1 (1883): 40–52; and Newcomb, *Sidelights on Astronomy and Kindred Fields of Popular Science: Essays and Addresses* (New York: Harper, 1906), 312–329. References in the present study are to the pamphlet published by Judd and Detweiler.

19. Newcomb, "Relation of Scientific Method," 3–7. Newcomb, "What Is a Liberal Education?", 435–436. See also Newcomb, "President Eliot on a Liberal Education," *Science* 3 (1884): 704–705.

20. Albert E. Moyer, "Physics Teaching and the Learning Theories of G. Stanley Hall and Edward L. Thorndike," *Physics Teacher* 19 (April 1981): 221.

21. John W. Myer, quoted in Gieryn, Bevins, and Zehr, "Professionalization of American Scientists," 394.

22. Newcomb, "Relation of Scientific Method," 7–8. Cf. Mill, "Book I: Of Names and Propositions," *System of Logic*, 11–106.

23. Newcomb, "Relation of Scientific Method," 7–8, 11.

24. Ibid., 11–12.

25. Ibid., 12, 14. Cf. Wright, "Philosophy of Herbert Spencer" (chap. 4, n. 37), 46–47. For a prior discussion of this point and Newcomb's earlier thoughts, see chap. 3, nn. 21, 22.

26. Newcomb, "Relation of Scientific Method," 14. Cf. Bertrand Russell, in *The Basic Writings of Bertrand Russell*, ed. Robert E. Egner and Lester E. Dennonn (New York: Simon & Schuster, 1961), 626.

27. Cf. Newcomb, "Relation of Scientific Method," 8, with Wright, "Speculative Dynamics" (see chap. 4, n. 38), 387–388. See also Mill (who quotes Whewell), *System of Logic*, 404–405.

28. Newcomb, "Relation of Scientific Method," 3–4, 10, 14. See also the prefatory comments by Karl Pearson, Bertrand Russell, and James R. Newman in William Clifford's posthumous book, *The Common Sense of the Exact Sciences* (1885; reprint, New York: Alfred A. Knopf, 1946), v–lxvi. Mill, *System of Logic*, 172–174.

29. Newcomb, "Relation of Scientific Method," 8–10.

30. Cf. Newcomb, "Relation of Scientific Method," 13–14, with Chauncey Wright, "Evolution of Self-Consciousness" (see chap. 1, n. 17), 244–246. See also Mill, *System of Logic*, 196–197, 521–526.

CHAPTER 7: POLITICAL ECONOMICS

1. Vincent P. DeSantis, "The Gilded Age in American History," *Hayes Historical Journal: A Journal of the Gilded Age* 7 (Winter 1988): 38–39. See also Irwin Unger, *The Greenback Era: A Social and Political History of American Finance, 1865–1879* (Princeton, N.J.: Princeton University Press, 1964); Haskell, *Emergence of Professional Social Science*; and Furner, *Advocacy and Objectivity*.

2. Newcomb, *Reminiscences*, 339–408. Campbell, "Simon Newcomb," 14. Archibald, "Simon Newcomb," *Science*, 871–878. Archibald, "Simon Newcomb, Bibliography," 58–65. Henry Adams to Newcomb, 5 Nov. 1873,

SNP, Box 14; see also Newcomb to Charles Norton, 6 Oct. 1866, Norton Papers, Houghton Library, Harvard University, Item bMS Am 1088 (4940). Harris to Newcomb, 15 May 1886, SNP, Box 25. Eliot to Newcomb, 4 Aug. 1879, 20 Aug. 1879, 28 Nov. 1879, and 1 Dec. 1879, SNP, Box 21. Hawkins, *Pioneer*, 185. For a general discussion of Newcomb's economic views, see Loretta M. Dunphy, "Simon Newcomb: His Contributions to Economic Thought" (Ph.D. diss., Catholic University of America, 1956). See also Paul F. Boller's chapter "The Old Political Economy and the New," in his *American Thought in Transition: The Impact of Evolutionary Naturalism, 1865–1900* (Chicago: Rand McNally & Co., 1969), 70–93; T. W. Hutchison's chapter "Simon Newcomb and Irving Fisher," in his *A Review of Economic Doctrines: 1870–1929* (Oxford: Clarendon Press, 1962), 269–278; Joseph Dorfman, *The Economic Mind in American Civilization* (New York: Viking Press, 1949), 3: 82–87.

3. Laughlin to Newcomb, 30 Nov. 1884 (along with 1884 "List of Members" [of Political Economy Club]), 26 Sept. 1885, 31 Dec. 1887, SNP, Box 29; an 1887 list of members is also in SNP, Box 136. Dorfman, *Economic Mind*, 272. A. W. Coats, "The Political Economy Club: A Neglected Episode in American Economic Thought," *American Economic Review* 51 (Sept. 1961): 624–637.

4. Haskell, *Emergence of Professional Social Science*, 236; Furner, *Advocacy and Objectivity*, 323.

5. McCloskey, *Rhetoric of Economics*, 20–35. For an evaluation of McCloskey's book, see A. W. Coats, "Economic Rhetoric: The Social and Historical Context," in *The Consequences of Economic Rhetoric*, ed. Arjo Klamer, Donald N. McCloskey, and Robert M. Solow (Cambridge: Cambridge University Press, 1988), 64–84. See also McCloskey, *If You're So Smart: The Narrative of Economic Expertise* (Chicago: University of Chicago Press, 1990), 1–9, 56–69; McCloskey, "What's the Science in Social Science? Two Theorems in Economic Meta-Science," paper read at the Conference on Science and Rhetoric, April 1991, Blacksburg, Virginia, pp. 1–25.

6. Philip Mirowski, *Against Mechanism: Protecting Economics from Science* (Totowa, N.J.: Rowman & Littlefield, 1988), 7, 141–143, 159. Mirowski, *More Heat Than Light: Economics as Social Physics: Physics as Nature's Economics* (Cambridge: Cambridge University Press, 1989), 196–197, 356–358, 396.

7. Margaret Schabas, *A World Ruled by Number: William Stanley Jevons and the Rise of Mathematical Economics* (Princeton, N.J.: Princeton University Press, 1990), 3, 139.

8. Newcomb, *Critical Examination of Our Financial Policy*, 12–19, 23–24, 45–46.

9. Schabas, "Alfred Marshall, W. Stanley Jevons, and the Mathematization of Economics," *Isis* 80 (1989): 60–63. Schabas, *World Ruled by Number*, 4, 140.

10. Mirowski, *Against Mechanism*, 31. Mirowski, *More Heat Than Light*, 217–219, 254–257. See also Robert B. Ekelund and Robert F. Hébert, *A*

History of Economic Theory and Method, 2d ed. (New York: McGraw-Hill Book Co., 1983), 312–315.

11. S[imon] N[ewcomb], review of *The Theory of Political Economy*, by W. Stanley Jevons, *North American Review* 114 (1872): 435–440. Schabas, "Mathematization of Economics," 63. Schabas, *World Ruled by Number*, 98, 126; as Schabas documents on pp. 54–79, Jevons also wrote extensively on logic and scientific method. Craufurd D. W. Goodwin, "Marginalism Moves to the New World," in *The Marginal Revolution in Economics*, ed. R. D. Collison Black, A. W. Coats, and Craufurd D. W. Goodwin (Durham, N.C.: Duke University Press, 1973), 292–295. Dorfman, *Economic Mind*, 82–87.

12. Newcomb, "The Method and Province of Political Economy" (see chap. 3, n. 24), 242, 260–265. For a draft of this review, see Newcomb, MS, "Methods and Objects of the Study of Political Economy," SNP, Box 110. For a response, ca. late 1880s, to J. E. Cairnes's objection to mathematics, see Newcomb, MS, "Mathematics in Political Economy," SNP, Box 95. On the moral nature of political economy, see also Newcomb, "The Labor Question," *North American Review* 111 (July 1870): 131–132. Cf. John Stuart Mill, "On the Definition of Political Economy and on the Method of Investigation Proper to It," *Westminster Review* (Oct. 1836), reprinted in *Mill's Philosophy of Scientific Method*, ed. Nagel, pp. 407–440.

13. Newcomb, "Method and Province of Political Economy," 248.

14. Newcomb, "The Standard of Value," *North American Review* 129 (Sept. 1879): 223–237; Newcomb, "The Silver Conference and the Silver Question," *International Review* 6 (Mar. 1879): 309–333. Dorfman, *Economic Mind*, 85–87.

15. Newcomb, *Principles of Political Economy*, iii, 3–13, 47, 202, 210–214, 338; see also Newcomb's definition of "value" on pp. 62–64, 199–203. Cf. Newcomb's dual definition of "value" with Henry Margenau's belief "that every accepted scientific measurable quantity [should] have at least two definitions, one formal and one instrumental"; for Margenau's statement, see "Interpretations and Misinterpretations of Operationalism," in *The Validation of Scientific Theories*, ed. Philipp G. Frank (New York: Collier Books, 1961), 46. For an introduction to Mill's *Principles of Political Economy* (1848), see Ekelund and Hébert, *History*, 148–175.

16. Archibald, "Simon Newcomb, Bibliography," 59. John Maynard Keynes's 1930 statement is quoted in Hutchison, *Review of Economic Doctrines*, 269 n. 1. Fisher, "Obituary: Simon Newcomb," 741–744. Philip Mirowski, "How Not to Do Things with Metaphors: Paul Samuelson and the Science of Neoclassical Economics," *Studies in History and Philosophy of Science* 20 (1989): 177. Mirowski, *Against Mechanism*, 31–32. Mirowski, *More Heat Than Light*, 222–223. See also Mary Morgan, *The History of Econometric Ideas* (Cambridge: Cambridge University Press, 1990). See also Goodwin, "Marginalism," 292–294.

17. Hutchison, *Review of Economic Doctrines*, 251, 269–278. Dorfman, *Economic Mind*, 83–87. For Peirce's views on the presumptuous tone of the *Principles*, see Eisele, "Peirce–Newcomb Correspondence," 413; reprinted in Eisele, *Studies*, 58.

18. Edmund J. James, "Newcomb's Political Economy," *Science* 6 (Nov. 1885): 470–471.

19. For the religious connection, see A. W. Coats, "The First Two Decades of the American Economic Association," *American Economic Review* 50 (Sept. 1960): 555–572.

20. Newcomb to Gilman, 3 May, 14 May, 1 June, 4 June 1884, Daniel Coit Gilman Papers Ms. 1, Special Collections, Milton S. Eisenhower Library, Johns Hopkins University. See also Hawkins, *Pioneer*, 179–180; Furner, *Advocacy and Objectivity*, 60–62. An anonymous reviewer in the *Nation* also attacked Ely's published speech; see "Notes," *Nation* 38 (24 July 1884): 74.

21. Furner, *Advocacy and Objectivity*, 59–63. Haskell, *Emergence of Professional Social Science*, 184, 187 n. 48.

22. Boller, *American Thought*, 84–85. Dorfman, *Economic Mind*, 87–98, 205–212. Coats, "First Two Decades," 555–572. Hawkins, *Pioneer*, 181. Haskell, *Emergence of Professional Social Science*, 185–187. Newcomb, review of *An Introduction to Political Economy* and *Outlines of Economics*, by Richard T. Ely, *Journal of Political Economy* 3 (Dec. 1894): 106. Furner, *Advocacy and Objectivity*, 62, 64, 72–73, 118–119.

23. Newcomb, *Reminiscences*, 404–406. W. G. Sumner to Newcomb, 19 Jan. 1886, SNP, Box 41. Franklin, "Letters to the Editor: Newcomb's 'Political Economy,'" *Science* 6 (Dec. 1885): 495–496. Newcomb, "Letters to the Editor: Newcomb's 'Political Economy,'" *Science* 6 (Dec. 1885): 495.

24. "Newcomb's Political Economy," review of *Principles of Political Economy*, by Newcomb, *Nation* 42 (14 Jan. 1886): 38–39.

25. Newcomb, "The Let-Alone Principle," *North American Review* 110 (1870): 1–2. Newcomb, *Principles of Political Economy*, 443–458. Fisher, "Obituary: Simon Newcomb," 642.

26. Newcomb, "Let-Alone Principle," 7–10. Newcomb, "Method and Province of Political Economy," 254. Newcomb, *Reminiscences*, 399–402.

27. Unger, *Greenback Era*, 129–130. Coe to Newcomb, 30 Jan. 1866, 11 Sept. 1866, Coe to McCulloch, 11 Sept. 1866, SNP, Box 20. McCulloch to Newcomb, 12 Dec. 1867, 15 Jan. 1879, 14 Dec. 1887, SNP, Box 30. Newcomb, *Reminiscences*, 243–246, 399, 402. McCulloch, *Men and Measures of Half a Century: Sketches and Comments* (New York: Charles Scribner's Sons, 1888), 262, 266. For Newcomb's mixed appraisal of McCulloch's performance as Comptroller of the Currency during the Civil War, see the anonymous article, "Our Financial Future," *North American Review* 102 (Jan. 1866): 127–132.

28. Newcomb, Diary for 1866, 7 July 1866, SNP, Box 1. Newcomb, MS, "Bill for Restoring Specie Payments," SNP, Box 140. Boller, *American Thought*, 78. Garfield to Newcomb, 3 April 1872, SNP, Box 22. Newcomb, *Reminiscences*, 167–170, 349–363.

29. Newcomb, "Method and Province of Political Economy," 249, 254. Newcomb, review of *Principles of Social Science*, by H. C. Carey, *North American Review* 103 (1866): 573–580. MS drafts of this review are in SNP, Box 109. See also: Diary for 1865, 8 Aug. 1865, 9 Oct. 1865, and 14 Oct.

1865, SNP, Box 1; and Newcomb, *Reminiscences*, 400–401. Cf. Mill, *System of Logic*, 550–554; and Mill, *Autobiography*, 96–97.

30. Newcomb, review of *Principles*, by Carey, p. 574. See also Steven Lukes, *Individualism* (London: Blackwell, 1973).

31. For parallels in religion, see Gieryn, Bevins, and Zehr, "Professionalization of American Scientists," 392–409.

32. Newcomb to Gilman, 3 May, 14 May, 1 June, 4 June 1884, Daniel Coit Gilman Papers Ms. 1, Special Collections, Milton S. Eisenhower Library, Johns Hopkins University. See also Benjamin G. Rader, *Academic Mind and Reform: The Influence of Richard T. Ely in American Life* (Lexington: University of Kentucky Press, 1966), 32.

33. Newcomb, "The Two Schools of Political Economy," *Princeton Review* 14 (Nov. 1884): 292–294. See also Newcomb, "Mathematics in Political Economy."

34. Newcomb, "Two Schools," 295–297.

35. Ibid., 299–301.

36. Laughlin to Newcomb, 30 Nov. 1884, 26 Sept. 1885, SNP, Box 29. Dorfman, *Economic Mind*, 271–275. Coats, "Political Economy Club," 627–632.

37. Newcomb, *Principles of Political Economy*, v, 14–31.

38. Ibid., 32–33.

39. Ibid., 33–40.

40. Rader, *Academic Mind and Reform*, 33. Furner, *Advocacy and Objectivity*, 63 n. 4, 83. Hawkins, *Pioneer*, 181–182. Albert Shaw, "Recent Economic Works," *The Dial* 6 (Dec. 1885): 210–211.

41. "Comment and Criticism," *Science* 7 (April 1886): 361. W. G. Sumner to Newcomb, a brief note with no date and an apparent follow-up letter dated 19 Jan. 1886, SNP, Box 41. Furner, *Advocacy and Objectivity*, 92.

42. Newcomb, "Aspects of the Economic Discussion," *Science* 7 (June 1886): 538–542; reprinted in *Science Economic Discussion* (New York: Science Co., 1886), 57–67. For a new-school reply, see Richard T. Ely, "The Economic Discussion in Science," *Science* 8 (July 1886): 3–6.

43. Newcomb, "Can Economists Agree upon the Basis of Their Teachings?" *Science* 8 (July 1886): 25–26.

44. [Newcomb], "Dr. Ely on the Labor Movement," review of *The Labor Movement in America*, by Richard T. Ely, *Nation* 43 (7 Oct. 1886): 293–294. (Newcomb is identified as the author in Archibald, "Simon Newcomb, Bibliography," 60.) See also Newcomb's anonymous note, "An Economist's Advice to the Knights of Labor," *Nation* 43 (8 April 1886): 292–293.

45. Rader, *Academic Mind and Reform*, 67–71. Hawkins, *Pioneer*, 182–184. Albert Shaw, "Seven Books for Citizens," *Dial* 7 (Nov. 1886): 149–150. W. G. Sumner to Newcomb, 26 Oct. 1886, SNP, Box 41.

46. Dorfman, *Economic Mind*, 208–209. Boller, *American Thought*, 86. Ekelund and Hébert, *History*, 171, 402. Haskell, *Emergence of Professional Social Science*, 184–185.

47. Newcomb, MS, "Rough Draft of Lecture University Penn. Feb. 1888," SNP, Box 88, pp. 3–4, 16–19, 41–42. Newcomb, *Reminiscences*, 405.

48. [Newcomb], "The Economists and the Public," *Nation* 52 (25 June 1891): 510–511.
49. Newcomb, "Correspondence: New-School Political Economists," *Nation* 53 (9 July 1891): 27.
50. Newcomb, "The Problem of Economic Education," *Quarterly Journal of Economics* 7 (July 1893): 375–399.
51. Rader, *Academic Mind and Reform*, 106–158. Laughlin remained allied with Newcomb; see Laughlin, "The Study of Political Economy in the United States," *Journal of Political Economy* 1 (Dec. 1892): 1–19.
52. Walker to Newcomb, 22 Dec. 1888, SNP, Box 42; quoted in Coats, "Political Economy Club," 632–633. Newcomb, review of *An Introduction to Political Economy* and *Outlines of Economics*, by Ely, *Journal of Political Economy* 3 (Dec. 1894): 106–111.
53. Newcomb, *Reminiscences*, 404–408.

CHAPTER 8: RELIGION

1. Simon Newcomb to Mary Newcomb, 21 Aug. and 23 Aug. 1878, SNP, Box 9. See also "The Scientific Association, the Addresses of Professors Thurston and Newcomb,—Thomas A. Edison," *Christian Herald*, Sept. 1878, p. 1.
2. In addition to the *Proceedings of the AAAS* 27 (1878): 1–28, and the AAAS pamphlet (Salem, Mass.: Publ. at the Salem Press, 1878), it was reprinted with minor variations in: *Independent* 5 Sept. 1878, pp. 5–8; *Kansas City Review* 2 (Sept.–Oct. 1878): 356–367, 392–396; *Popular Science Monthly, Supplement*, no. 18 (Oct. 1878): 481–493; and the London *Journal of Science*, n.s., 1 (1879): 64–89. References in the present study are to the version published in the *Proceedings* under the title "The Course of Nature." For Newcomb's own copies, see SNP, Box 87.
3. Frank M. Turner, "Public Science in Britain, 1880–1919," 591. For Thomas Huxley's views, see his "Lectures on Evolution [New York, 1876]," in *Science and Hebrew Tradition: Essays* (New York: Appleton, 1896), 46–138. For John Tyndall's views, see his "On Prayer as a Form of Physical Energy" and "The Belfast Address," in *Fragments of Science*, 6th ed. (New York: Appleton, 1897), 40–45, 135–201. See also W. H. Brock, N. D. McMillan, and R. C. Mollan, eds., *John Tyndall: Essays on a Natural Philosopher* (Dublin: Royal Dublin Society, 1981); and Houston Peterson, *Huxley: Prophet of Science* (London: Longmans, Green, 1932). For earlier tensions between science and religion, see Ronald L. Numbers, *Creation by Natural Law: Laplaces's Nebular Hypothesis in American Thought* (Seattle: University of Washington Press, 1977).
4. Bozeman, *Protestants in an Age of Science*, 81–86, 164; Hovenkamp, *Science and Religion*, 48–49. See also Bruce, *Launching of Modern American Science*, 119–127.
5. James Moore, *The Post-Darwinian Controversies: A Study of the Protestant Struggle to Come to Terms with Darwin in Great Britain and*

America 1870–1900 (Cambridge: Cambridge University Press, 1979), ix, 9, 218, 329–330, 340–341. For general discussions of late nineteenth-century science and religion, see A. Hunter Dupree, "Christianity and the Scientific Community in the Age of Darwin," and Frederick Gregory, "The Impact of Darwinian Evolution on Protestant Theology in the Nineteenth Century," in *God and Nature: Historical Essays on the Encounters between Christianity and Science*, ed. David C. Lindberg and Ronald L. Numbers (Berkeley, Los Angeles, London: University of California Press, 1986), 351–368, 369–390. See also Dupree's chapter "A Theist in the Age of Darwin," in his *Asa Gray*, 355–383; Boller, *American Thought*, 1–46; Russett, *Darwin in America*, 24–45; and Edward J. Pfeifer, "United States," in *The Comparative Reception of Darwinism*, ed. Thomas F. Glick (Austin: University of Texas Press, 1972), 168–206. For earlier analyses, see John Dewey, "The Influence of Darwinism on Philosophy" (1909), reprinted in Perry Miller, ed., *American Thought: Civil War to World War I* (New York, Holt, Rinehart & Winston, 1954), 214–225; Stow Persons, "Evolution and Theology in America," in *Evolutionary Thought in America*, ed. Persons (New Haven, Conn.: Yale University Press, 1950), 422–453.

6. Neal C. Gillespie, *Charles Darwin*, 13, 152–153.

7. Gieryn, Bevins, and Zehr, "Professionalization of American Scientists," 392–407. Gieryn and his coauthors tie demarcation using complementary images of science and religion to the Scopes "Monkey Trial" of 1925 and demarcation using the exclusion of religion from science to the McLean "Creation-Science" trial in 1981–82. Newcomb combines both tacks, advocating the complementarity of science and religion but the exclusion of natural theology.

8. Newcomb, "Course of Nature," 2–3. Cf. Newcomb, "Exact Science" (see chap. 6, n. 3), 293.

9. Newcomb, "Course of Nature," 2–3.

10. For the quotations from Wright and Peirce, see the third section of chap. 4.

11. Newcomb, "Course of Nature"; the phrasing quoted first is from the *Proceedings of the AAAS* 27 (1878): 3; the phrasing quoted second is from *Popular Science Monthly, Supplement*, no. 18 (Oct. 1878): 482. Cf. to Wright's "genetic" analysis of concepts, as reported by Madden, *Chauncey Wright and the Foundations of Pragmatism*, 91–92.

12. Newcomb, "Course of Nature," 3–4.

13. Ibid., 5–7, 19–20.

14. Ibid., 8–18.

15. Ibid., 20–28.

16. Wiener, *Evolution and the Founders of Pragmatism*, 48–65; Madden, *Chauncey Wright and the Foundations of Pragmatism*, 92–93.

17. In addition to the articles cited in subsequent notes, see: "The Scientific Association," *Christian Herald*, Sept. 1878, p. 1; D. D. Whedon, "Have Final Causes Disappeared," *Methodist Quarterly Review*, reprinted in *Independent*, 19 June 1879, pp. 5–6; and Newcomb, "Notes on Open Letters," *Sunday School Times*, 8 March 1879, n. pag.

18. Frank L. Mott, *A History of American Magazines: 1850–1865* (Cambridge, Mass.: Harvard University Press, 1938), 367–379.

19. A. Hunter Dupree later verified Gray's identity; see Dupree, *Asa Gray*, 370–376.

20. Asa Gray's letters and articles concerning Newcomb appeared under the pseudonym of "Country Reader" in the *Independent* on 19 Sept. 1878, pp. 16–17; 3 Oct. 1878, pp. 16–17; 10 Oct. 1878, p. 1; 7 Nov. 1878, pp. 1–3, 16; and 21 Nov. 1878, p. 15. Moore includes Gray among "Christian Darwinians"; see Moore, *Post-Darwinian Controversies*, 269–280. Dupree, *Asa Gray*, 370–371.

21. Newcomb's replies to Asa Gray ("Country Reader") appeared in the *Independent* on 3 Oct. 1878, pp. 4–5; 24 Oct. 1878, pp. 1–3; and 14 Nov. 1878, p. 1.

22. Porter's articles concerning Newcomb appeared initially under the pseudonym of "Another Country Reader" and later under his own name in the *Independent* on 12 Dec. 1878, pp. 1–2; 9 Jan. 1879, pp. 1–2; and 16 Jan. 1879, pp. 4–5. See also articles in the *Independent* by J. W. Andrews on 3 Oct. 1878, pp. 2–3, and William H. Ward on 26 Dec. 1878, pp. 3–4.

23. James McCosh, "Final Cause, M. Janet, and Professor Newcomb," *Princeton Review* 3 (1879): 370–371. Katherine R. Sopka, "John Tyndall: International Popularizer of Science," in *John Tyndall*, ed. Brock, McMillan, and Mollan, 198. J. David Hoeveler, Jr., *James McCosh and the Scottish Intellectual Tradition* (Princeton, N.J.: Princeton University Press, 1981), 180–211, 274–279, 355.

24. James McCosh, "Final Cause," pp. 367–369, 387. Moore numbers McCosh among "Christian Darwinists"; see Moore, *Post-Darwinian Controversies*, 245–251.

25. Cf. Newcomb, "Law and Design in Nature," *North American Review* 128 (May 1879): 538, with Mill, *System of Logic*, 183–210, and Wright, "Evolution of Self-Consciousness" (see chap. 1, n. 17), 244–245.

26. Newcomb, "Law and Design in Nature," 539–540, 542.

27. Noah Porter, Joseph Cook, James F. Clarke, and James McCosh, "Law and Design in Nature," *North American Review* 128 (May 1879): 546–547, 559–562. For Newcomb's personal, annotated copy of "Law and Design in Nature," see SNP, Box 94.

28. Newcomb, "Evolution and Theology" (see chap. 1, n. 16), 658, 661.

29. Ibid., 662. Newcomb followed this "rejoinder" with a similar but specific reply to Porter; see Newcomb, "The Attitude of Theology toward Science," *Independent*, 17 July 1879, p. 3, and 24 July 1879, p. 3. For Newcomb's views on the immortality of the soul, see Newcomb et al., "What Does Science Say," *The Christian Register* 66 (April 1887): 209–215.

30. Newcomb, "What Is the Course of Nature?" *Independent*, 24 Oct. 1878, p. 2.

31. Newcomb, unsigned note with the heading "Here is something for you to burn up when read," SNP, Box 12; for the dating of the note and the drafts of Newcomb's religious autobiography, see chap. 2, n. 4.

32. Newcomb, Diary for 1859, 2 Oct. 1859; Diary for 1860, 8 Jan. 1860; and Diary for 1861, 13 Oct. 1861; SNP, Box 1.

33. Newcomb, "A Religious Autobiography" (see chap. 2, n. 4), 16–27; Newcomb, "Development of My Religious Views" (see chap. 2, n. 4), 10; Newcomb, unsigned note with heading, "Here is something for you to burn up when read." See also, "Address of Mr. E. M. Gallaudet," in "Simon Newcomb: Memorial Addresses," *Bulletin of the Philosophical Society of Washington* 15 (1910): 164–165; in the note that Anita Newcomb McGee inserted into the MS "Development of My Religious Views," she mentioned that she showed the MS to Gallaudet before his memorial address. The Newcomb family Bible is in SNP, Box 126.

34. Newcomb, "Notes on Open Letters," *Sunday School Times*, 8 March 1879, n. pag.; copy in folder labeled "Speeches: 1874–1910," SNP, Box 87.

35. Newcomb, "Religious Autobiography" (see chap. 2, n. 4), 8; Newcomb, "Development of My Religious Views," 11; Newcomb, unsigned note with heading, "Here is something for you to burn up when read."

36. Newcomb, "Religious Autobiography," 28–30; Newcomb, "Development of My Religious Views," 12.

37. An Evolutionist [Newcomb], "An Advertisement for a New Religion" (see chap. 3, n. 31), 44–60. (Newcomb is identified as the author in Archibald, "Simon Newcomb, Bibliography," 23, 62.) One scientific doctor who answered the call for a new religion, independent of Newcomb, was Huxley; this Darwinian spokesman, like Newcomb, had appropriated church imagery to portray science as a religion replacing Christianity. See Dupree, "Christianity and the Scientific Community in the Age of Darwin," 363–366. For the views of Comte and Mill, see Cashdollar, *Transformation of Theology*, 12–15, 143, 160–161, 223–224, 410.

38. For attempts to reconcile evolution and theology, see Gregory, "Impact of Darwinian Evolution on Protestant Theology in the Nineteenth Century," 378–383, 387–388.

39. [Newcomb], "The Religion of To-day," *North American Review* 129 (Dec. 1879): 552–569. For Newcomb's annotated copy of this article, see SNP, Box 89.

CHAPTER 9: PHYSICS AND MATHEMATICS

1. Arthur L. Norberg, "Simon Newcomb's Role in the Astronomical Revolution of the Early Nineteen Hundreds" in *Sky with Ocean Joined*, ed. Dick and Doggett, 76, 84–85. Schaffer, "Astronomers Mark Time," 133–137.

2. Newcomb, *Reminiscences*, 147. Newcomb, MS, "Measuring the Velocity of Light," SNP, Box 104, in folder titled "Reminiscences of an Astronomer [Unpublished Portion]." See also Reingold, ed., *Science in Nineteenth-Century America*, 275–306.

3. Josiah Royce, "Introduction," in *Science and Hypothesis*, by Henri Poincaré, trans. George Halsted (New York: Science Press, 1905), xvi. Henry Adams, *The Education of Henry Adams*, ed. Ernest Samuels (1918; reprint, Boston: Houghton Mifflin Co., 1974), 450. Stallo's book and the initial

reviews of it are discussed in Moyer, *American Physics*, 3–32, 91–96, 114–118.

4. "Speculative Science," *International Review* 12 (1882): 334–341. Mott, *American Magazines*, 35.

5. Percy Bridgman, ed., "Introduction," in *Concepts and Theories of Modern Physics*, by John B. Stallo, 3d ed. (1888; reprint, Cambridge, Mass.: Harvard University Press, 1960), viii, xxi, xvi. See, e.g., Stallo, *Concepts and Theories of Modern Physics*, 4, 216. Von Joachim Thiele, "Karl Pearson, Ernst Mach, John B. Stallo: Briefe aus den Jahren 1897 bis 1904," *Isis* 60 (1969): 535–542. Albert E. Moyer, "P. W. Bridgman's Operational Perspective on Physics: Part I, Origins and Development," *Studies in the History and Philosophy of Science* 22 (1991): 243–249.

6. Newcomb, "Speculative Science," 336, 341.

7. Newcomb, "Course of Nature" (see chap. 3, n. 30), 20. Newcomb, "Investigation of the Dynamical Theory of Gases," *Proceedings of the American Academy of Arts and Sciences* 5 (1861): 112–114. Newcomb to William Bartlett, 7 Dec. 1862, SNP, Box 4. James C. Maxwell to Newcomb, 31 May 1879, SNP, Box 31. [Newcomb], "Note on Maxwell," *Nation* 29 (Dec. 1879): 403–404. Silvanus P. Thompson, *The Life of Lord Kelvin*, 2d ed. (1910; reprint, New York: Chelsea Publishing Co., 1976), 2: 674. Crosbie Smith and M. Norton Wise, *Energy and Empire: A Biographical Study of Lord Kelvin* (Cambridge: Cambridge University Press, 1989), 577. For an overview of atomo-mechanical physics, see Moyer, *American Physics*, 3–118. Cf. Henry, *Scientific Writings* 1: 298, 305 and 2: 89–98.

8. Henry, *Scientific Writings* 1: 259, 298–300 and 2: 34, 310–311. Newcomb, "Course of Nature," 7.

9. Newcomb, "Speculative Science," 337–338. On Newcomb's amenability to hypotheses in physics, see also his "The Philosophy of Hyper-Space," *Science*, n.s., 7 (Jan. 1898): 1–7.

10. Newcomb, "Speculative Science," 334–337.

11. For a similar distinction, see G. Schlesinger, "Operationalism," *Encyclopedia of Philosophy*.

12. Percy Bridgman, *The Logic of Modern Physics* (New York: Macmillan, 1927), 4, 6, 10. See also Moyer, *American Physics*, 142–166, and Moyer, "P. W. Bridgman's Operational Perspective on Physics."

13. Newcomb, "Speculative Science," 339.

14. Ibid., 339–340.

15. Peirce, "How to Make Our Ideas Clear" (see chap. 4, n. 17), 265–267; see also pp. 270–271; Madden, *Chauncey Wright and the Foundations of Pragmatism*, 80.

16. See, e.g., pp. 386 and 393 of Wright's "Speculative Dynamics" (see chap. 4, n. 38).

17. Newcomb, *Reminiscences*, 199–214. Cf. Newcomb, "Prof. Newcomb Discusses the Philosophy of Currency: What Gives Money Its Value," *Baltimore Sun*, Supplement, 18 June 1891; Newcomb, "Has the Standard Gold Dollar Appreciated?" *Journal of Political Economy* 1 (Sept. 1893): 503–504.

18. John B. Stallo, "Speculative Science," *Popular Science Monthly* 21 (1882), 146–152. See also Stallo, "Introduction to the Second Edition," in *Concepts and Theories of Modern Physics*, ed. Bridgman, 3d ed. Stallo's rebuttals are discussed in Moyer, *American Physics*, 26–32.

19. Stallo, "Speculative Science," 148–149, 163. Mill used a similar analogy between words and coins; see *System of Logic*, 431–432.

20. Newcomb, MS, "On the Fundamental Concepts of Physics," SNP, Box 94. An abstract of the MS, as read on 14 April 1888, appears in "Proceedings of the Philosophical Society of Washington and of Its Mathematical Section, 1888–1891," *Bulletin of the Philosophical Society of Washington* 10 (1892): 514. This MS is dated and discussed in James E. Beichler, "Ether/Or: Hyperspace Models of the Ether in America," in *The Michelson Era in American Science*, ed. Stanley Goldberg and Roger Stuewer, 206.

21. Newcomb, "The Units of Mass and Force," *Science* 2 (Oct. 1883): 493–494; "On the Definitions of the Terms 'Energy' and 'Work,'" *Philosophical Magazine*, 5th ser., 27 (Feb. 1889): 115–117; "Suggested Nomenclature of Radiant Energy," *Nature* 49 (Nov. 1893): 100; cf. Newcomb, MS, "On the Use of the Word Light in Physics," SNP, Box 95. See also Newcomb, "The Fundamental Definitions and Propositions of Geometry . . . ," *Nature* 21 (Jan. 1880): 293–295.

22. Archibald, "Simon Newcomb, Bibliography," sections 2 and 5.

23. "Correspondence," *Nation* 48 (1889): 488, 504–505, 524.

24. Newcomb to Charles Peirce, 9 March 1892, reprinted in Eisele, "Peirce–Newcomb Correspondence," 424–425; see also Eisele's comments on pp. 423–424. Beichler, "Ether/Or," 206–223.

25. Newcomb, "Philosophy of Hyper-Space," 7. Woodward, "Address of Mr. R. S. Woodward" (see pref., n. 2), 19.

26. Edward A. Krug, *The Shaping of the American High School* (New York: Harper & Row, 1964), 66. Theodore R. Sizer, *Secondary Schools at the Turn of the Century* (New Haven, Conn.: Yale University Press, 1964), xi. See also Albert E. Moyer, "Edwin Hall and the Emergence of the Laboratory in Teaching Physics," in *Physics History from AAPT Journals*, ed. Melba N. Phillips (College Park, Md.: American Association of Physics Teachers, 1985), 195.

27. Newcomb, *Reminiscences*, 7, 49, 56; "Formative Influences" (see chap. 2, n. 1), 183–184.

28. Untitled MS in folder titled "On Education," SNP, Box 114; a later note that Newcomb attached to the MS indicates that it was part of his 1854 unpublished "Essay on Human Happiness" (see chap. 2, n. 9).

29. Cf. Henry, *Scientific Writings* 1: 339. For an overview of early theories of learning, including developmental theories, see Moyer, "Physics Teaching," 221–228.

30. Archibald, "Simon Newcomb, Bibliography," 55–56. Campbell, "Simon Newcomb," 13. Newcomb, MS, "Mathematics in Education," pp. 1–2, SNP, Box 95. Newcomb, *Algebra for Schools and Colleges*, Newcomb's Mathematical Series (New York: Henry Holt and Co., 1881), iii. Newcomb to Alexander Agassiz, 15 May 1881, SNP, Box 14. For Newcomb's commitment

to "practical work" in astronomy teaching, see Newcomb to Gilman, 29 March 1886, Daniel Coit Gilman Papers Ms. 1, Special Collections, Milton S. Eisenhower Library, Johns Hopkins University. See also Bruce, *Launching of Modern American Science*, 349.

31. Newcomb, "The Teaching of Mathematics, I: Elementary Subjects," *Educational Review* 4 (Oct. 1892): 278. See also Newcomb, "Mathematical Teaching, II," 6 (Nov. 1893): 332–341.

32. Newcomb, "Teaching of Mathematics, I," 278–279.

33. Newcomb, MS, "Mathematical Education," pp. 17–27, SNP, Box 111. Mill, *System of Logic*, 148–161.

34. Newcomb, "Methods of Teaching Arithmetic," *Educational Review* 31 (April 1906): 339, 342.

35. Charles Eliot to Newcomb, 26 Nov. 1892, SNP, Box 21. National Educational Association, *Report of the Committee on Secondary School Studies* (1893; reprint, New York: Arno, 1969), 11, 104.

36. Newcomb et al., "Mathematics," in NEA, *Report*, 105–111.

37. NEA, *Report*, 18, 23–25, 50, 52. Charles Eliot had himself been a prime mover in bringing laboratory instruction to introductory chemistry and physics courses at Harvard. Also, he was a stanch proponent of teaching scientific method. In an early draft of the report, he had even claimed, "The scientific method of inquiry is the same for all subjects of human knowledge, however widely the several fields in which the method is applied may seem to differ." Eliot quoted in Sizer, *Secondary Schools*, 140; see also pp. 142, 192, 197. Krug, *Shaping*, 204–205. Moyer, "Edwin Hall," 193–196.

38. Harris to Hoke Smith, 8 Dec. 1893, reprinted in NEA, *Report*, i–ii.

CHAPTER 10: MENTAL AND PSYCHICAL SCIENCES

1. Newcomb, "Modern Scientific Materialism," *Independent*, 9 Dec. 1880, p. 1; 23 Dec. 1880, p. 1; 30 Dec. 1880, p. 3; 13 Jan. 1881, p. 3; and 27 Jan. 1881, pp. 1–2; esp. 9 Dec. 1880, p. 1, cols. 1–3.

2. Newcomb, "Materialism," 23 Dec. 1880, p. 1.

3. Ibid., p. 1.

4. Ibid., 13 Jan. 1881, p. 3; 27 Jan. 1881, pp. 2–3.

5. William James to Newcomb, 22 Oct. 1890, SNP, Box 28. Newcomb's prior letter to James is not in the James Collection at the Houghton Library, Harvard University.

6. For the history of late nineteenth-century psychical research, see Janet Oppenheim, *The Other World: Spiritualism and Psychical Research in England, 1850–1914* (Cambridge: Cambridge University Press, 1985); Seymour H. Mauskopf and Michael R. McVaugh, eds., *The Elusive Science: Origins of Experimental Psychical Research* (Baltimore: Johns Hopkins University Press, 1980); and Ivor Grattan-Guinness, *Psychical Research: A Guide to Its History, Principles and Practices* (Wellingborough, Northamptonshire: Aquarian Press, 1982).

7. Seymour H. Mauskopf, "The History of the American Society for Psychical Research: An Interpretation," *Journal of the American Society for*

Psychical Research 83 (Jan. 1989): 7–29. Eugene Taylor, "Psychotherapy, Harvard, and the American Society for Psychical Research: 1884–1889," in *Proceedings of Presented Papers: The Parapsychological Association 28th Annual Convention* (1985), 319–346. Arthur. S. Berger, "The Early History of the ASPR: Origins to 1907," *Journal of the American Society for Psychical Research* 79 (Jan. 1985): 43–53. Molly Noonan, "Science and the Psychical Research Movement," Ph.D. diss., University of Pennsylvania, 1977. More specifically, see "Formation of the Society," "Circular No. 1," "Circular No. 2," "Extracts from the Secretaries' Records," and "List of Members," *Proceedings of the American Society for Psychical Research* 1 (1885–89): 1–2, 2–4, 5–6, 49, 52–53. See also Newcomb, *Reminiscences*, 410; "Psychical Research in America," *Science* 4 (17 Oct. 1884): 369–370; Newcomb, "Psychic Force," *Science* 4 (17 Oct. 1884): 372; and Moyer, *American Physics in Transition*, 46–47.

8. William James to Thomas Davidson, 1 Feb. 1885, *Letters of William James*, 249–250. Not everyone found Newcomb's election as president to be appropriate; Newcomb recalled that naturalist and ethnologist John Wesley Powell found the choice "to be ridiculous in the highest degree." Newcomb, *Reminiscences*, 410–411.

9. Mauskopf, "History of the American Society for Psychical Research," 10–11.

10. Newcomb, "Address of the President," *Proceedings of the American Society for Psychical Research* 1 (1885–89): 66–67; Newcomb, "Modern Occultism," *The Nineteenth Century and After* 65 (Jan. 1909): 127–128; and Newcomb, *Reminiscences*, 408–409.

11. Newcomb, "Address of the President," 83.

12. Newcomb, "Psychic Force," 372–373. For a summary of Newcomb's experience with probability, see the "General Preface" to the MS "Theory of Probability," ca. 1906, SNP, Box 108.

13. Newcomb, "Psychic Force," 373.

14. Ibid., 373–374.

15. Oppenheim, *Other World*, 199–203.

16. Edmund Gurney, "Letters to the Editor: Psychical Research," *Science* 4 (5 Dec. 1884), 509–510; followed by Newcomb's response, pp. 510–511. Though Gurney's letter was published in early Dec., it is dated 4 Nov. Newcomb, "Can Ghosts Be Investigated," *Science* 4 (12 Dec. 1884): 525–527.

17. Ian Hacking, "Telepathy: Origins of Randomization in Experimental Design," *Isis* 79 (1988): 431–445. Charles Peirce and Joseph Jastrow, "On Small Differences of Sensation," *Memoirs of the National Academy of Sciences* 3 (1885): 79. Newcomb, "Psychic Force," 373. Charles Peirce, "Criticism on 'Phantasms of the Living': An Examination of an Argument of Messrs. Gurney, Myers, and Podmore," *Proceedings of the American Society for Psychical Research* 1 (1887–1889): 156–157; Charles Peirce, "Mr. Peirce's Rejoinder," 181–182. Edmund Gurney, "M. Richet's Recent Researches in Thought-Transference," *Proceedings of the Society for Psychical Research* 2 (1884): 239–257.

18. For the enthusiasm of psychical researchers in England, see Oppenheim, *Other World*.

19. G. Stanley Hall, "Introduction," in *Studies in Spiritism*, by Amy E. Tanner (New York: Appleton, 1910), xv. Newcomb, "The Georgia Wonder-Girl and Her Lessons," *Science* 5 (6 Feb. 1885), 106–108; Newcomb, *Reminiscences*, 412–416.

20. Henry, *Scientific Writings* 2: 515–519. For the British use of method to fault psychical research, see Oppenheim, *Other World*, 64, 201, 326–330.

21. Newcomb, "Address of the President," 63–69. The ASPR also issued the address as a separate pamphlet, "Annual Address of the President of the American Society for Psychical Research: January 12, 1886"; for a copy, see SNP, Box 87.

22. Newcomb, "Address of the President," 68–73. Kragh, *Historiography of Science*, 9–10.

23. Newcomb, "Address of the President," 73–86. Taylor, "Psychotherapy," 321. See also Eugene Taylor, "Experimental Psychology and Psychical Research at Harvard, 1872–1910," paper presented at the annual meeting of the History of Science Society, as part of a session on "The American Society for Psychical Research: Origin, Context, and Form," Gainesville, Florida, Oct. 1989, pp. 1–26.

24. Newcomb to Hall, 11 Jan. 1886, Records Collection, Archives of the American Society for Psychical Research (hereafter cited as ASPR); Hall to Newcomb, 28 Jan. 1886 and 1 Feb. 1886, SNP, Box 25. For drafts of Newcomb's address as originally read and as revised for publication, see SNP, Box 87.

25. "Comment and Criticism" and "Recent Psychical Researches," *Science* 7 (29 Jan. 1886): 89, 91–92.

26. Edwin Hall to Newcomb, 28 Jan. 1886, SNP, Box 25; William James to Newcomb, 12 Feb. 1886 and 7 July [1886], SNP, Box 28. William James, "Professor Newcomb's Address Before the American Society for Psychical Research," *Science* 7 (5 Feb. 1886): 123; Newcomb, "Professor Newcomb's Address Before the American Society for Psychical Research," *Science* 7 (12 Feb. 1886): 145–146. Sidgwick to Newcomb, 3 Feb. 1886, SNP, Box 39.

27. Newcomb to Edwin Hall, 11 Jan. 1886, ASPR; Hall to Newcomb, 28 Jan. 1886, SNP, Box 25; E. G. Gardiner to Newcomb, 27 Jan. 1887, SNP, Box 145; Newcomb to Hodgson, 28 Oct. 1887, ASPR. See also various lists of officers in *Proceedings of the American Society for Psychical Research* 1 (1885–89): 1, 57, 278, 571.

28. For British parallels, see Oppenheim, *Other World*, 202, 328.

CHAPTER 11: LATER YEARS

1. "Notes," *The Observatory* 21 (1898): 357. See also Archibald, "Simon Newcomb," *Science*, 873, 875–876. Marc Rothenberg, "Organization and

Control: Professionals and Amateurs in American Astronomy, 1899–1918," *Social Studies of Science* 11 (1981): 309–311.

2. Paul Forman, John L. Heilbron, and Spencer Weart, "Physics *circa* 1900: Personnel, Funding, and Productivity of the Academic Establishments," *Historical Studies in the Physical Sciences* 5 (1975): 5. Stephen Brush, "Looking Up," 43–48. Kevles, "Physics, Mathematics, and Chemistry Communities: A Comparative Analysis," in *Organization of Knowledge*, ed. Oleson and Voss, 139–172.

3. Shapley to Josepha Whitney, 11 April 1935, reprinted in "A Brief Sketch of Simon Newcomb: Astronomer, Mathematician, and Economist," report prepared for Newcomb's election to the Hall of Fame of New York University, ca. 1935, p. 3; copy with annotations by Anita Newcomb McGee, SNP, Box 125.

4. Newcomb, *Reminiscences*, 229–233. Campbell, "Simon Newcomb," 9. Norberg, "Newcomb's Role in the Astronomical Revolution," in *Sky with Ocean Joined*, ed. Dick and Doggett, 75–86.

5. Newcomb, "A New Determination of the Precessional Motion," *Astronomical Journal* 17 (11 June 1897): 161–167; Lewis Boss, "Note on Professor Newcomb's Determination of the Constant of Precession and on the Paris Conference of 1896," *Astronomical Journal* 18 (11 Aug. 1897): 9–12; Newcomb, "Reasons for the Adoption of New Values of the Precessional Motions: A Reply to the Remarks of Boss," *Astronomical Journal* 18 (27 Sept. 1897): 33–35; Boss, "The Paris Conference and the Precessional Motion," *Astronomical Journal* 18 (4 Jan. 1898): 113–118; Newcomb, "Remarks on the Precessional Motion: A Rejoinder," *Astronomical Journal* 18 (2 Feb. 1898): 137–139. See also Boss, "The Precessional Motion and the Paris Conference," *Astronomical Journal* 18 (19 Mar. 1898): 169–176; Newcomb, "Remarks on Prof. Boss's Third Paper on the Precessional Motion," *Astronomical Journal* 19 (14 Apr. 1898): 2–3; Boss, "Note on the Foregoing Communication," *Astronomical Journal* 19 (14 Apr. 1898): 4–5.

6. Newcomb, "On the Variation of Personal Equation with the Magnitude of the Star Observed," *Astronomical Journal* 16 (21 Mar. 1896): 65–67; Truman Safford, "On the Various Forms of Personal Equation in Meridian Transits," *Monthly Notices of the Royal Astronomical Society* 57 (1897): 504–514; Safford, "Personal Equation: Chronographic and Eye-and-Ear," *The Observatory* 20 (1897): 386–387. For more on the personal equation debate, see the various notes in *The Observatory* 20 (1897): 331–332, 353, 359–360, 387–388, 392–393, and also vol. 21 (1898): 179, 207–208. See also Safford, "The Two-Method Personal Equations of the Greenwich Observers," *Astronomical Journal* 18 (22 Jan. 1898): 129–130; Safford, "A Form of Personal Equation," *Astronomical Journal* 19 (25 Apr. 1898): 13–14. The British side of the controversy is discussed fully in Schaffer, "Astronomers Mark Time," 131–137.

7. The Newcomb-Boss debate offers a rich source for scholars trying to penetrate the micro-methods used implicitly by working scientists—what Larry Laudan calls "descriptive methodology" (see chap. 1, n. 3).

8. Norberg, "Newcomb's Role in the Astronomical Revolution," 85–86.

9. Archibald, "Simon Newcomb," *Science*, 875–878. [Charles Peirce], review of *Reminiscences of an Astronomer* (see chap. 4, n. 36), 237. Alvord, "Address of the Retiring President," 50.

10. Newcomb, "Reminiscences of an Astronomer," *Atlantic Monthly* 82 (1898): 242–253, 384–393, 519–526. [Charles Peirce], review of *Reminiscences*, 237. For the original MS of *Reminiscences*, see SNP, Box 113.

11. Newcomb, *His Wisdom the Defender: A Story* (1900; reprint, New York: Arno, 1975), 12, 44, 87, 110, 122, 150–151, 177, 316–317. Archibald, "Simon Newcomb, Bibliography," 67, entry 110. For the original MS of the novel, dated 1899, see SNP, Box 101. Judging by a rejection letter that he received in 1860, Newcomb had tried creative writing at an early date; see John Bartlett to Newcomb, 12 Nov. 1860, SNP, Box 16.

12. Newcomb, "Is the Airship Coming?" *McClure's Magazine* 17 (1901): 432–435. Newcomb, "The Outlook for the Flying Machine," *Independent* 55 (1903): 2509–2512, reprinted in Newcomb, *Sidelights on Astronomy*, 330–345; also reprinted in Byron E. Wall, ed., *Science in Society: Classical and Contemporary Readings* (Toronto: Wall & Thomson, 1989).

13. Newcomb, MS, "The Creed of Naturalism," pp. 1–16, SNP, Box 91; the date 1896 has been penciled next to the title of this typed MS.

14. Newcomb, MS, "The Scientific Idea," pp. 4–5, SNP, Box 87.

15. All of these writings, for example, are typed; such documents first appeared in Newcomb's papers in January 1883, when he acquired a typewriter.

16. Newcomb, MS, "Logic," SNP, Box 109; MS, "The Epistemology of Science," p. 10, SNP, Box 94. These and similar manuscripts offer fruitful historical digging for scholars interested in the nuances of Newcomb's philosophy of science.

17. Untitled MS in folder on "Induction," pp. 1–21, SNP, Box 113.

18. Newcomb, MS, "Theory of Probable Inference" (see chap. 3, n. 25), pp. 8, 14. "Winding Up of the Author's Scientific Work," p. 10, in folder labeled "Reminiscences of an Astronomer [Unpublished Portions]," SNP, Box 104; this is a chapter that Newcomb intended for a revised edition of his autobiography. For another unpolished discussion of "inductive logic" and the "statistical method," see Newcomb, MS, "Lecture," in the folder labeled "Boston Lecture 1903," SNP, Box 88.

19. Arthur Lovejoy, review of *Congress of Arts and Science, Universal Exposition, St. Louis, 1904*, ed. Howard J. Rogers, *Science* 23 (1906): 655. Albert E. Moyer, "Foreword," in *Physics for a New Century: Papers Presented at the 1904 St. Louis Congress*, ed. Katherine R. Sopka (New York: Tomash Publishers and American Institute of Physics, 1986), xiii–xx.

20. Newcomb, "The Coming International Congress of Arts and Science at St. Louis, September 19–24," *Popular Science Monthly* 65 (1904): 466–468. [Charles Peirce], review of *Congress of Arts and Science*, ed. Howard J. Rogers, *Nation* 82 (June 1906): 476; Charles Peirce was later identified by Daniel C. Haskell as the author of this anonymous review; see Haskell, *The Nation, Volumes 1–105, New York, 1865–1917: Indexes of Titles and Contributors*, 2 vols. (New York: New York Public Library, 1951–1953).

Newcomb, "The Evolution of the Scientific Investigator," reprinted in *Physics for a New Century*, ed. Sopka, p. 13. For Newcomb's continued experiential emphasis in technical writings, see Newcomb, "Philosophy of Hyper-Space" (see chap. 9, n. 9), 4–5, 7.

21. Newcomb to Peirce, 25 Jan. 1904, Charles S. Peirce Papers, Houghton Library, Harvard University, Section L 314. Cf. Peirce to Newcomb, 15 Jan. 1904, SNP, Box 34.

22. Thorndike to Newcomb, 6 Feb. 1907, p. 6, SNP, Box 41. In a personal letter of 11 July 1990, David M. Ment (Head, Special Collections, Milbank Memorial Library, Teachers College Columbia University) reports that Columbia has been unable to locate any main body of Thorndike's papers, including Newcomb's letters to him. See also Moyer, "Physics Teaching," 225–227.

23. Newcomb, "The Organization of Scientific Research," *North American Review* 182 (Jan. 1906): 32–43; reprinted in Newcomb, *Sidelights on Astronomy*, 165–181.

24. Newcomb, "Conditions Which Discourage Scientific Work" (see chap. 5, n. 8), 145–158. Cf. Newcomb, "Science and the Government," *North American Review* 170 (1900): 666–678.

25. Newcomb, "Aspects of American Astronomy," *Science* 6 (12 Nov. 1897): 718–719.

26. Newcomb, "Conditions Which Discourage Scientific Work," 145–158. Nathan Reingold and Ida H. Reingold, eds., *Science in America: A Documentary History, 1900–1939* (Chicago: University of Chicago Press, 1982), 3. Plotkin, "Astronomers versus the Navy," 387, 393–399.

27. Newcomb, "The Basis of Economics as an Exact Science," *Science* 21 (1905): 447–449. For the decline in Newcomb's publications, see Archibald, "Simon Newcomb, Bibliography," 58–60. For the so-called Progressive era, see Kevles, *Physicists*, 71–74, 94, and Burnham, *Science in America*, 252–254. Burnham quotes (p. 253) an 1898 statement by W. J. McGee as evidence of the impact of science and scientific method on the American social fabric; McGee was Newcomb's son-in-law.

28. Mauskopf, "History of the American Society for Psychical Research," 14–15. Berger, "Early History of the ASPR," 51–57.

29. Newcomb, "Modern Occultism" (see chap. 10, n. 10), 126–139; reprinted in *Living Age* 260 (Feb. 1909): 387–398. (For typed drafts of this article, see Box 93, SNP.) Sir Oliver Lodge, "The Attitude of Science to the Unusual: A Reply to Professor Newcomb," *The Nineteenth Century and After* 65 (Feb. 1909): 206–222; reprinted in *Living Age* 260 (March 1909): 707–719. James Hyslop, "Professor Newcomb and Occultism," *Journal of the American Society for Psychical Research* 3 (May 1909): 255–289. For prior tensions, see: Newcomb, "Has Telepathy Been Established?" *The Independent* 51 (June 1899): 1730–1733; Newcomb, *Reminiscences*, 410–416; and the exchange of letters between Lodge and Newcomb in the British *Journal of the Society for Psychical Research* (Jan. 1904): 169–170, and (May 1904): 243.

30. James Hyslop, "Professor Newcomb and Occultism," 257, 262.

31. For positivism, see Neal Gillespie, *Charles Darwin*. For the Progressive era, see Kevles, *Physicists*, 71–74, 97, and Burnham, *Science in America*, 252–254.

32. Neal Gillespie, *Charles Darwin*, 8–11, 53, 155–156. See also Dupree, "Christianity and the Scientific Community in the Age of Darwin" (see chap. 8, n. 5), 360–361, 365. Newcomb, "Development of My Religious Views" (see chap. 2, n. 4), 11. William James, *The Varieties of Religious Experience: A Study in Human Nature: Being the Gifford Lectures on Natural Religion Delivered at Edinburgh in 1901–02*, rev. ed. (1902; reprint, New York: Longmans, Green, & Co., 1909), 491.

33. Newcomb, "Development of My Religious Views," 1–15. Recall from chap. 2, n. 4, that Newcomb's daughter inserted a note into this MS stating: "This 'chapter' on SN's religious views was written within a few weeks of his death, as a contribution to a revised edition of his autobiography." For the circumstances surrounding Newcomb's death, see the files of extensive newspaper obituaries, SNP, Boxes 125, 126; see esp. "Simon Newcomb Dies in Washington," *New York Times*, 12 July 1909.

34. Newcomb, *Popular Astronomy* (New York: Harper & Brothers, 1878), 491–492.

35. Newcomb, "Development of My Religious Views," 13–15.

36. Putnam, "Address of Mr. Herbert Putnam" (see chap. 5, n. 1), 157–158.

CHAPTER 12: NEWCOMB AND AMERICAN PRAGMATISM

1. David A. Hollinger, "The Problem of Pragmatism in American History," *Journal of American History* 67 (1980): 88–93, 106; reprinted in Hollinger, *In the American Province* (Bloomington: Indiana University Press, 1985). For an abstruse criticism of Hollinger's thesis, see Peter T. Manicas, "Pragmatic Philosophy of Science and the Charge of Scientism," *Transactions of the Charles S. Peirce Society* 29 (Spring 1988): 179–222.

2. Hollinger, "Pragmatism," 88–93. See also Bruce, *Launching of Modern American Science*, 74.

3. Hollinger, "Pragmatism," 93–97. For Peirce's view, see Peirce, "How to Make Our Ideas Clear" (chap. 4, n. 17), 273. For a comparison of Peirce, James, and Dewey regarding their ideas on the relationship between science and philosophy, see also Wilson, *Science, Community, and the Transformation of American Philosophy*, 58–75.

4. Hollinger, "Pragmatism," 97–99.

5. Ibid., 99–100, 105. For a discussion from a rhetorical perspective of Dewey's use of scientific method, see James P. Zappen, "Scientific Rhetoric in the Nineteenth and Early Twentieth Centuries: Herbert Spencer, Thomas H. Huxley, and John Dewey," in *Textual Dynamics of the Profession: Historical and Contemporary Studies of Writing in Academic and Other Professional Communities*, ed. Charles Bazerman and James Paradis (Madison: University of Wisconsin Press, 1991), 145–167.

6. Hollinger, "Pragmatism," 93, 100–101.

7. Thomas Haskell judges collective, consensual inquiry of the type advocated by Peirce to be essential to the professionalization of American social scientists. See Haskell, *Emergence of Professional Social Science*, 237–239.

8. Chauncey Wright, "A Physical Theory of the Universe," *North American Review* (July 1864): 9n; "The Genesis of Species," *North American Review* (July 1871), 131–132; and "Evolution of Self-Consciousness," *North American Review* (April 1873), 243–245; all reprinted in *Philosophical Discussions*, ed. Charles E. Norton (1877; reprint, New York: Burt Franklin, 1971). See also Madden, *Chauncey Wright and the Foundations of Pragmatism*, 70–72, 83–87. Charles Peirce, "How to Make Our Ideas Clear," 267.

9. Madden, *Chauncey Wright and the Foundations of Pragmatism*, 43–50. Hollinger, "Pragmatism," 98.

10. Albert Moyer, "John Dewey on Physics Teaching," *The Physics Teacher* 20 (1982): 173–175.

11. Madden, *Chauncey Wright and the Foundations of Pragmatism*, 80–81. For a complimentary commentary on Madden's views on Wright, see Giuffrida, "Philosophical Thought of Chauncey Wright," 33–64.

12. Madden, *Chauncey Wright and the Foundations of Pragmatism*, 81.

13. Ibid., 77.

14. Ibid., 81. For a similar distinction, see G. Schlesinger, "Operationalism," *Encyclopedia of Philosophy*.

15. Madden, *Chauncey Wright and the Foundations of Pragmatism*, 81; includes quotation from William James. Cf. Charles Peirce, "How to Make Our Ideas Clear," 265.

16. Madden, *Chauncey Wright*, 62–63, 124–125; see also Madden, *Chauncey Wright and the Foundations of Pragmatism*, 16. Laudan, *Science and Values*, 81–82; see also Laudan, "Why Was the Logic of Discovery Abandoned?" in his *Science and Hypothesis*, 181–191. John Dewey, "An Empirical Survey of Empiricisms," *Studies in the History of Ideas* 3 (1935), reprinted in *On Experience, Nature, and Freedom*, ed. Richard J. Bernstein (New York: Bobbs-Merrill Co., 1960), 86.

17. Wright, *Chauncey Wright and the Foundations of Pragmatism*, 78.

18. Charles Peirce, "How to Make Our Ideas Clear," 273–274.

19. Newcomb, "Course of Nature" (see chap. 3, n. 30), 6–7.

20. Hollinger, "Pragmatism," 97.

21. William James, "What Pragmatism Means," from *Pragmatism: A New Name for Some Old Ways of Thinking* (1907), reprinted in *The American Pragmatists*, eds. Milton Konvitz and Gail Kennedy (New York: Meridian Books, 1960), 30.

22. This is not to say that other late nineteenth-century thinkers did nothing to usher in pragmatism; since William James proclaimed the twenty-year gap, scholars have located various Americans, for example within the legal profession, who also aided in filling the gap. See Max H. Fisch,

"Introduction," in Christian J. W. Kloesel, ed., *Writings of Charles S. Peirce*, xxxi–xxxvii; see esp. p. xxxvii n. 9.

23. For Chauncey Wright's influence on William James, see Madden, *Chauncey Wright*, 125–132. For Charles Peirce's influence on James, and for a discussion of developments before 1898, see Max H. Fisch, "American Pragmatism Before and After 1898," in *American Philosophy from Edwards to Quine*, ed. Robert W. Shahan and Kenneth R. Merrill (Norman: University of Oklahoma Press, 1977), 78–92.

CHAPTER 13: PRAGMATISM AND METHODOLOGICAL RHETORIC

1. Hollinger, "Pragmatism," 92–93.
2. Ibid., 93, 100–101.
3. Yeo, "Scientific Method and the Rhetoric of Science," in *Politics and Rhetoric*, ed. Schuster and Yeo, 262–263, 287–289. Of course, Yeo recognizes that the British proclamations and exchanges about method involved a spectrum of participants and took place on different levels of sophistication. Accordingly, he does not claim that there was full consensus in early nineteenth-century Britain on the unity, accessibility, and transferability of scientific method; nevertheless, influential commentators on science did endorse the three characterizations. Neither does he maintain that the three characterizations are uniformly applicable to the entire century; as we will see, the assumption of method's accessibility faded somewhat in Britain. Also, Yeo would not deny that symbiotic relationships existed between British and non-British commentators, especially Comte.
4. William James, *Pragmatism: A New Name for Some Old Ways of Thinking* (New York: Longmans, Green, & Co., 1907), v.
5. Yeo, "Scientific Method and the Rhetoric of Science," in *Politics and Rhetoric*, ed. Schuster and Yeo, 263–264, 274–275, 282–285. See also Schuster and Yeo, *Politics and Rhetoric*, ix. Mill is quoted in Yeo, "Reviewing Herschel's *Discourse*," review of *A Preliminary Discourse on the Study of Natural Philosophy*, by John. F. Herschel with a new foreword by Arthur Fine, *Studies in History and Philosophy of Science* 20 (Dec. 1989): 551.
6. Laudan, *Science and Values*, 83. See also Laudan, "Why Was the Logic of Discovery Abandoned?" in his *Science and Hypothesis*, 188.
7. Yeo, "Scientific Method and the Rhetoric of Science," in *Politics and Rhetoric*, ed. Schuster and Yeo, 260, 274–275.
8. Ibid., 263–264, 282–283.
9. Ibid., 264–273, 282–287.
10. Bozeman, *Protestants in an Age of Science*, xiii, 5–8, 24–26. Daniels, *American Science in the Age of Jackson*, 63–69, 86. See also Hovenkamp, *Science and Religion in America*, 23–36; Kevles, *Physicists*, 7; and Bruce, *Launching of Modern American Science*, 4, 68–69. For qualifications of Daniels's views on method, see Greene, "An Earlier Day," 638–640; see also Dupree, review of *American Science*, by Daniels, 281.

11. Bruce, *Launching of Modern American Science*, 4, 68–69. Bozeman, *Protestants in an Age of Science*, 26, 64–70. Daniels, *American Science in the Age of Jackson*, 69–85, 86.

12. Hollinger relates the timing of pragmatism to its ability to fill an intellectual void and to reinforce social values in an era of cultural change; see Hollinger, "Pragmatism," 104–105.

13. Newcomb, "The Place of Astronomy among the Sciences," *Sidereal Messenger* 7 (1888): 69–70, quoted in Lawrence Badash, "The Completeness of Nineteenth-Century Science," *Isis* 63 (1972): 53.

14. Moyer, *American Physics*.

15. Reingold, "American Indifference to Basic Research: A Reappraisal," in *Nineteenth-Century American Science: A Reappraisal*, ed. George H. Daniels (Evanston: Northwestern University Press, 1972), 50. Bruce, *Launching of Modern American Science*, 353. See also Kevles, *Physicists*, 25–74, and Kevles, Sturchio, and Carroll, "The Sciences in America," 27–32. For moves toward professionalization among antebellum American scientists, especially astronomers, see Mary Ann James, *Elites in Conflict*.

16. Yeo, "Scientific Method and the Rhetoric of Science," in *Politics and Rhetoric*, ed. Schuster and Yeo, 272–273, 283. Cf. Laudan, *Science and Values*, 81–81.

Select Bibliography

For an annotated bibliography of Newcomb's writings and early articles about him, see the list that Raymond C. Archibald prepared in 1924 for the National Academy of Sciences (the ninth item below). Of Newcomb's writings cited in the present volume, the following bibliography includes only his main books. For his many shorter works, see either this volume's notes or Archibald's list.

Adams, Henry. *The Education of Henry Adams*. Edited by Ernest Samuels. 1918. Reprint. Boston: Houghton Mifflin Co., 1974.
―――. *Henry Adams and His Friends: A Collection of His Unpublished Letters*. Edited by Harold D. Cater. 1947. Reprint. New York: Octagon Books, 1970.
―――. *Letters of Henry Adams, 1851–1891*. Edited by Worthington C. Ford. Boston: Houghton Mifflin Co., 1930.
Adams, Todd L. "The Commonsense Tradition in America: E. H. Madden's Interpretations." *Transactions of the Charles S. Peirce Society* 24 (Winter 1988): 1–31.
Alvord, William. "Address of the Retiring President of the Society, in Awarding the Bruce Medal to Professor Simon Newcomb." *Publications of the Astronomical Society of the Pacific* 10 (April 1898): 49–58.
Appleman, Philip, ed. *Darwin: A Norton Critical Edition*. 2d ed. New York: W. W. Norton & Co., 1979.
Archibald, Raymond C. "Bibliography of the Life and Works of Simon Newcomb." *Transactions of the Royal Society of Canada* 11, sect. 3 (1905): 79–110.
―――. "Simon Newcomb." *Science*, n.s. 44 (1916): 871–878.

———. "Simon Newcomb, 1835–1909, Bibliography of His Life and Work." *Memoirs of the National Academy of Sciences* 17, 1st mem. (1924): 19–69.

Aronowitz, Stanley. *Science as Power: Discourse and Ideology in Modern Science*. Minneapolis: University of Minnesota Press, 1988.

Badash, Lawrence. "The Completeness of Nineteenth-Century Science." *Isis* 63 (1972): 48–58.

Beichler, James E. "Ether/Or: Hyperspace Models of the Ether in America." In *The Michelson Era in American Science: 1870–1930*, edited by Stanley Goldberg and Roger H. Stuewer, 206–223. AIP Conference Proceedings, no. 179. New York: American Institute of Physics, 1988.

Benjamin, Andrew E., Geoffrey N. Cantor, and John R. R. Christie, eds. *The Figural and the Literal: Problems of Language in the History of Science and Philosophy, 1630–1800*. Manchester: Manchester University Press, 1987.

Benjamin, Marcus. "Simon Newcomb, Astronomer (1835–1909)." In *Leading American Men of Science*, edited by David Starr Jordan, 362–389. New York: Henry Holt & Company, 1910.

Berger, Arthur S. "The Early History of the ASPR: Origins to 1907." *Journal of the American Society for Psychical Research* 79 (Jan. 1985): 39–60.

Black, R. D. Collison, A. W. Coats, and Craufurd D. W. Goodwin, eds. *The Marginal Revolution in Economics: Interpretation and Evaluation*. Durham, N.C.: Duke University Press, 1973.

Boller, Paul F., Jr. *American Thought in Transition: The Impact of Evolutionary Naturalism, 1865–1900*. The Rand McNally Series on the History of American Thought and Culture, edited by David D. Van Tassel. Chicago: Rand McNally & Co., 1969.

Bozeman, Theodore D. *Protestants in an Age of Science: The Baconian Ideal and Antebellum American Religious Thought*. Chapel Hill: University of North Carolina Press, 1977.

Brasch, Frederick E. "Einstein's Appreciation of Simon Newcomb." *Science* 69 (1929): 248–249.

Brock, W. H., N. D. McMillan, and R. C. Mollan, eds. *John Tyndall: Essays on a Natural Philosopher*. Dublin: Royal Dublin Society, 1981.

Brooke, John H. Review of *The Politics and Rhetoric of Scientific Method: Historical Studies*, edited by John A. Schuster and Richard R. Yeo. *Isis* 78 (1987): 93–94.

Bruce, Robert V. *The Launching of Modern American Science: 1846–1876*. The Impact of the Civil War Series. New York: Alfred A. Knopf, 1987.

Brush, Stephen G. "Looking Up: The Rise of Astronomy in America." *American Studies* 20 (1979): 41–67.

Burnham, John C. *How Superstition Won and Science Lost: Popularizing Science and Health in the United States*. New Brunswick, N.J.: Rutgers University Press, 1987.

———, ed. *Science in America: Historical Selections*. New York: Holt, Rinehart & Winston, 1971.

Campbell, W. W. "Simon Newcomb." *Memoirs of the National Academy of Sciences* 17, 1st mem. (1924): 1–18.

Select Bibliography

Cashdollar, Charles D. *The Transformation of Theology, 1830–1890, Positivism and Protestant Thought in Britain and America.* Princeton, N.J.: Princeton University Press, 1989.

Cayley, Arthur. "Address Delivered by the President on Presenting the Gold Medal of the Society to Professor Simon Newcomb." *Monthly Notices of the Royal Astronomical Society* 34 (1874): 224–233. Reprinted in *The Collected Mathematical Papers of Arthur Cayley.* Vol. 9, pp. 176–184. Cambridge: Cambridge University Press, 1896.

Clifford, William. *The Common Sense of the Exact Sciences.* 1885. Reprint. New York: Alfred A. Knopf, 1946.

Coats, A. W. "Economic Rhetoric: The Social and Historical Context." In *The Consequences of Economic Rhetoric*, edited by Arjo Klamer, Donald N. McCloskey, and Robert M. Solow, 64–84. Cambridge: Cambridge University Press, 1988.

———. "The First Two Decades of the American Economic Association." *American Economic Review* 50 (September 1960): 555–574.

———. "The Political Economy Club: A Neglected Episode in American Economic Thought." *American Economic Review* 51 (September 1961): 624–637.

Cobbett, William. *A Grammar of the English Language.* The 1818 New York 1st ed., revised in 1819, 1820, and 1823. Reprinted in *Costerus*, edited by Charles C. Nickerson and John W. Osborne, n.s. vol. 39. Amsterdam: Rodopi, 1983.

Comte, Auguste. *Cours de philosophie positive.* Paris, 1830. In *Introduction to Positive Philosophy.* Translated by Frederick Ferré. Indianapolis: Bobbs-Merrill Co., 1970.

Conkin, Paul. *Puritans and Pragmatists.* New York: Dodd, Mead & Co., 1968.

Cooter, Roger. *The Cultural Meaning of Popular Science: Phrenology and the Organization of Consent in Nineteenth-Century Britain.* Cambridge: Cambridge University Press, 1984.

Daniels, George H. *American Science in the Age of Jackson.* New York: Columbia University Press, 1968.

———, ed. *Nineteenth-Century American Science: A Reappraisal.* Evanston, Ill.: Northwestern University Press, 1972.

DeSantis, Vincent P. "The Gilded Age in American History." *Hayes Historical Journal: A Journal of the Gilded Age* 7 (Winter 1988): 38–57.

Dewey, John. *On Experience, Nature, and Freedom: Representative Selections.* Edited by Richard J. Bernstein. New York: Bobbs-Merrill Co., 1960.

Dick, Steven J. "How the U.S. Naval Observatory Began, 1830–65." In *Sky with Ocean Joined: Proceedings of the Sesquicentennial Symposia of the U.S. Naval Observatory*, edited by Steven J. Dick and LeRoy E. Doggett, 166–181. Washington, D.C.: U.S. Naval Observatory, 1983.

Dorfman, Joseph. *The Economic Mind in American Civilization.* Vol. 3, 1865–1918. New York: Viking Press, 1959.

Dreyer, J. L. E., and H. H. Turner, eds. *History of the Royal Astronomical Society, 1820–1920.* 1923. Reprint. Oxford: Blackwell Scientific Publications, 1987.

Dunphy, Loretta M. "Simon Newcomb: His Contributions to Economic Thought." Ph.D. diss., Catholic University of America, 1956.

Dupree, A. Hunter. *Asa Gray: 1810–1888.* Cambridge, Mass.: Harvard University Press, 1959.

———. Review of *American Science in the Age of Jackson,* by George H. Daniels. *American Historical Review* 74 (October 1968): 281.

———. *Science in the Federal Government: A History of Policies and Activities to 1940.* Cambridge, Mass.: Harvard University Press, 1957.

Edwards, Paul, and Arthur Pap, eds. *A Modern Introduction to Philosophy: Readings from Classical and Contemporary Sources.* New York: Free Press, 1965.

Eisele, Carolyn. "Benjamin Peirce." *Dictionary of Scientific Biography.* 14 vols. Edited by Charles C. Gillespie. New York: Charles Scribner's Sons, 1970–76.

———. "The Charles S. Peirce–Simon Newcomb Correspondence." *Proceedings of the American Philosophical Society* 101 (1957): 409–433.

———. *Studies in the Scientific and Mathematical Philosophy of Charles S. Peirce.* Studies in Philosophy, edited by R. M. Martin, no. 29. The Hague: Mouton Publishers, 1979.

———, ed. *Historical Perspectives on Peirce's Logic of Science: A History of Science.* 2 parts. Berlin: Mouton Publishers, 1985.

Ekelund, Robert B., Jr., and Robert F. Hébert. *A History of Economic Theory and Method.* 2d ed. New York: McGraw-Hill Book Co., 1983.

Ely, Richard T. *Ground under Our Feet: An Autobiography.* New York: Macmillan Co., 1938.

Fisher, Irving. "Obituary: Simon Newcomb." *The Economic Journal* 19 (Dec. 1909): 641–644.

Flack, J. Kirkpatrick. *Desideratum in Washington: The Intellectual Community in the Capital City, 1870–1900.* Cambridge, Mass.: Schenkman, 1975.

Forman, Paul. "Independence, Not Transcendence, for the Historian of Science." *Isis* 82 (1991): 71–86.

Forman, Paul, John L. Heilbron, and Spencer Weart. "Physics *circa* 1900: Personnel, Funding, and Productivity of the Academic Establishments." *Historical Studies in the Physical Sciences* 5 (1975): 1–185.

Frank, Philipp G., ed. *The Validation of Scientific Theories.* New York: Collier Books, 1961.

Furner, Mary O. *Advocacy and Objectivity: A Crisis in the Professionalization of American Social Science, 1865–1905.* Lexington: University Press of Kentucky, 1975.

Garforth, Francis W., ed. *John Stuart Mill on Education.* New York: Teachers College Press, Columbia University, 1971.

Select Bibliography

Gieryn, Thomas F., George M. Bevins, and Stephen C. Zehr. "Professionalization of American Scientists: Public Science in the Creation/Evolution Trials." *American Sociological Review* 50 (1985): 392–409.

Gillespie, Neal C. *Charles Darwin and the Problem of Creation.* Chicago: University of Chicago Press, 1979.

Gilman, Daniel C. *The Launching of a University.* New York: Dodd, Mead, 1906.

Giuffrida, Robert, Jr. "The Philosophical Thought of Chauncey Wright: Edward Madden's Contribution to Wright Scholarship." *Transactions of the Charles S. Peirce Society* 24 (Winter 1988): 33–64.

Glick, Thomas F., ed. *The Comparative Reception of Darwinism.* Austin: University of Texas Press, 1974.

Golinski, Jan. "The Theory of Practice and the Practice of Theory: Sociological Approaches in the History of Science." *Isis* 81 (1990): 492–505.

Grattan-Guinness, Ivor. *Psychical Research: A Guide to Its History, Principles and Practices.* Wellingborough, Northamptonshire: Aquarian Press, 1982.

Greene, John C. *American Science in the Age of Jefferson.* Ames: Iowa State University Press, 1984.

––––––. "An Earlier Day in American Science." Review of *American Science in the Age of Jackson,* by George Daniels, and *Science in Nineteenth-Century America,* by Nathan Reingold. *Science* 160 (10 May 1968): 638–640.

Hacking, Ian. "Telepathy: Origins of Randomization in Experimental Design." *Isis* 79 (1988): 427–451.

Hall, G. Stanley. "Introduction." In *Studies in Spiritism,* by Amy E. Tanner. New York: Appleton, 1910.

Harman, P. M. *Energy, Force, and Matter: The Conceptual Development of Nineteenth-Century Physics.* Cambridge History of Science, edited by George Basalla and William Coleman. Cambridge: Cambridge University Press, 1982.

Haskell, Daniel C. *The Nation, Volumes 1–105, New York, 1865–1917: Indexes of Titles and Contributors.* 2 vols. New York: New York Public Library, 1951–1953.

Haskell, Thomas L. *The Emergence of Professional Social Science: The American Social Science Association and the Nineteenth-Century Crisis of Authority.* Urbana: University of Illinois Press, 1977.

Hawkins, Hugh. *Pioneer: A History of the Johns Hopkins University, 1874–1889.* Ithaca, N.Y.: Cornell University Press, 1960.

Heimert, Alan, and Andrew Delbanco, eds. *The Puritans in America: A Narrative Anthology.* Cambridge, Mass.: Harvard University Press, 1985.

Henry, Joseph. *Scientific Writings of Joseph Henry.* 2 vols. Washington, D.C.: Smithsonian Institution, 1886. These 2 vols. were issued as vol. 30 of *Smithsonian Miscellaneous Collections.* Washington, D.C.: Smithsonian Institution, 1887.

Hoeveler, J. David, Jr. *James McCosh and the Scottish Intellectual Tradition.* Princeton, N.J.: Princeton University Press, 1981.

Hollinger, David A. "The Problem of Pragmatism in American History." *Journal of American History* 67 (1980): 88–107. Reprinted in Hollinger, *In the American Province: Studies in the History and Historiography of Ideas.* Bloomington: Indiana University Press, 1985.

Hollinger, David A., and Charles Capper, eds. *The American Intellectual Tradition: A Source Book.* Vol. 2, *1865 to the Present.* Oxford: Oxford University Press, 1989.

Hovenkamp, Herbert. *Science and Religion in America: 1800–1860.* Philadelphia: University of Pennsylvania Press, 1978.

Howe, Frederic C. *The Confessions of a Reformer.* New York: Charles Scribner's Sons, 1925.

Hull, David L. *Darwin and His Critics: The Reception of Darwin's Theory of Evolution by the Scientific Community.* Cambridge, Mass.: Harvard University Press, 1973.

Hutchison, T. W. *A Review of Economic Doctrines: 1870–1929.* Oxford: Clarendon Press, 1962.

Huxley, Thomas. *Science and Hebrew Tradition: Essays.* New York: Appleton, 1896.

James, Mary Ann. *Elites in Conflict: The Antebellum Clash over the Dudley Observatory.* New Brunswick, N.J.: Rutgers University Press, 1987.

James, William. *Letters of William James.* Edited by Henry James. Boston: Atlantic Monthly Press, 1920.

———. *Pragmatism: A New Name for Some Old Ways of Thinking.* New York: Longmans, Green, & Co., 1907.

———. *The Varieties of Religious Experience: A Study in Human Nature: Being the Gifford Lectures on Natural Religion Delivered at Edinburgh in 1901–1902.* Rev. ed. 1902. Reprint. New York: Longmans, Green, & Co., 1909.

Janiczek, Paul M. "Remarks on the Transit of Venus Expedition of 1874." In *Sky with Ocean Joined: Proceedings of the Sesquicentennial Symposia of the U.S. Naval Observatory,* edited by Steven J. Dick and LeRoy E. Doggett, 52–73. Washington, D.C.: U.S. Naval Observatory, 1983.

Jarrell, Richard. *The Cold Light of Dawn: A History of Canadian Astronomy.* Toronto: University of Toronto Press, 1988.

Jones, Bessie Z., and Lyle G. Boyd. *Harvard College Observatory: The First Four Directorships, 1839–1919.* Cambridge, Mass.: Harvard University Press, 1971.

Ketner, Kenneth L., et al. *A Comprehensive Bibliography and Index of the Published Works of Charles Sanders Peirce with a Bibliography of Secondary Sources.* Greenwich, Conn.: Johnson Associates, 1977.

Kevles, Daniel J. *The Physicists: The History of a Scientific Community in Modern America.* New York: Alfred A. Knopf, 1978. 2d ed. Cambridge, Mass.: Harvard University Press, 1987.

———. "Physics and National Power, 1870–1930." In *The Michelson Era in American Science: 1870–1930,* edited by Stanley Goldberg and Roger H. Stuewer, 248–257. AIP Conference Proceedings, no. 179. New York: American Institute of Physics, 1988.

Select Bibliography

———. "The Physics, Mathematics, and Chemistry Communities: A Comparative Analysis." In *The Organization of Knowledge in Modern America, 1860–1920*, edited by Alexandra Oleson and John Voss. Baltimore: Johns Hopkins University Press, 1979.

Kevles, Daniel J., and Carolyn Harding. "The Physics, Mathematics, and Chemistry Communities in America, 1870–1915: A Statistical Survey." *Social Science Working Paper: No. 136*. Pasadena: California Institute of Technology, 1977.

Kevles, Daniel J., Jeffrey L. Sturchio, and P. Thomas Carroll, "The Sciences in America, Circa 1880." *Science* 209 (1980): 27–32.

Knight, David. *The Age of Science: The Scientific World-View in the Nineteenth Century*. Oxford: Basil Blackwell Publisher, 1986.

Kohlstedt, Sally Gregory. *The Formation of the American Scientific Community: The American Association for the Advancement of Science, 1848–60*. Urbana: University of Illinois Press, 1976.

Kohlstedt, Sally Gregory, and Margaret W. Rossiter, eds. *Historical Writing on American Science: Perspectives and Prospects*. Baltimore: Johns Hopkins University Press, 1986. Originally publ. in 1985 by the History of Science Society as vol. 1, 2d ser., of *Osiris: A Research Journal Devoted to the History of Science and Its Cultural Influences*.

Konvitz, Milton, and Gail Kennedy, eds. *The American Pragmatists*. New York: Meridian Books, 1960.

Kragh, Helge. *An Introduction to the Historiography of Science*. Cambridge: Cambridge University Press, 1987.

Krug, Edward A. *The Shaping of the American High School*. New York: Harper & Row, 1964.

Kuhn, Thomas S. "Objectivity, Value Judgment, and Theory Choice." In *The Essential Tension: Selected Studies in Scientific Tradition and Change*. Chicago: University of Chicago Press, 1977.

Kuklick, Bruce. *The Rise of American Philosophy: Cambridge, Massachusetts, 1860–1930*. New Haven, Conn.: Yale University Press, 1977.

Lankford, John. "Amateurs versus Professionals: The Controversy over Telescope Size in Late Victorian Science." *Isis* 72 (1981): 11–28.

Laudan, Larry. *Science and Hypothesis: Historical Essays on Scientific Methodology*. University of Western Ontario Series in Philosophy of Science, no. 19. Dordrecht, Holland: D. Reidel Publishing Co., 1981.

———. *Science and Values: The Aims of Science and Their Role in Scientific Debate*. Pittsburgh Studies in Philosophy and History of Science, no. 11. Berkeley, Los Angeles, London: University of California Press, 1984.

———. "Some Meta-Methodological Preliminaries." Paper presented at the Workshop on Scientific Methodology, Blacksburg, Virginia, April 1984. Pp. 1–24.

Lindberg, David C., and Ronald L. Numbers, eds. *God and Nature: Historical Essays on the Encounter between Christianity and Science*. Berkeley, Los Angeles, London: University of California Press, 1986.

Lukes, Steven. *Individualism*. London: Blackwell, 1973.

McCloskey, Donald N. *If You're So Smart: The Narrative of Economic Expertise.* Chicago: University of Chicago Press, 1990.

———. *The Rhetoric of Economics.* Madison: University of Wisconsin Press, 1985.

———. "What's the Science in Social Science? Two Theorems in Economic Meta-Science." Paper presented at the Conference on Science and Rhetoric, Blacksburg, Virginia, April 1991. Pp. 1–25.

McCulloch, Hugh. *Men and Measures of Half a Century: Sketches and Comments.* New York: Charles Scribner's Sons, 1888.

McKeon, Richard. *Rhetoric: Essays in Invention and Discovery.* Edited by Mark Backman. Woodbridge, Conn.: Ox Bow Press, 1987.

Madden, Edward H. *Chauncey Wright.* The Great American Thinkers Series. New York: Twayne, 1964.

———. *Chauncey Wright and the Foundations of Pragmatism.* Seattle: University of Washington Press, 1963.

———, ed. *Theories of Scientific Method: The Renaissance through the Nineteenth Century.* Seattle: University of Washington Press, 1960. Reprinted in Classics in the History and Philosophy of Science, edited by Roger Hahn, vol. 2. New York: Gordon & Breach Science Publishers, 1989.

Madden, Edward H., and Peter H. Hare. Review of *The Rise of American Philosophy*, by Bruce Kuklick. *Transactions of the Charles S. Peirce Society* 14 (1978): 53–72.

Manicas, Peter T. "Pragmatic Philosophy of Science and the Charge of Scientism." *Transactions of the Charles S. Peirce Society* 29 (Spring 1988): 179–222.

Marsden, Brian G. "Simon Newcomb." *Dictionary of Scientific Biography.* 14 vols. Edited by Charles C. Gillespie. New York: Charles Scribner's Sons, 1970–76.

Martin, Ronald E. *American Literature and the Universe of Force.* Durham, N.C.: Duke University Press, 1981.

Mauskopf, Seymour H. "The History of the American Society for Psychical Research: An Interpretation." *Journal of the American Society for Psychical Research* 83 (January 1989): 7–29.

Mauskopf, Seymour H., and Michael R. McVaugh, eds. *The Elusive Science: Origins of Experimental Psychical Research.* Baltimore: Johns Hopkins University Press, 1980.

Merrick, Sara Newcomb. "John and Simon Newcomb: The Story of a Father and Son." *McClure's Magazine* 35 (October 1910): 677–687.

Mill, John Stuart. *Autobiography.* Edited by Jack Stillinger. London: Oxford University Press, 1971.

———. *A System of Logic, Ratiocinative and Inductive; Being a Connected View of the Principles of Evidence and the Methods of Scientific Investigation.* New York: Harper & Brothers, Publishers, 1858.

———. *A System of Logic, Ratiocinative and Inductive; Being a Connected View of the Principles of Evidence and the Methods of Scientific Investigation.* In Collected Works of John Stuart Mill, edited by J. M.

Robson. Vols. 7 and 8. Toronto: University of Toronto Press, 1973 and 1974.

Miller, Howard S. *Dollars for Research: Science and Its Patrons in Nineteenth-Century America*. Seattle: University of Washington Press, 1970.

Miller, Perry, ed. *American Thought: Civil War to World War I*. New York: Holt, Rinehart & Winston, 1954.

Miller, Perry, and Thomas H. Johnson, eds. *The Puritans*. Rev. ed. 2 vols. New York: Harper & Row, 1963.

Mirowski, Philip. *Against Mechanism: Protecting Economics from Science*. Totowa, N.J.: Rowman & Littlefield, 1988.

———. "How Not to Do Things with Metaphors: Paul Samuelson and the Science of Neoclassical Economics." *Studies in History and Philosophy of Science* 20 (1989): 175–191.

———. *More Heat Than Light: Economics as Social Physics: Physics as Nature's Economics*. Cambridge: Cambridge University Press, 1989.

———. "Physics and the Marginalist Revolution." *Cambridge Journal of Economics* 8 (December 1984): 361–379.

Moore, James R. *The Post-Darwinian Controversies: A Study of the Protestant Struggle to Come to Terms with Darwin in Great Britain and America 1870–1900*. Cambridge: Cambridge University Press, 1979.

Morgan, Mary. *The History of Econometric Ideas*. Historical Perspectives on Modern Economics Series. Cambridge: Cambridge University Press, 1990.

Morison, Samuel E. *Three Centuries of Harvard, 1636–1936*. Cambridge, Mass.: Harvard University Press, 1936.

Morris, Charles W. *The Pragmatic Movement in American Philosophy*. New York: George Braziller, 1970.

Mott, Frank L. *A History of American Magazines: 1850–1865*. Cambridge, Mass.: Harvard University Press, 1938.

Moyer, Albert E. "American Physics in 1887." In *The Michelson Era in American Science: 1870–1930*, edited by Stanley Goldberg and Roger H. Stuewer, 102–110. AIP Conference Proceedings, no. 179. New York: American Institute of Physics, 1988.

———. *American Physics in Transition: A History of Conceptual Change in the Late Nineteenth Century*. History of Modern Physics Series, edited by Gerald Holton and Katherine Sopka. Los Angeles: Tomash, 1983.

———. "P. W. Bridgman's Operational Perspective on Physics: Part I, Origins and Development." *Studies in the History and Philosophy of Science* 22 (1991): 237-258.

———. "John Dewey on Physics Teaching." *The Physics Teacher* 20 (1982): 173–175.

———. "Edwin Hall and the Emergence of the Laboratory in Teaching Physics." In *Physics History from AAPT Journals*, edited by Melba N. Phillips, 191–198. College Park, Md.: American Association of Physics Teachers, 1985.

———. "History of Physics." In *Historical Writing on American Science: Perspectives and Prospects*, edited by Sally Gregory Kohlstedt and

Margaret W. Rossiter, 163–182. Baltimore: Johns Hopkins University Press, 1986.

———. "Physics Teaching and the Learning Theories of G. Stanley Hall and Edward L. Thorndike." *The Physics Teacher* 19 (April 1981): 221–228.

———. "A Puritan of Science." Review of *Science and Cultural Crisis*, by Maila Walter. *Science* 251 (1991): 815.

Myers, Greg. "Writing, Readings, and the History of Science." Review of *The Figural and the Literal: Problems of Language in the History of Science and Philosophy, 1630–1800*, edited by Andrew E. Benjamin, Geoffrey N. Cantor, and John R. R. Christie. *Studies in History and Philosophy of Science* 20 (1989): 271–284.

Nagel, Ernest, ed. *John Stuart Mill's Philosophy of Scientific Method*. New York: Hafner, 1950.

National Educational Association. *Report of the Committee on Secondary School Studies*. 1893. Reprint. New York: Arno Press, 1969.

Nelson, John S. "Approaches, Opportunities and Priorities in the Rhetoric of Political Inquiry: A Critical Synthesis." *Social Epistemology: A Journal of Knowledge, Culture and Policy* 2 (1988): 21–42.

Newcomb, Simon. *A Critical Examination of Our Financial Policy during the Southern Rebellion*. 1865. Reprint. New York: Greenwood, 1969.

———. *His Wisdom the Defender: A Story*. 1900. Reprint. New York: Arno Press, 1975.

———. *Popular Astronomy*. New York: Harper & Brothers, 1878.

———. *Principles of Political Economy*. 1886. Reprint. New York: A. M. Kelley, 1966.

———. *The Reminiscences of an Astronomer*. Boston: Houghton Mifflin Co., 1903.

———. *Sidelights on Astronomy and Kindred Fields of Popular Science: Essays and Addresses*. New York: Harper, 1906.

Noonan, Molly. "Science and the Psychical Research Movement." Ph.D. diss., University of Pennsylvania, 1977.

Norberg, Arthur L. "Simon Newcomb and Nineteenth-Century Positional Astronomy." Ph.D. diss., Univ. of Wisconsin-Madison, 1974.

———. "Simon Newcomb's Early Astronomical Career." *Isis* 69 (1978): 209–225.

———. "Simon Newcomb's Role in the Astronomical Revolution of the Early Nineteen Hundreds." In *Sky with Ocean Joined: Proceedings of the Sesquicentennial Symposia of the U.S. Naval Observatory*, edited by Steven J. Dick and LeRoy E. Doggett, 74–88. Washington, D.C.: U.S. Naval Observatory, 1983.

Norton, Charles E., ed. *Philosophical Discussions by Chauncey Wright*. New York: Henry Holt, 1877.

Numbers, Ronald L. *Creation by Natural Law: Laplace's Nebular Hypothesis in American Thought*. Seattle: University of Washington Press, 1977.

Olson, Richard. *Scottish Philosophy and British Physics 1750–1880: A Study in the Foundations of the Victorian Scientific Style*. Princeton, N.J.: Princeton University Press, 1975.

Oppenheim, Janet. *The Other World: Spiritualism and Psychical Research in England, 1850–1914*. Cambridge: Cambridge University Press, 1985.

Osterbrock, Donald E., John R. Gustafson, and W. J. Shiloh Unruh. *Eye on the Sky: Lick Observatory's First Century*. Berkeley, Los Angeles, London: University of California Press, 1988.

Peirce, Charles S. *Charles S. Peirce: Selected Writings*. Edited by Philip P. Wiener. New York: Dover Publications, 1960.

——— . *Philosophical Writings of Peirce*. Edited by Justus Buchler. 1940. Reprint. New York: Dover Publications, 1955.

——— . *Writings of Charles S. Peirce: A Chronological Edition*. Vol. 1, 1857–1866, edited by Max H. Fisch. Vol. 2, 1867–1871, edited by Edward C. Moore. Vol. 3, 1872–1888, edited by Christian J. W. Kloesel. Bloomington: Indiana University Press, 1982, 1984, 1986.

Persons, Stow, ed. *Evolutionary Thought in America*. New Haven, Conn.: Yale University Press, 1950.

Peterson, Houston. *Huxley: Prophet of Science*. London: Longmans, Green, 1932.

Plotkin, Howard. "Astronomers versus the Navy: The Revolt of American Astronomers over the Management of the United States Naval Observatory, 1877–1902." *Proceedings of the American Philosophical Society* 122 (1978): 385–399.

Porter, Theodore M. *The Rise of Statistical Thinking, 1820–1900*. Princeton, N.J.: Princeton University Press, 1986.

"Presentation of the Newcomb Library." *The City College Quarterly* 6 (1910): 233–243.

"Professor Newcomb's Library to Be Sold." *Publishers' Weekly*, no. 1956 (24 July 1909): 217.

Pycior, Helena M. "Benjamin Peirce's *Linear Associative Algebra*." *Isis* 70 (1979): 537–551.

Rader, Benjamin G. *The Academic Mind and Reform: The Influence of Richard T. Ely in American Life*. Lexington: University of Kentucky Press, 1966.

Reilly, Francis E. *Charles Peirce's Theory of Scientific Method*. New York: Fordham University Press, 1970.

Reingold, Nathan. "American Indifference to Basic Research: A Reappraisal." In *Nineteenth-Century American Science: A Reappraisal*, edited by George H. Daniels. Evanston: Northwestern University Press, 1972.

——— . "Joseph Henry." *Dictionary of Scientific Biography*. 14 vols. Edited by Charles C. Gillespie. New York: Charles Scribner's Sons, 1970–76.

——— , ed. *Science in Nineteenth-Century America: A Documentary History*. New York: Hill & Wang, 1964.

Reingold, Nathan, and Ida H. Reingold, eds. *Science in America: A Documentary History, 1900–1939*. Chicago: University of Chicago Press, 1982.

Rothenberg, Marc. "Organization and Control: Professionals and Amateurs in American Astronomy, 1899–1918." *Social Studies of Science* 11 (1981): 305–325.

Royce, Josiah. "Introduction." In *Science and Hypothesis*, by Henri Poincaré. Translated by George Halsted. New York: Science Press, 1905.
Russell, Bertrand. *The Basic Writings of Bertrand Russell*. Edited by Robert E. Egner and Lester E. Dennonn. New York: Simon & Schuster, 1961.
Russett, Cynthia Eagle. *Darwin in America: The Intellectual Response, 1865–1912*. San Francisco: W. H. Freeman & Co., 1976.
Santayana, George. *The Last Puritan: A Memoir in the Form of a Novel*. New York: Charles Scribner's Sons, 1935.
Say, Jean-Baptiste. *A Treatise on Political Economy; or the Production, Distribution, and Consumption of Wealth*. Translated from the 4th ed. of the French by C. R. Prinsep; new American ed., edited by Clement C. Biddle. Philadelphia: Grigg & Elliot, 1844.
Schabas, Margaret. "Alfred Marshall, W. Stanley Jevons, and the Mathematization of Economics." *Isis* 80 (1989): 60–73.
———. *A World Ruled by Number: William Stanley Jevons and the Rise of Mathematical Economics*. Princeton, N.J.: Princeton University Press, 1990.
Schaffer, Simon. "Astronomers Mark Time: Discipline and the Personal Equation." *Science in Context* 2 (1988): 115–145.
Schuster, John A., and Richard R. Yeo, eds. *The Politics and Rhetoric of Scientific Method: Historical Studies*. Australasian Studies in History and Philosophy of Science, edited by R. W. Home, vol. 4. Dordrecht, Holland: D. Reidel Publishing Co., 1986.
Shahan, Robert W., and Kenneth R. Merrill, eds. *American Philosophy from Edwards to Quine*. Norman: University of Oklahoma Press, 1977.
Shapin, Steven. Review of *The Figural and the Literal: Problems of Language in the History of Science and Philosophy, 1630–1800*, edited by Andrew E. Benjamin, Geoffrey N. Cantor, and John R. R. Christie. *Isis* 79 (1988): 127–128.
Sharlin, Harold I. *Lord Kelvin: The Dynamic Victorian*. University Park: Pennsylvania State University Press, 1979.
Simons, Herbert W., ed. *The Rhetorical Turn: Invention and Persuasion in the Conduct of Inquiry*. Chicago: University of Chicago Press, 1990.
Sizer, Theodore R. *Secondary Schools at the Turn of the Century*. New Haven, Conn.: Yale University Press, 1964.
Smith, Crosbie, and M. Norton Wise. *Energy and Empire: A Biographical Study of Lord Kelvin*. Cambridge: Cambridge University Press, 1989.
Sopka, Katherine R., ed. *Physics for a New Century: Papers Presented at the 1904 St. Louis Congress*. New York: Tomash Publishers and the American Institute of Physics, 1986.
Stallo, John B. *Concepts and Theories of Modern Physics*. Edited by Percy W. Bridgman. 3d ed. 1888. Reprint. Cambridge, Mass.: Harvard University Press, 1960.
Tanner, Amy E. *Studies in Spiritism*. New York: Appleton, 1910.
Taylor, Eugene. "Experimental Psychology and Psychical Research at Harvard, 1872–1910." Paper presented at the annual meeting of the

History of Science Society, as part of a session on "The American Society for Psychical Research: Origin, Context, and Form," Gainesville, Florida, October 1989.

——. "Psychotherapy, Harvard, and the American Society for Psychical Research: 1884–1889." In *Proceedings of Presented Papers: The Parapsychological Association 28th Annual Convention*, 319–346. 1985.

Thiele, Von Joachim. "Karl Pearson, Ernst Mach, John B. Stallo: Briefe aus den Jahren 1897 bis 1904." *Isis* 60 (1969): 535–542.

Thompson, Silvanus P. *The Life of Lord Kelvin*. 2 vols. 2d ed. 1910. Reprint. New York: Chelsea Publishing Co., 1976.

Turner, Frank M. "Public Science in Britain, 1880–1919." *Isis* 71 (1980): 589–608.

Unger, Irwin. *The Greenback Era: A Social and Political History of American Finance, 1865–1879*. Princeton, N.J.: Princeton University Press, 1964.

Wall, Byron E., ed. *Science in Society: Classical and Contemporary Readings*. Toronto: Wall & Thomson, 1989.

Walsh, Harold T. "Whewell and Mill on Induction." *Philosophy of Science* 29 (1962): 279–284.

Walter, Maila L. *Science and Cultural Crisis: An Intellectual Biography of Percy Williams Bridgman (1882–1961)*. Stanford, Calif.: Stanford University Press, 1990.

Wead, Charles K., et al. *Simon Newcomb: Memorial Addresses*. Read before the Philosophical Society of Washington, December 4, 1909. *Bulletin of the Philosophical Society of Washington* 15 (1910): 133–167.

Wiener, Philip P. *Evolution and the Founders of Pragmatism*. Cambridge, Mass.: Harvard University Press, 1949.

Wiener, Philip P., and Frederick H. Young, eds. *Studies in the Philosophy of Charles Sanders Peirce*. Cambridge, Mass.: Harvard University Press, 1952.

Wilson, Daniel J. *Science, Community, and the Transformation of American Philosophy, 1860–1930*. Chicago: University of Chicago Press, 1990.

Winnik, Herbert C. "The Role of Personality in the Science and the Social Attitudes of Five American Men of Science, 1876–1916." Ph.D. diss., University of Wisconsin, 1968.

Wright, Chauncey. *Letters of Chauncey Wright*. Edited by James B. Thayer. 1878. Reprint. New York: Burt Franklin, 1971.

——. *Philosophical Discussions*. Edited by Charles E. Norton. 1877. Reprint. New York: Burt Franklin, 1971.

Wright, Helen. *James Lick's Monument: The Saga of Captain Richard Floyd and the Building of the Lick Observatory*. Cambridge: Cambridge University Press, 1987.

Yeo, Richard R. "Reviewing Herschel's *Discourse*." Review of *A Preliminary Discourse on the Study of Natural Philosophy*, by John. F. Herschel, with a new foreword by Arthur Fine. *Studies in History and Philosophy of Science* 20 (Dec. 1989): 541–552.

Zappen, James P. "Scientific Rhetoric in the Nineteenth and Early Twentieth Centuries: Herbert Spencer, Thomas H. Huxley, and John Dewey." In *Textual Dynamics of the Profession: Historical and Contemporary Studies of Writing in Academic and Other Professional Communities*, edited by Charles Bazerman and James Paradis, 145–167. Madison: University of Wisconsin Press, 1991.

Index

Adams, Charles Francis, 99
Adams, Henry, 80, 148. See also under *North American Review*
Adams, Henry C., 99
Adams, John, 71
Agassiz, Alexander, 59, 68, 70
Agassiz, Louis, 31, 39, 68, 81, 186
Airy, Sir George, 74–75
Almanac Office. See Nautical Almanac Office
Alvord, William: appraisal of SN, 81, 186
American Academy of Arts and Sciences: and C. Peirce, 59; and debate between Gray and Agassiz, 39, 53: *Memoirs* of, 32, 59; and SN, 43, 46
American Association for the Advancement of Education, 160
American Association for the Advancement of Science (AAAS), 84, 126, 170–171, 196; *Proceedings* of, 128; and SN as president, 78, 128–129
American Economic Association (AEA), 108–109, 116, 120, 123, 124–125, 126
American Ephemeris and Nautical Almanac, 30, 75
American Journal of Mathematics: SN as editor, 61, 79, 160
American Journal of Science and Arts, 27, 83, 86
American Mathematical Society: SN as president, 157

American Society for Psychical Research (ASPR), 169–171, 180, 197; *Proceedings* of, 174, 175, 180; and SN as president, 78, 169–171, 174–175, 179–180, 181–182, 197
Archibald, Raymond C.: appraisal of SN, 81
Arthur, Chester, 111
Astronomical and Astrophysical Society of America (later American Astronomical Society): SN as president, 183
Astronomical Journal, 185
Astronomical Papers Prepared for the Use of the American Ephemeris and Nautical Almanac, 75
Astronomical Society of the Pacific, 81, 186
Astronomische Nachrichten, 32
Astronomy, 184; image in U.S. of, 76–77. See also under SN
Atkinson, Edward, 99
Atlantic Monthly, 186

Bachelard, Gaston, 6
Bacon, Francis, 10, 34, 47, 234
Baconianism: and Henry, 29; and influence on pragmatists, 228; and influence on rhetoric of method, 228–232, 234–236; in political economy, 27, 107, 113, 116; and Scottish Realism, 34–35, 48–49, 228; and SN, 10, 47–51, 113, 116.

293

See also Empiricism; Inductive reasoning
Baird, Spencer, 252 n. 4
Barker, George, 170
Barnes, Barry, 6
Barrett, William, 170
Bartlett, William, 39
Bell, Alexander Graham, 80, 112, 175
Boole, George, 34, 60
Boss, Lewis, 185
Bowditch, Henry, 170
Bowditch, Nathaniel, 28
Bowen, Francis, 34, 39, 42, 53
Bridgman, Percy W., 149, 151, 191, 217, 240 n. 16, 244 n. 25
British Association for the Advancement of Science, 129, 170
British Society for Psychical Research, 170, 171, 173–174, 177–178, 180–182; *Proceedings* of, 174
Bruce, Catherine Wolfe, 186
Bruce, Robert, 11, 231, 233
Bureau of Education, U.S., 158
Butler, Joseph, 33
Butler, Nicholas Murray, 161

Cairnes, John E., 102–103, 109, 116
Cambridge, Mass.: as science center, 19, 30, 58–59, 64, 90; and SN, 19, 29–30, 58–59, 143
Campbell, William W., 78
Carey, Henry, 112–113
Carnegie Institution, xv, 183
Cayley, Arthur, 71, 73
Civil War, 46; and economic and political legacy during Reconstruction, 98, 103, 110; and SN, 44, 54, 100, 110, 121, 197, 231, 233; and U.S. science, 11, 67, 90
Clifford, William, 40, 95, 190, 213
Coast and Geodetic Survey: and C. Peirce, 58–59, 60; and F. Hassler, 68; and SN, 29; and SN as possible superintendent, 68, 69, 71, 76
Cobbett, William, 26, 159
Coe, George S., 110–111
College of New Jersey (later Princeton), 136
Columbian University (later George Washington University), 79
Combe, George, 22–23, 240 n. 5
Comte, Auguste: general influence on SN, 10, 35, 51, 52, 65, 129, 222; positivism, 36–38; relation to W. James, 169; specific impact on middle-aged SN, 143–144, 167; specific impact on younger SN, 36–38, 47–49, 153, 216

Congress of Arts and Science, St. Louis: SN as president, 81, 192–193
Cope, Edward, 170
Cournot, A. A., 60

Dana, James D., 81, 86, 233
Darwin, Charles: general influence on SN, 35, 36, 51, 52, 65, 129, 222; relation to Comte, 37, 39; specific impact on middle-aged SN, 134, 141, 143, 149; specific impact on younger SN, 38–40; views on scientific inquiry, 38–39, 40. *See also* Evolutionary ideas
Davis, Charles Henry, 59
Davis, William Harper: appraisal of SN, 81
Delaunay, Charles, 72
Dewey, John: pedagogy, 161; pragmatism's cultural aspects, 207–213 *passim*, 221; pragmatism's emergence, 222; pragmatism's philosophical aspects, 206, 217, 219, 220, 221; ties between pragmatism and rhetoric of method, 225–227, 228, 237, 274 n. 5
Dial, 119, 122
Draper, John, 25
Dudley Observatory, 185
Dunbar, Charles F., 99
Dutch Academy of Science, 73

Economics. *See* Political economics
Education: in colleges and universities, 11, 83–84, 158, 184; and Committee of Ten (NEA), 158, 164–165; in secondary schools, 158; and theories of learning and instruction, 92. *See also* SN, pedagogy; *and names of schools*
Educational Review, 161–164
Edwards, Jonathan, 55
Einstein, Albert: appraisal of SN, xi, 76; general theory of relativity, 76
Eliot, Charles W.: advocacy of scientific method, 268 n. 37; chair of Committee of Ten, 158, 163; relation with SN, 69, 80, 99, 163
Ely, Richard T., 99, 116; controversy with SN, 107–109, 113–114, 119, 120–122, 124, 126, 223
Empiricism: and SN, 10, 47–51, 157, 159, 162, 176. *See also* Baconianism; Hume; Inductive reasoning; Mill, empiricism and induction
Enlightenment, the, 23, 228
Europe, science in, 83–84, 88, 183–185, 195

Index

Evolutionary ideas: Darwinian, 36, 38–40, 87, 129; developmental, 137; Lamarckian, 129; and nebular hypothesis, 128–129; and religion, 40, 128–129, 134–135, 137

Feyerabend, Paul, 6
Fisher, Irving, 105, 110
Fiske, John, 36, 37
Fowler, Orson, 22, 25
Franklin, Benjamin, 81, 186
Franklin, Fabian, 109
French Academy of Sciences, 69, 186

Galileo (Galileo Galilei), 234
Garfield, James A., 111–112
Geographical Society of New York, 84
Gibbs, J. Willard, 105
Gieryn, Thomas, 8
Gilded Age, xii, xiv, 9–10, 15; and Twain, 98
Gilliss, Capt. James, 46, 252 n. 3
Gilman, Daniel C., 61, 86; relation with SN, 79, 80, 87, 107, 175
Godkin, Edwin, 99
Gould, Benjamin, 46, 59
Gray, Asa, 53, 135, 170; controversy with SN, 135–136, 143, 145, 147, 199, 223; debate with L. Agassiz, 39
Greenwich Observatory, 75
Gurney, Edmund, 173–174

Haeckel, Ernst, 40
Hall, Edwin, 170, 180
Hall, G. Stanley, 170, 175
Halley's Comet, 20, 98
Hamilton, William, 41, 44–45, 52, 54, 57
Hansen, Peter, 62, 72
Harper publications: books, 77–78, 99, 105; *Harper's Magazine*, 77; *Harper's Weekly*, 99
Harris, William T., 165, 170; appraisal of SN, 99
Harvard College (later Harvard University): and Bowdoin Prize, 37; and SN's middle years, 69, 79, 99, 114; and SN's younger years, 31–35, 36, 58–59, 69; and W. James, 237. See also Lawrence Scientific School
Harvard Observatory, 30; and SN as possible director, 68, 69, 71, 76
Hassler, Ferdinand R., 68
Hassler, Mary C. See Newcomb, Mary C. (wife)
Henry, Joseph, 28–29, 78, 82, 135, 249 n. 13; appraisal of SN, 29–30, 69;

Philosophical Society of Washington, 60, 90, 176; relation to middle-aged SN, 68–69, 70, 72, 111, 149–150, 171, 222; relation to younger SN, 28–30, 35, 46, 48–51, 141, 160
Herschel, Sir John, 78; influence on rhetoric of method, 228–230
Hilgard, Julius E., 29
Historiography: internalist, 6, 207, 214; and scientific method, 6–7; Whig, 6, 207, 214
Hodgson, Richard, 181
Hollinger, David, 205–209, 210, 221–222, 225–226, 227, 277 n. 12
Holt, Henry, and Company, 160, 161
Howe, Frederic: appraisal of SN, 66
Hume, David, 32–33, 93, 216, 218
Huntington, Frederick D., 32
Huxley, Thomas, 40, 129, 135, 137, 141, 213, 229, 265 n. 37
Hypothetico-deductive reasoning, 10, 43, 218–219; and Henry, 29; in political economy, 107, 113–117; rhetorical deployment of, 234–236; and SN, 10, 49–51, 113–117, 192, 235–236
Hyslop, James, 197–199

Imperial Academy of Sciences, St. Petersburg, 80
Independent, 99, 135–136, 167, 171
Inductive reasoning, 10, 43, 218–219; and Henry, 29; in political economy, 27, 107, 113; rhetorical deployment of, 230–232, 234–236; and SN, 10, 45, 48–50, 113, 176, 178, 192, 235. See also Baconianism; Empiricism; Mill, empiricism and induction
International Review, 148
Intuitionism. See Hamilton; McCosh

James, Edmund J., 99, 106, 108, 116; criticism of SN, 106–107, 109, 119, 120, 137, 223; relation with SN, 123–124, 126
James, William: appraisal of SN, 78; appraisal of Wright, 58, 213; attitudes and views about science, 43, 199; pragmatism's cultural aspects, 207–213 *passim*, 221; pragmatism's emergence, 64, 222–223, 228; pragmatism's philosophical aspects, 206, 218–219, 220, 221; psychical research, 169–171, 179, 181, 182; relation with SN, 80, 169–171, 182; skirmish with SN, 181, 223; ties between pragmatism and rhetoric of

method, 225–227, 228, 237; mentioned, 193
Jevons, W. Stanley, 101–102, 105, 106
Johns Hopkins University: and C. Peirce, 61; and Dewey, 237; and Ely, 107, 126; and SN, 61, 78–79, 87, 99, 183. *See also* Gilman
Johnson, Andrew, 111
Journal of Political Economy, 126

Kevles, Daniel, 11
Keynes, John Maynard: appraisal of SN, 105
Knox, John Jay, 99
Koyré, Alexandre, 6
Kuhn, Thomas, 6

Langley, Samuel, 80, 187
Laplace, Pierre, 28, 31, 43
Latour, Bruno, 6
Laudan, Larry: perspective on scientific method, 5, 6
Laughlin, James Laurence, 116, 126; appraisal of SN, 99–100; relation to SN, 116, 119
Lawrence Scientific School, at Harvard: and SN, 31–32, 36, 58–59, 69
Lazzaroni, 46, 82
LeConte, John L., 85, 233
Leverrier, Urbane J. J., 62, 71, 75, 76
Lick, James, 79
Lick Observatory, 78, 79; SN as advisor to, 79
Lincoln, Abraham, 46, 141, 171
Literature and science, 187–188
Lockyer, Norman, 80
Lodge, Henry Cabot, 82
Lodge, Oliver, 170, 197

McCosh, James, 136; controversy with SN, 135, 136–138, 143, 145, 199, 223
McCulloch, Hugh, 110–111
McGee, Anita Newcomb. *See* Newcomb, Anita
McGee, W. J. (SN's son-in-law), 273 n. 27
Mach, Ernst, 149, 240 n. 16
Madden, Edward, 215–222
Marcet, Mrs. Jane, 24
Margenau, Henry, 259 n. 15
Maritime Provinces, 19, 20, 25, 158–159, 186
Materialism, scientific, 166. *See also under* SN
Mathematics, 184; and Conference on Mathematics (NEA), 158, 163–165. *See also under* SN

Maury, Matthew, 46
Maxwell, James C., 80, 149
Merrick, Sara Newcomb (SN's sister): appraisal of SN, 66–67, 80
Method, rhetorical use of, xii–xiii, xiv, 6–10, 15; on disciplinary level, 7, 9, 13–14, 147, 224–225; in economics, 100, 106, 108, 127; and emergence in U.S., 230–237; in Great Britain, 226, 227–231, 234–236, 276 n. 3; on internal level, 7, 8–9, 13–14, 148, 224–225; and "new rhetoric," 9; on private level, 7–8, 12–13, 146–147, 225; on public level, 7–8, 9, 11–12, 13–14, 146–148, 224–225; and ties to pragmatism, 210, 224, 226–230, 237. *See also under* SN
Method, scientific, xii, 3–6, 271 n. 7. *See also* Baconianism; Empiricism; Hypothetico-deductive reasoning; Inductive reasoning; Operationalism; Positivism; Scottish Realism; *and names of individual philosophers and scientists, including under* SN
Michelson, Albert A., 147–148, 155, 251 n. 31
Mill, John Stuart: compatibilism, 54–57, 96, 103, 212; debate with Whewell, 42–43, 50, 94, 162, 218–219; empiricism and induction, 36, 41, 45, 93, 218, 267 n. 19; general influence on SN, 10, 35, 36, 52, 65, 129; 222; influence on pragmatists, 228; influence on rhetoric of method, 228–230; opposition to Hamilton, 41, 44–45; political economy, 44, 45, 110, 112–113, 209; position on liberal education, 89; relation to Comte, 37, 144; specific impact on middle-aged SN, 54–57, 85, 137–138, 143, 162, 167, 219; specific impact on younger SN, 38, 41–44, 47–51, 216; ties to "old school," 102, 107, 109, 116, 125; visit with SN, 44–45, 72
Morley, Edward, 155

Nation: and C. Peirce, 186; and C. Peirce-SN controversy, 61; and Ely, 126; and SN, 77, 99, 109, 122, 124–125; and Wright, 63
National Academy of Sciences (NAS), 90, 173, 195; and SN as vice-president, 78
National Educational Association (NEA), 158, 163, 193
National Intelligencer, 28

Index

Nature, 156
Nautical Almanac Office: and SN, 29–32, 36–37, 52–54, 58–59, 72; and SN as superintendent, 68, 69–70, 71, 74–75, 76, 79, 126, 183; status of, xii, 196
Naval Academy, U.S., 46
Naval Observatory, 46, 196; status of, xii, 77; and SN, 46–47, 54, 60, 68–70, 71–74, 196; and SN's health, 67
Navy, U.S., and science, 29–30, 46, 74, 77, 81
Newcomb, Anita (daughter), 142, 160, 265 n. 33; appraisal of SN, 24
Newcomb, Anna Josepha (daughter), 142
Newcomb, Emily Kate (daughter), 142
Newcomb, Emily Prince (mother), 20, 21–22, 23, 33, 141, 142
Newcomb, John (father), 20–24, 26, 31, 33, 141, 158–159
Newcomb, Mary C. (wife), 68, 111, 141–143
Newcomb, Simon (abbreviated in Index as SN):
— astronomy: early exposures, 24, 28, 29–30, 30–31; expeditions, 33, 60, 72, 74, 112; international project for uniformity, 76, 183, 184–186; lunar, planetary, and stellar research, 32, 46–47, 68–69, 71–76, 183, 184, 199, 232; popularizations, 77–78; rhetorical use of method, 147, 157, 185–186; telescopes, 73–74, 79–80, 195–196
— biography during formative years: appointment to Almanac Office, 29–32; apprenticeship to herbal doctor, 20, 25–26; birth, childhood, and youth, 19–26; early writings, 27–28, 159–160; enrollment at Lawrence Scientific School, 31–32; family background, 20; first teaching positions, 26–28, 159; immigration to the U.S. and Maryland, 26; influential books, 19, 21, 22–25, 26–28, 31–34; interest in science and mathematics, 20–25, 27, 28–30; Washington contacts, 28–30. See also Lawrence Scientific School; Nautical Almanac Office
— biography during later years: autobiographical and fictional writings, 186–188, 199; death and funeral, 81, 199–201; honors and awards, 186, 192; retirement, 70, 126, 183, 186; university teaching, 183; views on flying machines, 187–188
— biography during middle years: appointment to Naval Observatory, 46–47; correspondence, 80, 255 n. 27; first trip to Europe, 44–45, 60, 72; honors and awards, 73, 79–80, 80–81, 114; marriage to Mary Hassler, 68; officer in professional societies, 78, 90, 99, 128–129; physical and personal characteristics, 66–67, 80; public image, 76–78, 81, 119, 137; salary, 161; superintendent of the Almanac Office, 69–70; ties to Republican party, 111; university teacher, 61, 78–79, 87, 99, 107, 160; U.S. citizenship, 47. See also Nautical Almanac Office; Naval Observatory
— business issues and language, xiii, 12, 94–95, 130
— historical appraisals of SN, xi–xii, 75–76, 105, 157, 186
— materialism, scientific, 166–169; rhetorical use of method, 166–169
— mathematics: chair of Conference on Mathematics, 158, 159, 161, 163–165; early exposures, 20–21, 23, 27–31; midcareer involvements, 160; non-Euclidean geometry, 150, 157, 162; rhetorical use of method, 147, 157; theories of learning and instruction, 159–165; theory of probability, 43–44, 172–174, 191
— method, rhetorical use of, xii–xiii, 8, 9–10, 11–15; central features, 146–148, 200–201, 205, 224–225, 234, 235–237; linkage to social progress, 82, 88–89, 91–92, 95, 96–97, 146; parallels with Dewey, James, C. Peirce, and Wright, 225–227; ties to pragmatism, 225–227, 228, 230. See also under areas of SN's interest
— method, scientific, xii, 3–4, 5–6, 9–11, 14–15; central features, 85, 91–92, 117, 130; characteristics of linguistic, empirical method, 10–12, 92–97, 130–132, 200–201, 240 n. 9; impact in U.S., 198–199; later reflections on linguistic, empirical method, 187–188, 188–194; parallels with Mill, C. Peirce, and Wright, 85, 94–96, 131, 137–138; roots of linguistic, empirical method, 24, 26, 27, 45, 51, 63, 65; textbook statement, 47–51, 115, 117, 176. See also Baconianism; Empiricism; Hypothetico-deductive reasoning; Inductive reasoning; Operationalism;

Positivism; Scottish Realism; *and names of philosophers and scientists*
—pedagogy: chair of Conference on Mathematics, 158, 159, 161, 163–165; early views, 27–28, 159–160; later reflections, 193–194; position on liberal education, 89, 92; relevance to scientific method, 158, 159, 161, 165; rhetorical use of method, xiii, 12; textbooks, 80, 160–161; theories of learning and instruction, 93, 159–165
—philosophy, 272 n. 16; axiological issues, 14–15; compatibilism, 54–57, 96, 103, 189, 212, 249 n. 16; epistemic issues, 14–15, 42–43, 93–94, 150, 159, 162, 219; involvements while in Cambridge, 34–35; ontological issues, 14–15, 150; rhetorical use of method, xiii, 12, 94, 95–97, 131–132
—phrenology, 22–23, 25
—physics: classical research program, 49, 61, 76, 148–153, 168–169, 187; first exposure, 24; involvement while in Cambridge, 46, 47, 147; midcareer interests and orientations, 147–150, 155–157; operational definitions, 151–153, 155–156; parallels with Mill, C. Peirce, and Wright, 152–153; rhetorical use of method, 12, 94–95, 147–155 *passim*, 156–157. *See also under* Stallo
—political economics: basic involvements, 60, 98–105, 184; controversy between old and new schools, 105–109, 113–126, 131, 137, 143, 147; first book, 54, 100, 110; first exposure, 27; later reflections, 196–197; liberalism, xiii, 12, 23, 109–113, 127, 146, 187–188; national issues and policies, 88, 91, 103–104, 110–112, 123–126, 153; operational definitions, 153; parallels with Mill, 101, 103, 104–105, 109–110, 112–113, 123; rhetorical use of method, xiii, 12, 82, 98–127 *passim*, 146–147, 176, 194, 196–197; scientific and mathematical rigor, 13–14, 44, 100–105, 108, 119, 161; university teaching, 79, 99. *See also names of individuals and organizations involved with political economics*
—pragmatism, 10–11, 14–15; cultural aspects, 209–213, 221–222, 224, 227; cultural parallels with Dewey, James, C. Peirce, and Wright, 210–213, 221–222; philosophical aspects, 214–222, 224; philosophical parallels with Dewey, James, C. Peirce, and Wright, 214–222 *passim;* role in emergence, 221–223, 237
—psychical research: early views, 171; later reflections, 197–198; midcareer interests and orientations, 169–182; moral imperative to investigate, 174–175, 182; parallels with C. Peirce, 173–174; rhetorical use of method, 13–14, 166–167, 171–173, 175–182, 194, 197–198. *See also names of individuals and organizations involved with psychical research*
—religion: exposure during childhood and youth, 21–22, 27, 141; involvements while in Cambridge, 32–34, 141; later reflections, 199–200; natural theology, 13, 33, 130, 139, 140, 147, 190; parallels with Comte, Mill, and Wright, 134, 143–144, 189; personal beliefs, 139–145; relation to evolutionary ideas, 40, 134, 138–139, 144, 149, 199; relation to science, 128–139, 140–141, 145; rhetorical use of method, xiii, 12, 132–141 *passim,* 145, 146–147, 188–191, 194, 199–200; skepticism, xiii, 12, 40, 141–143, 146, 199; truth, theory of, 132–133. *See also under* Gray; McCosh; Porter
—science in U.S., perceptions of: conceptual framework, 11–14, 82–83, 87–89, 148–151, 166, 232–234, 236–237; democratic doctrines, 82, 87–88; government scientists, 70, 74, 89, 161, 195–196; governmental and public support, 82–89, 129–130, 140, 146, 150, 184, 195–196; institutional framework, 11–14, 82–89, 91, 129–130, 145, 150, 184, 194–196, 232–237
—writings, select, 80–81; "Abstract Science in America, 1776–1876," 86–89, 91, 92; "Address of the President" (ASPR), 175–181; "Course of Nature," 128–135, 137–145 *passim,* 150, 168, 188–189; "Exact Science in America," 82–86, 130; *His Wisdom the Defender,* 187–188; "Modern Scientific Materialism," 166–169, 171; *Popular Astronomy,* 77–78, 79, 199–200; *Principles of Political Economy,* 80, 104–107, 109, 117–119, 122, 153; "Relation of Scientific Method to Social Progress," 90–97, 98, 105, 120, 125,

Index

128, 131; *Reminiscences of an Astronomer* (autobiography), 19, 44–45, 62, 70, 111, 126–127, 186–187; "Speculative Science," 148–155; "Two Schools of Political Economy," 107–108, 113–116, 117–119, 120
Newcomb, Thomas (younger brother), 23
Newton, Isaac, 27
Newtonian science. See Physics, and classical research programs
New York Observer, 40
New York Tribune, 77
North American Review: and SN, 57, 77, 109, 143, 194–195; and SN's interaction with editor H. Adams, 82, 86, 99; and symposium on "Law and Design in Nature," 137–139, 143, 199; mentioned, 45, 53, 54
Norton, Charles E., 57–58

Operationalism, 151–153, 191, 217

Paley, William, 33, 129
Paris Academy of Sciences. See French Academy of Sciences
Paris Observatory, 72, 75
Peabody, Andrew P., 32–33
Peirce, Benjamin, 30, 31; appraisal of SN, 69, 72; relation to middle-aged SN, 31, 39, 42, 46, 49–51, 59; relation to younger SN, 69, 71, 72–73; son C. Peirce, 59, 60
Peirce, Benjamin O., 170
Peirce, Charles Sanders: appraisal of SN, 62, 186–187, 193; appraisal of Wright, 52; early philosophical writings, 59–60; father B. Peirce, 59, 60; general influence on SN, 10–11, 51, 222; personal interaction with SN, 58–62, 156–157, 193; pragmatism's cultural aspects, 207–213 *passim*, 221; pragmatism's emergence, xiv, 6, 62–63, 64–65, 131, 222–223; pragmatism's philosophical aspects, 152, 191, 206, 214–218, 220–221; relation to W. James, 64, 222–223; relation to Wright, 58–59, 64–65; scholarly interests and activities, 60–62, 156–157, 173–174, 253 n. 6; specific impact on SN, 63, 65, 131, 191; ties between pragmatism and rhetoric of method, 225–227, 228, 229, 237
Perry, Arthur L., 99
Philadelphia, 86, 175

Philosophical Magazine, 156
Philosophical Society of Washington, 60, 90, 155–156, 200; and SN as president, 78, 90–91, 200–201
Phrenology, 22–23, 25. See also under SN
Physics, 183–184; and classical research programs, 11, 63, 148–155; and economics, 100, 101, 127. See also under SN
Pickering, Edward, 170
Political economics: liberalism in, 109–110, 209; and Mill, 44; national issues involving, 98, 103, 107, 110; old and new schools of, 106–108, 120–122; scientific and mathematical rigor in, 27, 100, 101, 113; and scientific method, 100, 105, 109. See also under SN
Political Economy Club of America, 99–100, 108, 116–117, 123, 126; and SN as president, 99–100
Popular Science Monthly: and C. Peirce, 64–65; and SN, 77, 128, 192; and Stallo, 154; mentioned, 81, 83
Porter, Noah, 135; controversy with SN, 135, 136, 137–138, 143, 145, 199, 223
Positivism, 198; opposition to, 136; and SN, 10, 129, 134. See also Comte
Powell, John Wesley, 269 n. 8
Pragmatism, xiv, 205–206, 210; cultural aspects of, 206–213, 221–222, 224, 225–230; emergence in U.S. of, 230–237, 275 n. 22; impact in U.S. of, 199; philosophical aspects of, 206, 207, 214–222, 224; roots of, 9–10, 43, 62–65, 222–223, 227–228; and ties to the rhetoric of method, 210, 224, 226–230, 237. See also names of pragmatic thinkers, including under SN
Princeton Review, 108, 113, 136
Progressive era, xii, 199, 209
Psychical research, 169–171, 180–182; and psychology, 180. See also under SN
Publications, U.S. scientific, 11, 83–84, 87, 184
Public science, 8, 13
Pulkovo Observatory, 79–80
Putnam, Herbert, 200
Putnam, Hilary, 6

Quarterly Journal of Economics, 125

Radicalism, philosophical, 41
Reid, Thomas, 34

Reingold, Nathan, 233
Religion: Baptist, 21; Calvinist, 21, 33, 135, 141, 240 n. 5; Congregationalist, 135; Episcopal, 21, 32; Methodist, 32; and natural theology, 13, 33, 129–130; Presbyterian, 141; Puritan, 21–22, 33–34; and relation to evolutionary ideas, 40, 129, 134–135, 137; and relation to science, 8, 119, 128–130, 199; and social reform, 107, 110; Swedenborgian, 32; Unitarian, 32–33, 34, 36. See also under SN
Rhetoric: and Cobbett, 26; and "new rhetoric," 9, 240 n. 8. See also Method, rhetorical use of
Ricardo, David, 107, 124–125
Robeson, George M., 69
Rowland, Henry, 61, 80
Royal Astronomical Society, 71, 73, 77, 82
Royal Society of London, 81, 114
Royce, Josiah, 148, 170
Rumford, Count, 186
Runkle, John, 52

Safford, Truman, 185
Santayana, George, 34
Say, Jean-Baptiste, 27
Schuster, John, and Richard Yeo: history of method, 9–10; rhetoric of scientific method, 6–9, 224–225, 227, 236; science as collective enterprise, 12–13, 225. See also Method, rhetorical use of; Yeo, Richard
Science: economic forum in, 120–121; psychical issues in, 171–176, 178, 180–181; and SN, 92, 109, 156, 196
Science, U.S. See United States, science in
Scientism, xii, xiii, 14, 149, 207
Scottish Realism, 34–35, 36, 48–49, 228
Scribner's Monthly, 77, 79
Seligman, Edwin, 120–121
Sensationism. See Mill
Shapley, Harlow: appraisal of SN, 184
Shaw, Albert, 119–120, 122
Sidgwick, Henry, 181
Silliman, Benjamin, Jr., 83, 85, 86, 233
Silliman's journal. See Popular Science Monthly
Smith, Adam, 107, 109–110, 125
Smithsonian Institution, 28, 90, 187; and SN as possible secretary, 68–69, 71, 76

Smithsonian Miscellaneous Collections, 90–91
Social sciences, 13; and professionalization, 11, 98, 275 n. 7; scientific method in, 100. See also Political economics
Societies, professional, 11, 83–84, 87, 184. See also names of societies
Spencer, Herbert, 40, 63, 229
Stallo, John, 148, 154; controversy with SN, 148–155, 156, 166, 168, 223
Stewart, Dugald, 34
Struve, Otto, 79, 80
Sumner, Charles, 111
Sumner, William Graham, 86, 99, 108; relation to SN, 108, 109, 120, 122–123
Sunday School Times, 142
Sylvester, James, 69, 79

Telescopes. See under SN, astronomy
Thayer, James B., 53, 55, 58
Thomson, William (Lord Kelvin): appraisal of SN, 66; relation to SN, 80, 149
Thorndike, Edward L., 193–194
Turner, Frank, 8
Twain, Mark, 98
Tyler, Samuel, 231
Tyndall, John, 40, 129, 135, 136, 137, 167–168

United States, science in, 67–68; conceptual framework of, 11, 233–234, 237; government support of, 11, 69, 70; institutional framework of, 11, 46, 129–130, 171, 183–184, 231, 233–237; public support of, 11. See also SN, science in U.S.
University of California, Berkeley, 78; and SN as possible president, 79
University of Pennsylvania, 106, 123–124, 126
Utilitarianism, British, 41

Walker, Francis A., 99, 124–125, 126
Walker, James, 34
Washington, D.C.: as science center, 19, 78, 90–91; and SN, 19, 28, 68–70, 141, 143, 196
Washington Philosophical Society. See Philosophical Society of Washington
Washington Scientific Club, 111
Wells, David A., 99
Whately, Richard, 34
Whewell, William, 42–43, 50, 178, 218–219; influence on rhetoric of method, 228–229

Index

White, Andrew, 170
Wiener, Philip, 252 n. 40
Winlock, Joseph: appraisal of SN, 30; relation with SN, 29–30
Woodward, Robert S.: appraisal of SN, xv
Wright, Chauncey: Darwin and evolution, 39, 53, 96; dialogue with SN on compatibilism, 54–57, 95–96; general influence on SN, 10–11, 51; personal interaction with SN, 31, 36, 45, 52–58; positivism, 15, 63; pragmatism's cultural aspects, 212–213, 221; pragmatism's emergence, xiv, 6, 62–65, 222–223; pragmatism's philosophical aspects, 43, 214–221; relation to C. Peirce, 58, 63–65; relation to James, 63, 223; specific impact on SN, 63, 65, 85, 131, 134, 137, 222; ties between pragmatism and rhetoric of method, 225–227, 228, 237
Wright, George Frederick, 135

Yeo, Richard, 227–230, 234–236, 276 n. 3. *See also* Schuster, John, and Richard Yeo
Yerkes Observatory, 195
Youmans, Edward L., 83, 85

Ziman, John, 6

Designer:	U.C. Press Staff
Compositor:	Auto-Graphics, Inc.
Text:	10/13 Sabon
Display:	Sabon
Printer:	Thomson-Shore, Inc.
Binder:	Thomson-Shore, Inc.